Research Notes in Mathematics

Submission of proposals for consideration
Suggestions for publication, in the form of outlines and representative samples, are invited by the Editorial Board for assessment. Intending authors should approach one of the main editors or another member of the Editorial Board, citing the relevant AMS subject classifications. Alternatively, outlines may be sent directly to one of the publisher's offices. Refereeing is by members of the board and other mathematical authorities in the topic concerned, throughout the world.

Preparation of accepted manuscripts
On acceptance of a proposal, the publisher will supply full instructions for the preparation of manuscripts in a form suitable for direct photo-lithographic reproduction. Specially printed grid sheets are provided and a contribution is offered by the publisher towards the cost of typing. Word processor output, subject to the publisher's approval, is also acceptable.

Illustrations should be prepared by the authors, ready for direct reproduction without further improvement. The use of hand-drawn symbols should be avoided wherever possible, in order to maintain maximum clarity of the text.

The publisher will be pleased to give any guidance necessary during the preparation of a typescript, and will be happy to answer any queries.

Important note
In order to avoid later retyping, intending authors are strongly urged not to begin final preparation of a typescript before receiving the publisher's guidelines and special paper. In this way it is hoped to preserve the uniform appearance of the series.

Advanced Publishing Program
Pitman Publishing Inc
1020 Plain Street
Marshfield, MA 02050, USA
(tel (617) 837 1331)

Advanced Publishing Program
Pitman Publishing Limited
128 Long Acre
London WC2E 9AN, UK
(tel 01-379 7383)

Titles in this series

1 Improperly posed boundary value problems
A Carasso and A P Stone
2 Lie algebras generated by finite dimensional ideals
I N Stewart
3 Bifurcation problems in nonlinear elasticity
R W Dickey
4 Partial differential equations in the complex domain
D L Colton
5 Quasilinear hyperbolic systems and waves
A Jeffrey
6 Solution of boundary value problems by the method of integral operators
D L Colton
7 Taylor expansions and catastrophes
T Poston and I N Stewart
8 Function theoretic methods in differential equations
R P Gilbert and R J Weinacht
9 Differential topology with a view to applications
D R J Chillingworth
10 Characteristic classes of foliations
H V Pittie
11 Stochastic integration and generalized martingales
A U Kussmaul
12 Zeta-functions: An introduction to algebraic geometry
A D Thomas
13 Explicit *a priori* inequalities with applications to boundary value problems
V G Sigillito
14 Nonlinear diffusion
W E Fitzgibbon III and H F Walker
15 Unsolved problems concerning lattice points
J Hammer
16 Edge-colourings of graphs
S Fiorini and R J Wilson
17 Nonlinear analysis and mechanics: Heriot-Watt Symposium Volume I
R J Knops
18 Actions of fine abelian groups
C Kosniowski
19 Closed graph theorems and webbed spaces
M De Wilde
20 Singular perturbation techniques applied to integro-differential equations
H Grabmüller
21 Retarded functional differential equations: A global point of view
S E A Mohammed
22 Multiparameter spectral theory in Hilbert space
B D Sleeman
24 Mathematical modelling techniques
R Aris
25 Singular points of smooth mappings
C G Gibson
26 Nonlinear evolution equations solvable by the spectral transform
F Calogero
27 Nonlinear analysis and mechanics: Heriot-Watt Symposium Volume II
R J Knops
28 Constructive functional analysis
D S Bridges
29 Elongational flows: Aspects of the behaviour of model elasticoviscous fluids
C J S Petrie
30 Nonlinear analysis and mechanics: Heriot-Watt Symposium Volume III
R J Knops
31 Fractional calculus and integral transforms of generalized functions
A C McBride
32 Complex manifold techniques in theoretical physics
D E Lerner and P D Sommers
33 Hilbert's third problem: scissors congruence
C-H Sah
34 Graph theory and combinatorics
R J Wilson
35 The Tricomi equation with applications to the theory of plane transonic flow
A R Manwell
36 Abstract differential equations
S D Zaidman
37 Advances in twistor theory
L P Hughston and R S Ward
38 Operator theory and functional analysis
I Erdelyi
39 Nonlinear analysis and mechanics: Heriot-Watt Symposium Volume IV
R J Knops
40 Singular systems of differential equations
S L Campbell
41 N-dimensional crystallography
R L E Schwarzenberger
42 Nonlinear partial differential equations in physical problems
D Graffi
43 Shifts and periodicity for right invertible operators
D Przeworska-Rolewicz
44 Rings with chain conditions
A W Chatters and C R Hajarnavis
45 Moduli, deformations and classifications of compact complex manifolds
D Sundararaman
46 Nonlinear problems of analysis in geometry and mechanics
M Atteia, D Bancel and I Gumowski
47 Algorithmic methods in optimal control
W A Gruver and E Sachs
48 Abstract Cauchy problems and functional differential equations
F Kappel and W Schappacher
49 Sequence spaces
W H Ruckle
50 Recent contributions to nonlinear partial differential equations
H Berestycki and H Brezis
51 Subnormal operators
J B Conway
52 Wave propagation in viscoelastic media
F Mainardi
53 Nonlinear partial differential equations and their applications: Collège de France Seminar. Volume I
H Brezis and J L Lions

54 Geometry of Coxeter groups
H Hiller
55 Cusps of Gauss mappings
T Banchoff, T Gaffney and C McCrory
56 An approach to algebraic K-theory
A J Berrick
57 Convex analysis and optimization
J-P Aubin and R B Vintner
58 Convex analysis with applications in the differentiation of convex functions
J R Giles
59 Weak and variational methods for moving boundary problems
C M Elliott and J R Ockendon
60 Nonlinear partial differential equations and their applications: Collège de France Seminar. Volume II
H Brezis and J L Lions
61 Singular systems of differential equations II
S L Campbell
62 Rates of convergence in the central limit theorem
Peter Hall
63 Solution of differential equations by means of one-parameter groups
J M Hill
64 Hankel operators on Hilbert space
S C Power
65 Schrödinger-type operators with continuous spectra
M S P Eastham and H Kalf
66 Recent applications of generalized inverses
S L Campbell
67 Riesz and Fredholm theory in Banach algebra
B A Barnes, G J Murphy, M R F Smyth and T T West
68 Evolution equations and their applications
F Kappel and W Schappacher
69 Generalized solutions of Hamilton-Jacobi equations
P L Lions
70 Nonlinear partial differential equations and their applications: Collège de France Seminar. Volume III
H Brezis and J L Lions
71 Spectral theory and wave operators for the Schrödinger equation
A M Berthier
72 Approximation of Hilbert space operators I
D A Herrero
73 Vector valued Nevanlinna Theory
H J W Ziegler
74 Instability, nonexistence and weighted energy methods in fluid dynamics and related theories
B Straughan
75 Local bifurcation and symmetry
A Vanderbauwhede
76 Clifford analysis
F Brackx, R Delanghe and F Sommen
77 Nonlinear equivalence, reduction of PDEs to ODEs and fast convergent numerical methods
E E Rosinger
78 Free boundary problems, theory and applications. Volume I
A Fasano and M Primicerio
79 Free boundary problems, theory and applications. Volume II
A Fasano and M Primicerio
80 Symplectic geometry
A Crumeyrolle and J Grifone
81 An algorithmic analysis of a communication model with retransmission of flawed messages
D M Lucantoni
82 Geometric games and their applications
W H Ruckle
83 Additive groups of rings
S Feigelstock
84 Nonlinear partial differential equations and their applications: Collège de France Seminar. Volume IV
H Brezis and J L Lions
85 Multiplicative functionals on topological algebras
T Husain
86 Hamilton-Jacobi equations in Hilbert spaces
V Barbu and G Da Prato
87 Harmonic maps with symmetry, harmonic morphisms and deformations of metrics
P Baird
88 Similarity solutions of nonlinear partial differential equations
L Dresner
89 Contributions to nonlinear partial differential equations
C Bardos, A Damlamian, J I Díaz and J Hernández
90 Banach and Hilbert spaces of vector-valued functions
J Burbea and P Masani
91 Control and observation of neutral systems
D Salamon
92 Banach bundles, Banach modules and automorphisms of C*-algebras
M J Dupré and R M Gillette
93 Nonlinear partial differential equations and their applications: Collège de France Seminar. Volume V
H Brezis and J L Lions
94 Computer algebra in applied mathematics: an introduction to MACSYMA
R H Rand
95 Advances in nonlinear waves. Volume I
L Debnath
96 FC-groups
M J Tomkinson
97 Topics in relaxation and ellipsoidal methods
M Akgül
98 Analogue of the group algebra for topological semigroups
H Dzinotyiweyi
99 Stochastic functional differential equations
S E A Mohammed
100 Optimal control of variational inequalities
V Barbu
101 Partial differential equations and dynamical systems
W E Fitzgibbon III
102 Approximation of Hilbert space operators. Volume II
C Apostol, L A Fialkow, D A Herrero and D Voiculescu
103 Nondiscrete induction and iterative processes
V Ptak and F-A Potra

104 Analytic functions – growth aspects
O P Juneja and G P Kapoor
105 Theory of Tikhonov regularization for Fredholm equations of the first kind
C W Groetsch
106 Nonlinear partial differential equations and free boundaries. Volume I
J I Díaz
107 Tight and taut immersions of manifolds
T E Cecil and P J Ryan
108 A layering method for viscous, incompressible L_p flows occupying R^n
A Douglis and E B Fabes
109 Nonlinear partial differential equations and their applications: Collège de France Seminar. Volume VI
H Brezis and J L Lions
110 Finite generalized quadrangles
S E Payne and J A Thas
111 Advances in nonlinear waves. Volume II
L Debnath
112 Topics in several complex variables
E Ramírez de Arellano and D Sundararaman
113 Differential equations, flow invariance and applications
N H Pavel
114 Geometrical combinatorics
F C Holroyd and R J Wilson
115 Generators of strongly continuous semigroups
J A van Casteren
116 Growth of algebras and Gelfand–Kirillov dimension
G R Krause and T H Lenagan
117 Theory of bases and cones
P K Kamthan and M Gupta
118 Linear groups and permutations
A R Camina and E A Whelan
119 General Wiener–Hopf factorization methods
F-O Speck
120 Free boundary problems: applications and theory, Volume III
A Bossavit, A Damlamian and M Fremond
121 Free boundary problems: applications and theory, Volume IV
A Bossavit, A Damlamian and M Fremond
122 Nonlinear partial differential equations and their applications: Collège de France Seminar. Volume VII
H Brezis and J. L. Lions
123 Geometric methods in operator algebras
H Araki and E G Effros
124 Infinite dimensional analysis–stochastic processes
S Albeverio
125 Ennio de Giorgi Colloquium
P Krée
126 Almost-periodic functions in abstract spaces
S Zaidman
127 Nonlinear variational problems
A Marino, L Modica, S Spagnolo and M Degiovanni
128 Second-order systems of partial differential equations in the plane
L K Hua, W Lin and C-Q Wu
129 Asymptotics of high-order ordinary differential equations
R B Paris and A D Wood
130 Stochastic differential equations
R Wu
131 Differential geometry
L A Cordero
132 Nonlinear differential equations
J K Hale and P Martinez-Amores
133 Approximation theory and applications
S P Singh
134 Near-rings and their links with groups
J D P Meldrum
135 Estimating eigenvalues with *a posteriori/a priori* inequalities
J R Kuttler and V G Sigillito
136 Regular semigroups as extensions
F J Pastijn and M Petrich
137 Representations of rank one Lie groups
D H Collingwood
138 Fractional calculus
G F Roach and A C McBride
139 Hamilton's principle in continuum mechanics
A Bedford
140 Numerical analysis
D F Griffiths and D A Watson

J I Díaz

Universidad Complutense, Madrid

Nonlinear partial differential equations and free boundaries

VOLUME I

Elliptic equations

π

Pitman Advanced Publishing Program

BOSTON · LONDON · MELBOURNE

PITMAN PUBLISHING LIMITED
128 Long Acre, London WC2E 9AN

A Longman Group Company

Jan 1986
Phys. Sci.

First published 1985

AMS Subject Classifications: (main) 35BXX, 35KXX, 35JXX
(subsidiary) 35R35, 35Q20, 47H15

ISSN 0743-0337

Library of Congress Cataloging in Publication Data

Díaz, J. I.
Nonlinear partial differential equations and free boundaries.

Bibliography: v. 1, p.
Includes index.
Contents: v. 1. Elliptic equations.
1. Differential equations, Partial. 2. Boundary value problems. I. Title.
QA377.D59 1985 515.3′53 84-28940
ISBN 0-273-08572-7

British Library Cataloguing in Publication Data

Díaz, J. I.
Nonlinear partial differential equations and free boundaries.—(Research notes in mathematics, ISSN 0743-0337; 106)
Vol. 1
1. Differential equations, Partial
2. Differential equations, Nonlinear
I. Title II. Series
515.3′53 QA374

ISBN 0-273-08572-7

Reproduced and printed by photolithography
in Great Britain by Biddles Ltd, Guildford

Contents

to Cuayi,

Vicente and Amanda

Preface

This work is an attempt to unify various known aspects and to present recent results on nonlinear partial differential equations giving rise to a free boundary, mainly defined by the boundary of the region where the solution vanishes identically. The material is organized in two volumes; the first is devoted to elliptic equations and a second, in preparation, will deal with the study of parabolic and hyperbolic equations.

I would like to express my sincere thanks to Haim Brezis, who encouraged me to write this book. Throughout the past ten years he and Philippe Benilan have given me the continuous present of their profound advice and friendship.

I don't forget the valuable help, during the preparation of this volume, of the many friends who provided material, read parts of the manuscript, made suggestions and also gave me their moral support and encouragement. In particular, I would like to mention C. Bandle, F. Bernis, L. Boccardo, C.M. Brauner, J. Carrillo, A. Damlamian, G. Diaz, J. Hernandez, M.A. Herrero, R. Jimenez, S. Kamin, B. Kawohl, R.Kersner, P.L. Lions, J.M. Morel, L.A. Peletier, M. Pierre, J.F. Rodrigues, I. Stakgold, J.L. Vazquez and L. Veron, among other friends. Also my deepest thanks go to my friend and colleague Alfonso Casal, who helped to improve my use of English. He and my brother Gregorio shared very closely the conception and evolution of the manuscript. My thanks also go to Josesteban Prieto who made a very good job of the figures, and to Pilar Aparicio for her patience, skill and effieiency in typing the original.

Finally, I would like to reflect here my sincere thanks to the organizers of Pitman Advanced Publishing Program, for their cooperation and understanding during the anxious and lengthy period of gestation of this volume.

Madrid
July, 1985

J.I. Díaz

Introduction

This work is devoted to the study of some partial differential equations whose nonlinear character gives rise to a free boundary: the boundary of an a priori unknown and positively measured region where the solution of the equation vanishes identically.

The formation of this free boundary holds under some adequate balance between two of the terms of the equation representing the different peculiarities of the phenomenon under consideration: diffusion, absorption, convection, evolution, etc. These particular balances take place neither in the case of linear equations nor in every nonlinear equation. The characterization of these special balances for several classes of nonlinear equations, important in the applications, is one of the main goals of the work. This first volume is dedicated to elliptic equations; a second volume dealing with parabolic and hyperbolic equations is now in preparation.

The present work is an attempt to unify various known aspects of a nonlinear PDE giving rise to such a free boundary and also to present recent results and methods.

Physical motivation

Although in this work the treatment will be of a mathematical character it seems interesting to mention, at least in this introduction, the physical motivation of the equations dealt with. The three following stationary problems form, perhaps, the main context in which the nonlinear equations in consideration appear.

(A) Reaction-diffusion problems. Consider, for instance, a single. irreversible, steady-state reaction taking place in a bounded domain, Ω in $\mathbb{R}^N$. The reactant being consumed in Ω is replaced through diffusion from the ambient region so that a steady state is possible. Problems of this type are discussed in detail in Aris [1]. A nonlinear system is obtained for the density u and the temperature T of the reactant. Upon eliminating T the system can be reduced to a scalar problem for the concentration

$$-\Delta u + \lambda f(u) = 0 \quad \text{in} \quad \Omega \tag{1}$$

$$u = 1 \quad \text{on} \quad \partial\Omega \tag{2}$$

where Δ is the Laplacian operator, λ is a positive constant (the Thiele modulus) and $f(u)$ is the ratio of the reaction rate at concentration u to the reaction rate at concentration unity. Clearly, the concentration u must be nonnegative. The given function $f(u)$ is defined for $u \geqslant 0$, is nonnegative and satisfies $f(0) = 0$ and $f(1) = 1$. Moreover, if the reaction is isothermal or endothermic, f turns out to be monotone increasing, which is not usually true for exothermic reactions. In any case, f may fail to be differentiable or even continuous at the origin. This is illustrated for isothermal reactions of the form $f(u) = u^q$, where q is called the <u>order of the reaction</u>. If $0 < q < 1$, f is continuous but not differentiable at the origin, whereas if $q = 0$, f is actually discontinuous at the origin (recall that $f(0)$ must be zero).

It turns out that the density af the reactant u may be zero in a closed interior region Ω_0 called a <u>dead core</u>. In such a set, no reaction takes place so that Ω_0 is wasted. For instance, in a catalyst pellet, one could just as well to dispense with the region Ω_0 and save that amount of catalyst. Such a dead core can only occur if the reaction rate remains high as the concentration decreases, for it may be then impossible for diffusion to draw reactant sufficiently fast from the exterior of Ω to reach the central part of Ω_0. One of the most important goals of this volume is to find out when a set where the solution vanishes exists, to give estimates of its size and location and to study its geometry. If, for instance, $f(u) = u^q$, $q > 0$, a particularization of the results in Chapter 1 shows that a dead core may only exist if and only if $0 < q < 1$ and λ is large enough. This also shows how for the equation (1) the existence of the free boundary given by $\partial(\Omega - \Omega_0) - \partial\Omega$ is ambiguous, and it takes place only for some adequate nonlinear terms f. (A complete characterization will be given in Chapter 1).

(B) <u>Non-Newtonian fluids</u>. The second order nonlinear elliptic equation

$$-\operatorname{div}(|\nabla u|^{p-2}\nabla u) + \lambda u = 0 \qquad p > 1, \ \lambda > 0, \tag{3}$$

where ∇u denotes the gradient of u and $|\nabla u|$ is the Euclidean norm in $\mathbb{R}^N$ of the vector ∇u, appears in the study of non-Newtonian fluids. Indeed, when studying the laws of motion of fluid media, Newtonian fluids are usually considered to be those for which the relation between the shear stress τ and the velocity gradient $\frac{du}{dx}$ (for simplicity we shall here restrict ourselves to the plane case) takes the form

$$\tau = \mu \frac{du}{dx} . \tag{4}$$

However, this approximation is satisfactory only for a limited number of actual fluid media. Dispersive media treated according to a continuum model do not obey the law given by (4). The motions of such non-Newtonian fluids are studied in rheology (see e.g. Astarita-Marruci [1]). Usually (4) is substituted by the power rheological law

$$\tau = \mu \left| \frac{du}{dx} \right|^{p-2} \frac{du}{dx} \quad , \quad p > 1 . \tag{5}$$

The quantities μ and p are the rheological characteristics of the medium. Media with $p > 2$ are called <u>dilatant fluids</u>, and those with $p < 2$ are called <u>pseudoplastics</u>. When $p = 2$ they are Newtonian fluids. The study of the non-Newtonian flow properties of such media having conductivity in electromagnetic fields leads to equations similar to (3). Consider, for instance, that the conducting fluid moves in a flat channel, $x = \pm L$ whose non-conducting walls move along the x axis with a velocity $\pm u_0$, (magnetohydrodynamic Couette flow). There is no pressure gradient, no electric field and the external magnetic field of induction is perpendicular to the walls. By normalizing, we arrive at the problem

$$-\frac{d}{dx}\left(\left|\frac{du}{dx}\right|^{p-2} \frac{du}{dx} \right) + \lambda u = 0 \quad , \quad \text{in} \quad \Omega = (-1,1) \tag{6}$$

$$u(\pm 1) = 1, \tag{7}$$

where λ is a positive constant (the generalized Hartmann number). The physics of the problem shows (see L.K.Martinson-K.B.Pavlov [1]) that for dilatant fluids $(p > 2)$, and only for these fluids, and for λ large enough, flow zones appear in which the fluid moves at velocity, which vanish over the channel cross-section. Again, the regions $\Omega_0 \equiv \{u = 0\}$, now called quasi-solid zones, may appear, for adequate values of p, and so the existence of the free boundary $\partial\Omega_0 - \partial\Omega$ is not automatically derived for any equation (6). The study of the free boundary in this particular situation is also contained in the general treatement made in Chapter 1.

(C) Nonlinear diffusion problems. The study of the steady-state of many different problems governed by a nonlinear diffusion in the presence of an absorption term leads to the equation

$$- \Delta\varphi(u) + f(u) = g(x) \quad \text{in} \quad \Omega \tag{8}$$

$$u = h(x) \quad \text{on} \quad \partial\Omega \tag{9}$$

where Ω is an open bounded set (eventually unbounded), φ and f are real continuous nondecreasing functions such that $\varphi(0) = f(0) = 0$ and g and h are given functions. Equation (8) is sometimes written in the so called Fickian diffusion form

$$- \operatorname{div}(k(u)\nabla u) + f(u) = g(x)$$

where function k is any primitive of φ. Some typical choices of φ in the applications are the following:

(C_1) Flows through porous media (slow diffusion problems). Via Darcy's law equation (8) holds for φ satisfying the additional assumptions $\varphi'(0) = 0$ and $\varphi'(u) > 0$ if $u \neq 0$ as, for instance, $\varphi(u) = |u|^{m-1}u$, $m > 1$ (see Bear [1]). This same type of φ also occurs for nonlinear heat conduction when the thermal conductivity depends on temperature (see Zeldovich-Raizer [1]), in the spread of certain biological populations (Gurtin-MacCamy [1] and Okubo [1]), in the spread of a thin drop of viscous fluid over a horizontal plane under gravity (Lacey-Ockendon-Taylor [1]), in solar prominences (Ames [1]) and in galactic civilizations (Newman-Sagan [1]).

(C_2) Plasma physics (fast diffusion problems). Certain mathematical models of a thermal evolution of a heated plasma lead to nonlinear operators as in (8). (See Berryman-Holland [1] or Kamin-Rosenau [1]). Now the natural assumptions on φ are $\varphi'(0) = +\infty$ and $\varphi'(u) > 0$ if $u \neq 0$ as for instance, $\varphi(u) = |u|^{m-1}u$ with $0 < m < 1$.

(C_3) Stefan-like problems. Equation(8) is also related to the classic as well as generalizations of the two phase Stefan problem (see references in the survey Niezgodgka [1]). Mainly, φ is taken such that $\varphi([0,a])=\{0\}$ and $\varphi'(u)>0$ for $u \notin (0,a)$ for some $a>0$.

As in the preceding examples, in the third context the existence of a subset Ω_0 where u vanishes has a physical meaning and it depends on the behaviour of the nonlinear terms φ and f at the origin.

A short summary

In spite of the general title of this work, by now the reader should be aware that we shall not study different types of free boundaries here, but only the one generated by the boundary of a subset where the solution of a nonlinear PDE vanishes.(dead core, quasi-solid zone, etc.). To be more precise, given any general function u defined on $\overline{\Omega}$, we introduce the notation

$$N(u) \equiv \text{null set of } u \equiv \{x \in \overline{\Omega} : u = 0\}$$

$$S(u) \equiv \text{support of } u \equiv \overline{\{x \in \Omega : u \neq 0\}}.$$

The free boundary under study is given by

$$\mathcal{F}(u) \equiv \text{free boundary} \equiv \partial S(u) \cap \partial N(u).$$

when u represents the solution of the nonlinear PDE in consideration. (Exceptionally, in some examples the critical value $u=0$, defining the free boundary, can be replaced by another constant value $u=c$, $c \neq 0$. See Theorem 1.14 in Chapter 1).

Although different types of nonlinear PDE's will be considered, for the sake of clarity in the exposition, a special emphasis will be put on the Dirichlet problem associated to a concrete quasilinear elliptic second order equation:

$$-\Delta_p u + f(u) = g(x) \quad \text{in} \quad \Omega \tag{QE}$$

$$u = h \quad \text{on} \quad \partial\Omega \ , \tag{DC}$$

where the operator Δ_p denotes the pseudo-Laplacian operator defined, for $p > 1$, by

$$\Delta_p u = \operatorname{div}(|\nabla u|^{p-2} \nabla u) = \sum_{i=1}^{N} \frac{\partial}{\partial x_i} (|\nabla u|^{p-2} \frac{\partial u}{\partial x_i})$$

Some remarks on the interest of the above formulation seem to be in order. First of all, we note that all the results for the (QE) equation are also true for the semilinear equation

$$- \Delta u + f(u) = g(x). \qquad \text{(SE)}$$

Indeed, when $p = 2$ the operator Δ_p coincides with the usual Laplacian operator. We also note that the nonlinear diffusion equation (8) can be reduced to the semilinear one (SE) when φ is strictly increasing (in that case u in (SE) must be substituted by $\tilde{u} = \varphi(u)$ and f by $\tilde{f} = f \circ \varphi^{-1}$). Although the semilinear equation (SE) is perhaps much more popular than the quasilinear one (QE), one important reason to choose (QE) as the general setting is that its generality is very useful in the applications. The operator Δ_p with $p \neq 2$ appears in many other contexts (besides that mentioned of non-Newtonian fluids). It is also used in some reaction-diffusion problems (see Aris [1] p.207) as well as in flow through porous media (for instance in flow through rock filled dams, Ahmed-Sunada [1] or Volker [1]). It also appears in nonlinear elasticity (e.g. Oden [1]), glaceology (Pelissier [1]),and petroleum extraction (Schoenaver [1]). Equation (QE) has also geometrical interest for $p \geqslant 2$ (see references in Uhlenbeck [1]).

We also remark that the term $g(x)$ is not only present in the equations for a systematic mathematical study but for its interest in particular formulations such as that associated with the Thomas-Fermi atomic model in which, in (SE), f is given by $f(u) = [u^+]^{3/2}$ (see Brezis [9]).

Before proceeding further it seems interesting to examine what may be the reasons for the existence of the free boundary $\mathcal{F}(u)$. Recall that in the case of linear equations the solution of an elliptic equation as (QE) with $p = 2$, $f(s) = s$, satisfying a Dirichlet condition (DC) corresponding to data, say $g \geqslant 0$ and $h \geqslant 0$, is such that $u > 0$ on Ω. This well-known fact can be proved in many different ways: strong maximum principle, Harnack inequality, unique continuation property, and so on. Thus, in some sense, the existence of the free boundary is a nonlinear typical phenomenon. However, it doesn't appear in every nonlinear equation (QE).

The analysis becomes clear if, for instance, we reformulate a semilinear equation (SE) as a nonlinear diffusion one. Assume, for instance, $f \in C^2$, $f' > 0$ and u satisfying (SE). Then $w = f(u)$ satisfies

$$- \Delta\varphi(w) + w = g$$

where $\varphi = f^{-1}$. The operator $-\Delta\varphi(w)$ has the peculiarity of being non-uniformly elliptic if $\varphi'(0) = 0$ (i.e. if the f of (SE) is such that $f'(0) = +\infty$), and degenerated in the sense that it loses its elliptic character around the set $\{w = 0\}$, where it is hyperbolic. Note that

$$\Delta\varphi(w) = \varphi'(w)\, \Delta w + \varphi''(w)\, |\nabla w|^2.$$

Analogously, the operator $\Delta_p u$ is not uniformly elliptic if $p > 2$, being degenerated around the set $\{\nabla u = 0\}$ (in particular in $\{u = 0\}$). Note that

$$\Delta_p u = |\nabla u|^{p-2}\, \Delta u + \sum_{i=1}^{N} \frac{\partial u}{\partial x_i} \frac{\partial}{\partial x_i} (|\nabla u|^{p-2}).$$

It turns out that this degenerated character of the associated operators is a necessary condition for the existence of $\mathcal{F}(u)$ (recall, for instance, that the strong maximum principle also holds for uniformly elliptic operators with Lipschitz perturbations; see Protter-Weinberger[1]). Nevertheless, this reason is not enough, and the existence of $\mathcal{F}(u)$ is characterized by the fulfilment of the two following conditions:

(a) Balance between the diffusion and absorption terms:
for the (QE), this is given for the convergence of the following improper integral

$$\int_{0^+} \frac{ds}{F(s)^{1/p}} < +\infty \quad , \qquad (H_1)$$

where $F(s) = \int_0^s f(t)dt$. Note that if in (QE) is $f(u) = |u|^{q-1}u$ with $q > 0$, then (H_1) holds if and only if $0 < q < p - 1$. In particular for (SE), p is equal to two, and it must be $0 < q < 1$, but in the case of a (QE) with linear perturbation $q = 1$, it holds if $p > 2$.

(b) Balance between the "sizes" of Ω and of the solution u:
this kind of condition (which already appears in many other free boundaries problems), says, roughly speaking, that the measure of

the set $N(g) \cup N(h|_{\partial\Omega})$ (respectively $\|u\|_{L^\infty}$) must be large enough (respec. small enough). If, for instance, (QE) contains $f(u)=|u|^{q-1}u$, $q > 0$, it is enough that the radius ρ of the largest ball contained in the set $N(g) \cup N(h|_{\partial\Omega})$ be such that

$$\rho \geqslant \left(\frac{K}{\|u\|_{L^\infty}} \right)^{\frac{p-1}{2}} \tag{H_2}$$

where K is a positive constant only dependent on the dimension of the space N and, perhaps, on the parameter λ when it is present in the equation as in (1) or (6). (Here the notation $N(g)$, $N(h|_{\partial\Omega})$ represents the null set of these functions). In fact an estimate of $\|u\|_{L^\infty}$ is only needed on compact subsets of $N(g)$, i.e. where the equation becomes homogeneous, which allows a large generality in the results with respect to g and h. Note that (H_2) is trivially satisfied if Ω is unbounded and the support of g and h are compact subsets of $\overline{\Omega}$.

The optimality of both conditions will be shown in Chapter 1 when the perturbation term f is assumed nondecreasing; Nevertheless,similar results hold in other circumstances which we shall consider in Chapter 2 and 3:

(a) Equation with a non-monotone perturbation term. Of particular interest in some exothermic reactions (Aris [1]) and population dynamics (Okubo [1]). In most of cases, f satisfies $f(u)u \geqslant 0$.

(b) The obstacle problem. In its strong or complementary formulation it can be stated as the problem of finding u which satisfies (DC) as well as the relations

$$u \geqslant \psi, \ -\Delta u \geqslant g \ \text{ and } \ (-\Delta u - g)(\psi - u) = 0 \ \text{ in } \ \Omega, \tag{OP}$$

where ψ is a function (called an obstacle) given a priori and the free boundary is now $F(\psi - u)$ (see, e.g. Duvaut-Lions [1], Friedman [3] and Kinderlehrer-Stampacchia[2]). This problem can be reformulated when $\psi \equiv 0$, as a multivalued equation

$$-\Delta u + \beta(u) \ni g \quad \text{in} \quad \Omega \tag{ME}$$

where β is an adequate maximal monotone graph of $\mathbb{R}^2$ (see Chapter 2 for definitions). The use of mutivalued equations (ME) is also of interest in the study of zero order reactions, as well as in nonlinear diffusion equations (8) in which φ is not strictly increasing.

(c) Singular equations. The study of some reaction-diffusion problems (the Langmuir-Hinshelwool chemical kinetics, Aris [1] p. 168) leads to formulations such as the following

$$- \Delta u + \lambda u^{-k} = 0 \quad \text{in} \quad \Omega$$

$$u = 1 \quad \text{on} \quad \partial\Omega$$

where, now, $\lambda > 0$ and $0 < k < 1$. Note that even in this case (H_1) holds.

In all the mentioned problems, the free boundary $\mathcal{F}(u)$ appears as an essential consequence of the diffusion-absorption balance.

This same balance can be also interpreted for other problems where $\mathcal{F}(u)$ may exist :

(d) Fully nonlinear equations of the form $F(x,u,Du,D^2u) = 0$, including, for instance, the Hamilton-Jacobi-Bellman equation (see Krilov [1] and Gilbarg-Trudinger [1]).

(e) Nonlinear systems in reaction-diffusion equations (see references in Remark 1.8 and Subsection 2.1b.

(f) Higher order quasilinear equations. Particular formulations occur in the theory of elastic bars and plates (see Langenbach [1]) and in optimal control problems (Berkovitz-Pollard [1] and Brunovsky-Mallet-Paret [1]).

To end the list of problems we shall consider, we also mention some nonlinear equations where the existence of the free boundary $\mathcal{F}(u)$ is derived from a different relation: the diffusion-convection balance (see Subsection 2.4a). We remark that an adequate convection-absorption may also be the reason for the existence of the free boundary $\mathcal{F}(u)$. Nevertheless, such a given balance is peculiar to some stationary first order equations, but of hyperbolic character. That's why its consideration will be postponed to the second volume of this work.

In the above equations the Dirichlet boundary conditions can be replaced by some others, even involving nonlinear terms. In some cases, the peculiar boundary conditions may be the cause of the formation of a free boundary

$F(u)$, now localized exclusively on $\partial\Omega$ (the Signorini problem, see Subsection 2.5b).

As we have already indicated, the study of the existence of the free boundary is, perhaps, the main goal of this book, in such a way that this volume has been organized according the two main general methods available to carry out such a study: the super and subsolutions method (Chapters 1 and 2) and the energy methods (Chapter 3). The first method is based on the construction of adequate super and subsolutions to which, jointly with the solution of the considered PDE, the comparison principle is applied. Due to the use of this principle (strongly related to the weak maximum principle) this method is specially useful in the study of second (and also first) order equations. Although the comparison principle implies automatically the uniqueness of the solution, by means of adequate arguments the method can be also applied to some nonmonotone problems and systems in which the uniqueness fails. The idea of using super and subsolutions to derive different conclusions for the solution is quite classical in the theory of PDE. Nevertheless, in order to prove the existence of the free boundary $F(u)$ for nonlinear PDE on bounded domains, such functions are taken locally as suitable interior (or boundary) barrier functions. Our treatment here develops and systematizes the author's memoir, Diaz [4] (see also Diaz-Hernandez [1]).

The energy methods rely on the idea of finding some ordinary differential inequalities satisfied by some energy functions involving the integral, over some subsets, of some differential expressions of the solutions suitably chosen. Such inequalities are obtained by means of different interpolation results. Energy methods are actually available for equations of any order, and even with no monotonicity assumptions on the nonlinear terms of the equation. For instance, they apply to the general second order quasilinear equation

$$- \operatorname{div} A(x,u,\nabla u) + B(x,u,\nabla u) = 0$$

When both methods can be applied to a particular class of equations, the energy methods are less general with respect to the nonlinearities (they must be majored and / or minored by power functions) and the estimates on the location of the free boundary are hardly explicit and less precise than the ones obtained by the super and subsolution method. A pioneering application of an energy method was given by Antoncev[2] when considering some second order parabolic equations. A more systematic treatment , including the consideration of second order elliptic or parabolic equations, was made in Diaz-Veron [2],[3].Finally,the general treatment of quasilinear elliptic equations of any order was given by Bernis [2],[3] by choosing very sharp energy domains and energy functions.

In some applications, it is desirable to avoid the existence of the free boundary $F(u)$. For instance, in chemical reactions, the existence of a dead core N(u) means that no reaction takes place there and the catalyst is wasted. The study of nonexistence of the free boundary (Section 1.2) is related to the optimality of the above mentioned balances (a) and (b), as well as to the positivity of solutions corresponding to nonnegative data g and h. When condition (H_1) fails, this result can be interpreted as a strong maximum principle in the sense of Hopf [1] and it is proved by means of the construction of adequate local subsolutions. The necessity of the balance (b) between the sizes of Ω and u , is proved via a "best" maximum principle in the sense of Payne [1] and it is based on sharp estimates of the gradient of u in terms of a function of u. This kind of gradient estimate is of great interest in the study of nonlinear equations giving rise to the free boundary $F(u)$.

In addition to the question of existence, other properties of the free boundary are also considered. For instance , with respect to its localization, several estimates are given (see subsection 1.1b and 1.1c). A curious property may occur when the behaviour of the data g and h is adequately flat near the boundary of its null set. If, for instance , we assume h = 0 on the boundary, in some cases the influence of the external perturbation g(x) is even strictly localized to its support and the support of the solution u coincides with that of the g. This property of nondiffusion of the support seems to be new in the present literature and can be considered as an elliptic version of the waiting time property , which is well-known for nonlinear parabolic equations.

Another property of the free boundary analyzed here is its geometry . This is done as an application of the theory of the symmetric rearrangement of a function in the sense of Schwarz (see Section 1.3). A special isoperimetric inequality is shown with respect to the free boundary : among all the domains with the same measure, the ball is the set for which the solution of the homogeneous equation and nonhomogeneous boundary condition, as e.g. (2), has a greater null set (dead core in terms of model (A)). This result is obtained here via the general inequaltiy

$$\int_{\Omega} f(u)dx \leq \int_{\Omega^*} f(v)dx,$$

where Ω^* is a ball of measure $|\Omega|$ and u and v are now solutions of (QE) on Ω and Ω^*, respectively, with homogeneous boundary conditions. This inequality is of a great interest, both from the mathematical point of view and from that of the applications. Indeed, it allows us to obtain a priori estimates on different norms of solutions and, in terms of chemical engineering literature, by means of the change u = 1 - U, it expresses that the effectiveness, defined by

$$e = \frac{1}{|\Omega|} \int_{\Omega} f(U)dx,$$

where U is a solution of (1), (2), is lowest for balls. Physically the effectiveness represents the ratio of the actual amount of reactant consumed per unit of time in Ω to the amount that would be consumed if the interior concentration were equal everywhere to the ambient concentration. Although such results are already well known in the literature (see references in Section 1.3) our formulation seems to be more general and the proofs are different.

For the sake of completeness we also have compiled some very sharp results on the free boundary, such as the Hausdorff measure estimates, the behaviour of the solution near the free boundary and so on. These results have their origins in the works of L.A. Caffarelli and H.W. Alt for variational inequalities and have been obtained recently by different authors (see references in Section 1.4) for the semilinear equation (1), (2).

The study of the existence and properties of the free boundary can be carried out independently of the general theory of nonlinear PDE which concerns the existence, uniqueness and regularity of the solution. Indeed, most of the results on the free boundary $F(u)$ are obtained for any solution u satisfying the equation in some weak sense independently of the way in which it may be obtained. Reciprocally, in general the existence of this free boundary bears no relation to any extra difficulty in proving the existence of weak solutions. Only the obtaining of the sharp regularity of weak solutions becomes much more difficult when the free boundary $F(u)$ exists. In that case some singularities on $F(u)$ can appear making it impossible, for instance, to obtain classical solutions even for very smooth data. Also sometimes the existence of $F(u)$ may cause the absence of uniqueness in some problems.

Due to this independence between the general theory and the study of $F(u)$ and in order not to distract the reader specially interested in the consideration of the free boundary, we have postponed the mention of the general theory to the last chapter, as a kind of long appendix. In this way, sometimes the reader is referred to this last chapter in order to find some more concrete details. Due to the vast literature on this general topic this last chapter is written in the form of a survey where only very few proofs are given but, in contrast, many references are indicated. It is clear tnat there we do not try to make a complete revision of the general theory. Our exposition is obviously motivated by the special nonlinear equations considered in the book, such as those already mentioned in this introduction. Some of the results in this chapter such as, e.g., the accretiveness of the considered operators, will be applied in the second volume when studying the associated evolution problems.

We conclude with a few words on the organization of the results and references within the text. We follow a number sequence for theorems, propositions and lemmas; only definitions and remarks are numbered separately, the mathematical expressions being numbered within each section. In general, the bibliographical sources of the results are given, with comments, at the ends of chapters. Open problems are also noted at the ends of some chapters.

NOTATION

Ω represents a proper open subset of $\mathbb{R}^N$ which is assumed of boundary $\partial\Omega$ smooth enough and (unless we indicate the contrary) bounded. Its measure is indicated by $|\Omega|$ or meas Ω and its characteristic function by $\mathbb{1}_\Omega$.

$B_R(x_0)$ denotes an open ball of radius R centered at x_0.

ω_N is the volume of the unit ball in $\mathbb{R}^N$, i.e. equal to $2\pi^{N/2}/N\Gamma(N/2)$.

The gradient of a function u is denoted by ∇u and, sometimes, also by $Du = (D_1u,\ldots,D_Nu)$, i.e. $D_iu = \partial u/\partial x_i$. The Hessian matrix of second derivatives is represented by $D^2u = [D_{ij}u]$, $D_{ij}u = \partial^2 u/\partial x_i \partial x_j$, $i,j = 1,2,\ldots,N$. Moreover, if u is assumed to be defined on $\bar{\Omega}$ we make systematic use of the sets

$$N(u) \equiv \text{null set of } u \equiv \{x \in \bar{\Omega} : u = 0\}$$

$$S(u) \equiv \text{support of } u \equiv \overline{\{x \in \Omega : u \neq 0\}}$$

$$\tilde{S}(u) \equiv \{x \in \bar{\Omega} : u(x) \neq 0\}.$$

Finally the *free boundary* under study is defined by

$$\mathcal{F}(u) \equiv \text{ free boundary } \equiv \partial S(u) \cap \partial N(u),$$

where u represents the solution of the nonlinear PDE in consideration. In the case of the obstacle problem (see Section 2.2) the free boundary is defined as the common boundary of the coincidence set $\{x \in \bar{\Omega} : u = \psi\}$ and the continuation set $\{x \in \bar{\Omega} : u > \psi\}$, where ψ represents here the obstacle.

In the context of the rearrangement results, $\mu(t)$, $\tilde{u}(s)$ and $u^*(x)$ represent the distribution function, the decreasing and the radially symmetric rearrangement of u, respectively, and Ω^* is a ball centered at the origin of the same measure as Ω. The order relation $\lesssim$ is also there introduced (see Subsection 1.3a).

The more common differential operators considered are the Laplace operator Δ as well as the pseudo-Laplacian operator Δ_p defined by

$$\Delta_p u = \operatorname{div}(|\nabla u|^{p-2}\nabla u) = \sum_{i=1}^{N} \frac{\partial}{\partial x_i}(|\nabla u|^{p-2}\frac{\partial u}{\partial x_i}),\ 1 < p < \infty .$$

In general, L represents a general second order elliptic operator (see (6.4) of Section 1.1). The symbol $\mathbb{L}$ represents the nonlinear ordinary differential operator defined in (3.1) of Section 1.1.

The main functional spaces used in the book are the usual Lebesgue and Sobolev spaces $L^p_{loc}(\Omega)$, $L^p(\Omega)$, $W^{m,p}(\Omega)$, $W^{m,p}_o(\Omega)$, $H^m(\Omega)$ and $H^m_o(\Omega)$ with $m \in \mathbb{R}$ and $1 < p < \infty$, as well as the spaces of test functions $\mathcal{D}(\Omega)$, the space of distributions $\mathcal{D}'(\Omega)$ and that of Hölder continuous functions $C^{\alpha}(\Omega)$, $0 < \alpha \leqq 1$. Occasionally, some other functional spaces are mentioned: the Orlicz and Orlicz-Sobolev spaces $L^a(\Omega)$ and $W^{1,a}(\Omega)$, respectively (see Subsection 4.1a); the space $V^{1,p,q}(\Omega)$ (see Subsection 4.1a); the space of bounded Radon measures $M(\Omega)$ (topological dual of the space $C^o(\bar{\Omega})$); and the Marcinkiewicz space (see Subsection 4.2a).

In general, the nonlinear perturbation of the PDE equation is represented by a real continuous function which is denoted by f. Associated to f we define $F(t) = \int_o^t f(s)ds$. Sometimes the perturbation term is a general maximal monotone graph β of $\mathbb{R}^2$ to which we associate its domain $D(\beta)$, range $R(\beta)$, sections $\beta^o, \beta^+, \beta^-$ as well as its "primitive" j (i.e., a convex l.s.c. and proper function of subdifferential $\partial j = \beta$). See definitions in Subsections 2.3a or 4.1b.

We also use very often the auxiliary functions ψ and ψ_μ as well as their inverses $\eta(r)$ and $\eta(r,\mu)$, introduced in Subsection 1.1a.

1 The free boundary in the Dirichlet problem for second order elliptic quasilinear equations

In this chapter, the existence and properties of the free boundary $F(u)$ given by the boundary of the support of the solutions of the Dirichlet problem for second order quasilinear equations is studied. The maximum and comparison principles are used in order to get sufficient and necessary conditions for the existence of this free boundary. In Section 1.1 some sufficient conditions are given via the construction of adequate local super and subsolutions, for which a systematic study of the case of symmetric solutions is previously done. As a by-product of the existence results, several estimates on the localization of $F(u)$ are given. In particular, it is shown that, under adequate assumptions, the support of the data is not diffused. Global super and subsolutions are also constructed.

The strict positivity of the solutions is considered in Section 1.2, showing in this way the optimality of the sufficient conditions. This is done in two different ways: studying for which nonlinearities the strong maximum principle holds, and by giving sharp estimates of the gradient of the solution u in terms of adequate functions of u, which shows the necessity of an adequate balance between the size of the domain and that of the solution.

Several applications of the symmetric rearrangement of a function are given in Section 1.3. As a consequence of the main inequality (Theorem 1.26) an isoperimetric inequality is obtained on the free boundary. As other applications, the symmetry of the solution and a general sufficient condition (without monotonicity) for the existence of $F(u)$ are also obtained.

The chapter ends with the consideration of the special case of semilinear equations. By means of a Harnack type inequality, it is possible to show two important conclusions: first, that near a point P where $u(P) > 0$, u is uniformly bounded away from 0, and second, that it is possible to control the growth of u away from $P \in \partial\{u > 0\}$. The optimal regularity of the solution can be obtained in this way. This program is also applied to the case of a semilinear equation with an x-dependent perturbation which appears in some biological problems. With some additional assumptions, it is proved that $\mathcal{F}(u)$ is Lipschitz and star-like about the origin, as well as the uniqueness of the solution. Other delicate results on the free boundary such as Hausdorff estimates, dependence with respect to a parameter, regularity and convexity are also collected .

1.1. ON THE EXISTENCE OF THE FREE BOUNDARY.

The main goal of this section is to give some sufficient conditions for the existence of the free boundary. For the sake of simplicity in the exposition, the results in this section will concern mainly the Dirichlet problem

$$-\Delta_p u + f(u) = g(x) \quad \text{in} \quad \Omega \tag{1}$$

$$u = h \quad \text{on} \quad \partial\Omega \tag{2}$$

where Ω is a regular open set (not necessarily bounded) of $\mathbb{R}^N$, $p > 1$ and the operator Δ_p represents the pseudo-Laplacian operator defined by

$$\Delta_p u = \operatorname{div}(|\nabla u|^{p-2}\,\nabla u) ,$$

where $|\xi| = (\xi_1^2 + \ldots + \xi_N^2)^{1/2}$ for every $\xi = (\xi_1, \ldots \xi_N)$. We remark that

$$\Delta_2 u = \operatorname{div}(\nabla u) = \sum_{i=1}^{N} \frac{\partial^2 u}{\partial x_i} = \Delta u.$$

Therefore, the equation (1) becomes semilinear if $p = 2$. Throughout this section we shall always assume the following condition on f:

$$f \text{ is a nondecreasing continuous real function such that } f(0) = 0 \tag{3}$$

(More general conditions will be considered in the next chapter).

As we said in the Introduction, we are interested in the study of the free boundary generated by the unkown boundary of the support of the solution of (1),(2). To be more precise, given any general function u defined on $\overline{\Omega}$, we introduce the notation

$$N(u) \equiv \text{null set of } u \equiv \{x \in \overline{\Omega} : u = 0\}$$

$$S(u) \equiv \text{support of } u \equiv \overline{\{x \in \Omega : u \neq 0\}}.$$

Now, if u is any solution of (1),(2), the associated free boundary is the a priori unknown subset of $\overline{\Omega}$ defined by

$$\mathcal{F}(u) \equiv \text{ free boundary } \equiv \partial S(u) \cap \partial N(u).$$

(Note that the sets $N(u)$ and $S(u)$ are well defined even for $u \in \mathcal{D}'(\Omega)$: Schwartz[1]).

A well-known fact of the general theory of nonlinear partial differential equations is that it is possible to associate different notions of solution to a problem such as (1),(2), according to the regularity of the data g and h as well as the boundedness or unboundedness of the domain Ω. A very natural setting corresponds to the case in which the solution u is sought in an energy space: in our case the Sobolev space $W^{1,p}(\Omega)$, defined as customary by

$$W^{1,p}(\Omega) = \{u \in L^p(\Omega) : \frac{\partial u}{\partial x_i} \in L^p(\Omega) , \quad i = 1...N \} .$$

In this case, the boundary condition (2) holds in the sense that $u - h \in W_0^{1,p}(\Omega)$ (the closure in $W^{1,p}(\Omega)$ of the set $C_0^\infty(\Omega)$ of the infinitely differentiable functions with compact support). With respect to g, it suffices for it to belong to the dual space $W^{-1,p'}(\Omega) = (W_0^{1,p}(\Omega))'$. Note that by the characterization of $W^{-1,p'}(\Omega)$ (see e.g. Adams[1]) it is enough that g may be written as

$$g = g_0 + \sum_{i=1}^{N} \frac{\partial g_i}{\partial x_i} \quad , \quad g_0, g_i \in L^{p'}(\Omega) \quad , \quad 1/p + 1/p' = 1.$$

Nevertheless, in many physical examples g does not belong to that dual space but it is merely in L^1, L^1_{loc} or g is even a measure. In spite of this generality , a study of the existence is possible, specially for the semilinear case, $p = 2$. A detailed exposition of all this very rich problem goes beyond the scope of this book. Nevertheless, in the last chapter, Chapter 4, we give a long survey of the general theory with many recent references.

The study we shall make of the free boundary has very few interconnections with the general theory of elliptic equations. Indeed, throughout this chapter we shall only use the two following properties of that general theory,

a) comparison principle : for $i = 1,2$, let u_i be any solution of (1), (2) corresponding to $g = g^i$ and $h = h^i$, $i = 1,2$, then if $g^1 \leq g^2$ in Ω and $h_1 \leq h_2$ on $\partial\Omega$ we have $u_1 \leq u_2$ in Ω.

b) local boundedness of the solution where the equation becomes homogeneous: for every compact subset $D \subset \mathrm{int}(N(g))$, any solution u of (1) (2) satisfies $u \in L^\infty(D)$.

We remark that a) implies trivially the uniqueness of the solution and that this holds under very general assumptions on g and h due to the monotonicity of f. For this reason and some regularity results, the property b) is also verified without any extra assumptions on g and h. In fact a stronger result holds: $u \in C^{1,\alpha}(D)$ for some $\alpha \in (0,1)$ and for any compact subset $D \subset \mathrm{int}(N(g))$. Here $C^{1,\alpha}$ denotes the usual space of functions with a Holder continuous gradient,

$$C^{1,\alpha}(D) = \{v \in C^1(D) : \exists\, C > 0 \; |\frac{\partial v}{\partial x_i}(x) - \frac{\partial v}{\partial x_i}(y)| \leq C|x-y|^\alpha \; \forall x,y \in D,\ \forall i = 1\ldots N\}$$

We also remark that under additional hypotheses on g and h it is possible to find global bounds on $\|u\|_{L^\infty(\Omega)}$. The easier case corresponds to when Ω is bounded, and $g,h \in L^\infty(\Omega)$. Then, it suffices to use a) and compare u with the constant $\pm M$, $M = f^{-1}(\|g\|_{L^\infty} + f(\|h\|_{L^\infty}))$. Other sharper results will be discussed later.

For the sake of completeness of the exposition we recall here two particular statements on a bounded domain Ω, and relative to the variational and L^1 frameworks.

Theorem 1.1. *Let* Ω *be a regular bounded open set*, $p > 1$, f *satisfying* (3) *and let* F *be the primitive of* f, $F(s) = \int_0^s f(t)dt$. *Consider* g *and* h *satisfying*

$$g \in W^{-1,p'}(\Omega)\ ,\quad i.e.,\ g = g_0 + \sum_{i=1}^{N} \frac{\partial g_i}{\partial x_i}\ ,\quad g_0, g_i \in L^{p'}(\Omega) \tag{4}$$

and

$$h \in W^{1,p}(\Omega) \quad \text{such that} \quad F(h(x)) \in L^1(\Omega). \tag{5}$$

Then there exists a unique $u \in W^{1,p}(\Omega)$ *verifying* (1),(2) *in the sense that* u *minimizes the functional*

$$J(u) = \int_\Omega [|\nabla u|^p + F(u) - g_0 u + \sum_{i=1}^{N} g_i \frac{\partial u}{\partial x_i}]dx$$

on the set $K = \{v : v-h \in W_0^{1,p}(\Omega)$ *such that* $F(v) \in L^1(\Omega)\}$. *If for* $i = 1,2,$ u_i *is the solution of* (1),(2) *corresponding to* $g = g^i$, $h = h^i$ *as above, then* $g^1 \leq g^2$ *in* Ω *and* $h^1 \leq h^2$ *on* $\partial\Omega$ *implies* $u_1 \leq u_2$ *in* Ω. *Finally,* $u \in C^{1,\alpha}(D)$ *for any compact set* $D \subset \mathrm{int}(N(g))$ *and* $\sup\{|u(x)| : x \in D\} \leq M$, *with*

$$M = C(||g||^{p'/p}_{W^{-1,p'}(\Omega)} + ||h||_{W^{1,p}(\Omega)} + ||F(h)||^{1/p}_{L^1(\Omega)}) \tag{6}$$

for some positive constant C, *depending only on* p, N, Ω *and* $d(D, \partial N(g))$. □

With respect to the L^1-framework, and for simplicity in the exposition we shall restrict ourselves to the semilinear equation

$$-\Delta u + f(u) = g(x) \quad \text{in} \quad \Omega, \tag{1*}$$

with the boundary condition (2). We have

Theorem 1.2. *Let* Ω *be a regular bounded open set and let* f *satisfy* (3). *Let* g *and* h *be such that*

$$g \in L^1(\Omega) \tag{4*}$$

$$h \in W^{1,1}(\Omega) \text{ (and so } h \in L^1(\partial\Omega)) \text{ such that } \Delta h \in L^1(\Omega). \tag{5*}$$

Then there exists a unique $u \in W^{1,1}(\Omega)$, *solution of* (1*),(2) *in the sense that the function* $w = u - h \in W_0^{1,1}(\Omega)$ *and satisfies* $f(w) \in L^1(\Omega)$ *and* $-\Delta w + f(w) = g + \Delta h$ *a.e. on* Ω. *If for* $i = 1,2$, u_i *is the solution of* (1*),(2) *corresponding to* $g = g_i$ *and* $h = h_i$ *as above then* $g_1 \leq g_2$ *in* Ω *and* $h_1 \leq h_2$ *on* $\partial\Omega$ *implies* $u_1 \leq u_2$ *in* Ω. *Finally,* $u \in C^{1,\alpha}(D)$ *for every compact set* $D \subset \mathrm{int}(N(g))$ *and* $\sup\{|u(x)| : x \in D\} \leq M$, *with*

$$M = C(||g||_{L^1(\Omega)} + ||h||_{L^1(\partial\Omega)}) \tag{6*}$$

for some positive constant C *only depending on* N, Ω *and* $d(D, \partial N(g))$. □

We emphasize that the above results are only particular statements: for instance, some results for the quasilinear equation (1) are also available in the L^1 setting (see Section 4.2) Except in the main result of this section (Theorem 1.9), in general, we shall state our theorems for the free boundary $F(u)$ when u is the solution of (1),(2) in the variational setting. Corresponding versions when u is the solution in the L^1-sense are left to the reader.

To obtain some qualitative information of the free boundary $F(u)$ we shall need in 1.1c. not only the local boundedness of u on int(N(g)) but on the whole domain Ω . Results of this nature are also presented in Chapter 4. They are based on additional hypotheses on g and h and/or on adequate growing conditions on f. A particular and easy statement, true for any f verifying (3), says that if

$$g \in L^s(\Omega) \quad \text{with} \quad s > N/p \quad \text{if} \quad 1 < p \leqslant N \quad \text{and} \quad s = 1 \quad \text{if} \quad p > N \tag{4**}$$

and

$$h \in W^{1,p}(\Omega) \cap L^\infty(\Omega) \tag{5**}$$

then for any $u \in W^{1,p}(\Omega)$ satisfying (1),(2) we have $u \in C^{1,\alpha}(D)$, for some $\alpha \in (0,1)$ and for any compact subset $D \subset \overline{\Omega}$. Moreover $\sup\{|u(x)| : x \in D\} \leqslant M$ with

$$M = f^{-1}(\|g\|_{L^\infty} + f(\|h\|_{L^\infty})) \text{ if } g \in L^\infty(\Omega) \text{ and } M = C(\|g\|_{L^s} + \|h\|_{L^\infty}) \tag{6**}$$

in general, for some positive constant C only depending on p and Ω .

The study of the free boundary we shall make is based on the preliminary systematic consideration of the particular case in which the solutions are radially symmetric. There the problem can be solved "almost" explicity and will provide useful comparison problems. This is given in subsection 1.1a where the one-dimensional case is also considered. Adequate radial solutions will be employed as local barrier functions in more general problems. The application of this local barrier as local super and subsolutions will be different according to whether they are applied around a point x_0 of the interior set of $N(g) \cup N(h|_{\partial\Omega})$ (<u>interior estimates</u>; subsection 1.1b) or in points x_0 of the boundary of $N(g) \cup N(h|_{\partial\Omega})$ (<u>boundary estimates</u>;subsection 1.1c)

Here we have used the notation introduced for the null set of a function. Finally, in subsection 1.1d we construct global super and subsolutions, useful to show the compactness of the support of the solution and of interest for many other purposes.

1.1a. One-dimensional and radially symmetric solutions.

The keystone in the study of the free boundary $\mathcal{F}(u)$ for the problem (1), (2), is the consideration of some simpler problems on balls, for which the symmetry of the solution takes the PDE into an ordinary differential equation.

The main results of this subsection concern the existence of the free boundary generated by the solution of

$$- \Delta_p u + f(u) = 0 \quad \text{in} \quad B_R(0) \tag{7}$$

$$u = k \quad \text{on} \quad \partial B_R(0) , \tag{8}$$

where k is a positive constant. Due to the uniqueness of the solution (see Theorem 1.1), it is easy to see that the solution u of (7),(8) must be radially symmetric (see, for instance, the subsection 1.3b). Then, writing $u = u(r)$, $r = |x|$, u must satisfy

$$- \frac{1}{r^{N-1}} \frac{d}{dr} \left(r^{N-1} \left|\frac{du}{dr}\right|^{p-2} \frac{du}{dr} \right) + f(u) = 0 \quad , \quad r \in (0,R) \tag{9}$$

or, equivalently,

$$-(|u'|^{p-2} u')' - \frac{(N-1)}{r} |u'|^{p-2} u' + f(u) = 0 \quad , \quad r \in (0,R) \tag{10}$$

where $u' = \frac{du}{dr}$. One of the main difficulties in the study of (9) comes from the fact that, if $N > 1$, the equation is not autonomous. That is why we shall start by studying the simpler case of the one-dimensional equation:

$$- a(u')' + f(u) = 0 \quad \text{in} \quad (-R,R) \tag{11}$$

$$u(\pm R) = k \tag{12}$$

It is clear that the one-dimensional version of (9) corresponds to the case of $a(s) = |s|^{p-2}s$. Here, we shall consider the general case in which a satisfies

$$a \in C^0(\mathbb{R}) \cap C^1(\mathbb{R} - \{0\}) \ , \ a'(s) > 0 \ \text{ if } s \neq 0 \text{ and } \ a(s) = -a(-s) \qquad (13).$$

We shall study the free boundary $F(u)$ for (11),(12) by giving necessary and sufficient conditions for the measure of the null set $N(u)$ to be positively measured. This leads in a natural way to the study of nontrivial solutions of the homogeneous Cauchy Problem

$$- a(u')' + f(u) = 0 \qquad (14)$$

$$u(0) = 0 \quad , \quad u'(0) = 0 \qquad (15)$$

By the results of the general theory (see Theorem 1.1 for $a(s) = |s|^{p-2} s$ and Chapter 4 in general) problem (11),(12) has a unique solution u. From (13) it is clear that $\tilde{u}(r) = u(-r)$ is also a solution. Then, u is an even function and $u'(0) = 0$. On the other hand, by the comparison principle, $u \geqslant 0$, and then, from the equation and (14), it follows that u is convex (strictly convex where $u > 0$). Then

$$u(0) = \min_{(-R,R)} u(r) \equiv m.$$

Thus, u can vanish in some region N of $(-R,R)$ if and only if $m = 0$. In this case there must exists an $\varepsilon \geqslant 0$ such that $N = [-\varepsilon,\varepsilon]$ where $u = 0$, $r\, u'(r) > 0$ on the set $(-R,R)-[\varepsilon,\varepsilon]$ and, hence, u must satisfy (14)(15).

The existence of a nontrivial solution of the Cauchy Problem (14),(15) is closely related to the question of the uniqueness of solutions. So, it is clear that $u \equiv 0$ is always a solution. Moreover if we assume for instance $a(s) = s$ and f Lipschitz continuous, it is the only solution, as we deduce from the standard uniqueness theorems for ordinary differential equations. Nevertheless, we shall prove that under suitable hypotheses on a and f we can obtain non-trivial solutions of (14)(15). First for $r \in \mathbb{R}$ we introduce the auxiliary functions

$$A(r) = \int_0^r a'(s) s \, ds \ , \qquad (16)$$

and

$$F(r) = \int_0^r f(s) \, ds \ . \qquad (17)$$

Our main hypothesis will be expressed by assuming the singular function $1/A^{-1}(F(r))$ integrable in $r = 0^+$. This also can be formulated in the following way: for $\tau > 0$ define the function $\psi : [0,\infty) \to [0,+\infty]$ by

$$\psi(\tau) = \int_0^{\tau} \frac{ds}{A^{-1}(F(s))} \tag{18}$$

Due to the monotonicity of A^{-1} and F, the hypothesis of $1/A^{-1}(F(r))$ being integrable in $r = 0^+$ i.e.

$$\psi(0^+) < +\infty \tag{19}$$

is equivalent to the condition

$$\psi(\tau) < +\infty \quad \text{for every} \quad 0 \leqslant \tau \leqslant \tau_\infty , \tag{20}$$

where $\tau_\infty \leqslant \infty$ is such that $\psi(\tau_\infty)=+\infty$. (By convention $\psi(+\infty) = +\infty$ if $\tau_\infty < +\infty$).

Remark 1.1. Condition (19) (or 20)) will play an important role in this chapter. In order to make clear the scope of such a condition we shall particularize a and f by taking

$$a(s) = |s|^{p-2} s \ , \ p > 1 \ , \quad f(s) = \lambda |s|^{q-1} s \ , \ q > 0, \ \lambda > 0 \tag{21}$$

which in the following will be called the case of homogeneous nonlinear terms. Some easy computations show that in this case, if $r > 0$, we have

$$A(r) = \frac{(p-1)}{p} r^p \quad , \quad F(r) = \frac{\lambda}{q+1} r^q \ , \tag{22}$$

and

$$\psi(0^+) < +\infty \ \text{if and only if} \ q < p-1 . \tag{23}$$

In addition, if $q < p-1$

$$\psi(r) = K r^{\frac{p-q-1}{p}} \quad , \quad K = \frac{p}{p-q-1} \left(\frac{(q+1)p}{(p-1)\lambda} \right)^{\frac{1}{p}} . \ \square \tag{24}$$

Now, we return to the Cauchy Problem.

Lemma 1.3. *Assume* (3) , (13) *and* (19). *Let* ψ *be given by* (18) *and define* $\eta = \psi^{-1}$ *on the interval* $[0,\psi(\infty))$. *Then, for every* $\tau \geq 0$ *, the function*

$$u_\tau(x) = \eta((x-\tau)^+) = \begin{cases} 0 & \text{if } 0 \leq x \leq \tau \\ \eta(x-\tau) & \text{if } \tau \leq x < \psi(\infty) + \tau \end{cases} \tag{25}$$

is a weak solution of the Cauchy Problem (14) (15).

Proof. We multiply the equation (14) by u' and integrate. Then

$$\int_0^x a(u'(s))'u'(s)ds = \int_0^x f(u(s))u'(s)ds.$$

Using the definitions of A and F we obtain

$$\int_0^x A(u'(s))'ds = \int_0^x F(u(s))'ds$$

and so

$$u'(x) = A^{-1}(F(u(x))). \tag{26}$$

The function $\eta = \psi^{-1}$ is defined on $[0,\psi(\infty))$, $\eta \in C^2((0,\psi(\infty)) \cap C^1([0,\psi(\infty))$ $\eta(0) = \eta'(0) = 0$ and, after some easy computations, we see that $u(x) = \eta(x)$ satisfies the equation (26). Finally, given $\tau > 0$ the function $u_\tau(x) = \eta((x-\tau)^+)$ is such that $u_\tau(0) = u_\tau'(0) = 0$, and satisfies $u_\tau \in C^2([0,\tau) \cap (\tau, \psi(\infty)+\tau)) \cap C^1[0,\psi(\infty)+\tau)$. Then, by the autonomous character of the equation, u_τ is also solution of (14),(15) on the intervals $(0,\tau)$ and $(\tau, \psi(\infty)+\tau)$. □

Let's go back to the one-dimensional boundary problem (11),(12). The following result gives a necessary and sufficient condition for the existence of the null set $N(u)$.

Theorem 1.4. *Let us assume* (3) *and* (13). *Moreover, if* ψ *is defined by* (18) , *we suppose that*

$$\psi(0+) < \infty \tag{27}$$

and

$$\psi(k) \leq R. \tag{28}$$

Then the solution of (11),(12) *is given by*

$$u(x) = \eta([|x| - R + \psi(k)]^+) = \begin{cases} 0 & \text{if } |x| \leq R - \psi(k) \\ \eta(|x| - R + \psi(k)) & \text{if } |x| > R - \psi(k), \end{cases} \quad (29)$$

where $\eta = \psi^{-1}$. *Finally, if* $R < \psi(k)$ *then* $u > 0$ *on* $[-R,R]$.

Proof. If $R \geq \psi(k)$ and we take $\tau = R - \psi(k)$, the function $u(x)$ is such that $u(x) = u_\tau(x)$ if $0 \leq x \leq R$, $u(x) = u_\tau(-x)$ if $-R \leq x \leq 0$, and satisfies the equation (11) in every point of $(-R,R)$ except, perhaps, in the points $\pm\tau$. Moreover,

$$u(\pm R) = \eta(\psi(k)) = k.$$

and, by uniqueness, it is the unique solution of (11),(12). To prove the second part of the statement, let u be the solution of (11),(12). Then $u'(0) = 0$ and $u(0) = m \geq 0$. As in the proof of Lemma 1.3 we have

$$\int_m^x A(u'(s))'ds = \int_m^x F(u(s))ds = \int_m^x F(u(s)) - F(m))'ds.$$

Therefore, if $x \in [0,R]$ we have $A(u') = F(u) - F(m)$ and so

$$x = \int_m^{u(x)} \frac{ds}{A^{-1}(F(s) - F(m))} .$$

In particular, if we define the function $\theta : \mathbb{R}^+ \times \mathbb{R}^+ \to \mathbb{R}^+$ by

$$\theta(r_1,r_2) = \int_{r_1}^{r_2} \frac{ds}{A^{-1}(F(s) - F(m))}$$

we know that $R = \theta(m,k)$. But $\theta(m,k) > \theta(M + \varepsilon,k)$ for every $\varepsilon > 0$. Indeed, by using a trivial change of variable and the strict monotonicity of A^{-1} we have

$$\theta(m,k) = \int_{m+\varepsilon}^{k+\varepsilon} \frac{ds}{A^{-1}(\int_{m+\varepsilon}^{s} f(t-\varepsilon)dt)} \geq \int_{m+\varepsilon}^{k+\varepsilon} \frac{ds}{A^{-1}(\int_{m+\varepsilon}^{s} F(t)dt)} = \theta(m+\varepsilon,k).$$

Then for $k > 0$ fixed, the function $\theta(\cdot,k)$ is a strictly decreasing isomorphism from $[0,k]$ into $[0, \psi(k)]$ (remark that $\psi(k) = \theta(0,k)$). Therefore, if $R < \psi(k)$, for k fixed there is a unique $m > 0$ satisfying $\theta(m,k) = R$ and so $u(x) \geq m > 0$ for every $x \in [-R,R]$. □

The above one-dimensional result already points out the nature of the two sufficient conditions for the existence of the free boundary:

a) a balance between the diffusion and absortion term, condition $\psi(0+) < +\infty$ i.e, $q < p - 1$ in the case of homogeneous nonlinearities (21), and

b) a balance between the sizes of the domain and of the solution, expressed here by $\psi(k) \leqslant R$, i.e.

$$R \geqslant \left(\frac{(p-1)(q+1)}{\lambda p} \right)^{1/p} \left(\frac{p}{p-1-q} \right) k^{\frac{p-1-q}{p}} \tag{30}$$

in the case of homogeneous nonlinearities.

Now we return to the radially symmetric boundary problem (7),(8). As in the one-dimensional case, the existence of the free boundary $\mathcal{F}(u)$, that is, the existence of nonempty null set $N(u)$, is related to the existence of nontrivial solutions of the Cauchy problem

$$\mathbb{L}(u) := -\frac{1}{r^{N-1}} \frac{d}{dr} \left[\left(r^{N-1} \left| \frac{du}{dr} \right|^{p-2} \frac{du}{dr} \right) \right] + f(u) = 0 \tag{31}$$

$$u(0) = 0 \quad , \quad u'(0) = 0 \tag{32}$$

(Again, note that if $N(u) \neq \emptyset$ then the solution of (7) satisfies (32)). To state the results, we need to modify slightly the definition of function ψ in (18) : given $\mu > 0$, we introduce the function $\psi_\mu : [0,\infty) \to [0,\infty]$ by

$$\psi_\mu(\tau) = \int_0^\tau \frac{ds}{A^{-1}(\mu F(s))} = \left(\frac{p-1}{p\mu} \right)^{1/p} \int_0^\tau \frac{ds}{F(s)^{1/p}} \tag{33}$$

(note that, now, $a(r) = |r|^{p-2} r$ and then A is given by (22)).

The main result of this subsection is the following

Theorem 1.5. *Assume* $N \geqslant 2$, $p > 1$, f *satisfying* (3) *and*

$$\int_{0^+} \frac{ds}{F(s)^{1/p}} < +\infty \tag{34}$$

For every $\mu > 0$ *let* $\eta(r, \mu) : [0,\psi_\mu(+\infty)) \to [0,+\infty)$ *be the function defined by* $\eta(\tau,\mu) = \psi_\mu^{-1}(\tau)$, *i.e.*

$$\left(\frac{p\mu}{p-1} \right)^{1/p} \tau = \int_0^{\eta(\tau,\mu)} \frac{ds}{F(s)^{1/p}} \tag{35}$$

then we have:

i) *If* $\mu \geq 1$ *then* $\mathbb{L}(\eta(r,\mu)) < 0$ *for* $r > 0$

ii) *If* $0 < \mu < 1/N$ *then* $\mathbb{L}(\eta(r,\mu)) > 0$, *and* $\mathbb{L}(\eta(r:1/N)) \geq 0$, *for* $r > 0$

iii) *For every* $\tau > 0$ *the function* $\eta_\tau(r) = \eta([r-\tau]^+,\mu)$ *satisfies* $\mathbb{L}(\eta_\tau(r)) < 0$ *(resp.* $\mathbb{L}(\eta_\tau(r)) > 0$*) if* $\mu \geq 1$ *(resp.* $\mu < 1/N$*),* *for* $r > \tau$ *and* $\mathbb{L}(\eta_\tau(r)) = 0$ *for* $0 < r < \tau$.

Proof. Let $\eta(r) = \eta(r,\mu)$ defined by (35). We have that $\eta \in C^2$, $\eta(0)=0$ and $\eta(r) > 0$ if $r > 0$. Differentiating in (35) we find that

$$\eta'(r) = \left(\frac{p\mu}{p-1} \right)^{1/p} F(\eta(r))^{1/p} . \tag{36}$$

So, $\eta'(0) = 0$, $\eta'(r) > 0$ if $r > 0$, and η is a convex bijection from $[0,\psi_\mu(\infty))$ onto $[0,\infty)$, i.e. $\eta(r,\mu) \to +\infty$ when $r \to \psi_\mu(\infty)$. Moreover,

$$(\eta'(r))^{p-1} = \left(\frac{p\mu}{p-1} \right)^{(p-1)/p} F(\eta(r))^{(p-1)/p}$$

and differentiating

$$((\eta'(r))^{p-1})' = \mu f(\eta(r)). \tag{37}$$

Then, if we apply the operator $\mathbb{L}$ of (31), we obtain

$$\mathbb{L}(\eta(r)) = (1-\mu)f(\eta(r)) - \frac{(N-1)}{r} \left(\frac{p\mu}{p-1}\right)^{(p-1)/p} F(\eta(r))^{(p-1)/p} . \tag{38}$$

From (40), the conclusion i) is clear. To prove ii), consider the function

$$\phi(r) = \left(\frac{p\mu}{p-1}\right)^{(p-1)/p} F(\eta(r))^{(p-1)/p} \tag{39}$$

Obviously, $\phi(0) = 0$ and $\phi(r) > 0$ if $r > 0$. Moreover, ϕ is convex because, from (37), $\phi'(r) = \mu f(\eta(r))$, which is a nondecreasing function of r. So, by elementary results

$$\phi(r) \leq \phi'(r)r.$$

Hence , if $\mu < 1/N$

$$\mathbb{L}(\eta(r,\mu)) \geq f(\eta(r))(1-\mu)-(N-1)\phi'(r) = (1-\mu N)f(\eta(r)) > 0.$$

Analogously, $\mathbb{L}(\eta(r,1/N)) \geq 0$ and ii) is proved. To conclude, let $\tau > 0$ and $r \geq \tau$. Making $s = r - \tau$ we have that

$$\mathbb{L}(\eta([r-\tau]^+;\mu) = -\frac{d}{ds}[(\frac{d\eta}{ds}(s))^{p-1}] - \frac{(N-1)}{(s+\tau)}[\frac{d\eta}{ds}(s)]^{p-1} + f(\eta(s)) \;,\; s \geq 0.$$

We also remark that $\mathbb{L}(\eta([r-\tau]^+,\mu)) \in C^\circ((0,\psi_\mu(+\infty)+\tau))$. Using again (37) we see that , if $\mu \geq 1$, then $\mathbb{L}(\eta_\tau(r)) < 0$ if $r > \tau$.

In a similar way from the convexity of ϕ we obtain that

$$\phi(s) \leq s\phi'(s) \leq (s+\tau)\phi'(s) \quad , \quad s \geq 0.$$

So, if $\mu < 1/N$

$$\mathbb{L}(\eta([r-\tau]^+,\mu) \geq f(\eta(s))(1-\mu N) > 0 \quad \text{if} \quad s = r-\tau > 0 \;,$$

and iii) follows.□

From Theorem 1.5 it is easy to show the existence of solutions.

<u>Theorem 1.5*</u>. *Let* p,f *and* $\eta(r,\mu)$ *as in the above theorem. Then, for every* $R \in (0,\psi_{1/N}(+\infty))$ *there exists a solution* $u_\tau(r)$ *of* (31),(32) *such that* $u_\tau(R) = \eta(R,1/N)$. *Moreover*

$$0 \leq u_\tau(r) \leq \eta(r,1/N) \quad \textit{for every} \quad r \in (0,R), \tag{40}$$

and $u_\tau(r) > 0 \quad \forall r \in [\tau,R]$, *being* $\tau \leq R(1-1/N^{1/p})$.

<u>Proof</u>. Let $u_\tau(x) = u_\tau(|x|)$ be the unique (radially symmetric) solution of the boundary value problem (7),(8) with $k = \eta(R,1/N)$. Using part ii) of Theorem 1.5 and the comparison principle we obtain that $0 \leq u_\tau(|x|) \leq \eta(|x|,1/N)$ for $|x| < R$. Finally, let $\tau_0 = R(1-1/N^{1/p})$. Then $u_\tau(R) = \eta(R-\tau_0,1)$ and so by Theorem 1.5, part i), and comparison results

we deduce that

$$\eta([|x|-\tau_0]^+,1) \leqslant u_\tau(|x|) \quad \text{for} \quad |x| \leqslant R$$

which ends the proof.□

Since those existing solutions of the Cauchy Problem (31),(32) are not explicit, when studying the formation and location of the free boundary $F(u)$ for solutions of (1),(2) it will be more useful to work with the explicit supersolution $\eta(r,1/N)$ instead of the solutions $u_\tau(r)$.

The lack of explictness in Theorem 1.5* may be overcome in the case of homogeneous nonlinearities where, in fact, a solution $u_0(r)$ of (31) (32) with $u_0(r) > 0$ if $r > 0$ can be exhibited. Indeed; if we assume $q < p - 1$ then

$$\eta(r,u) = K_\mu r^{\frac{p}{p-1-q}}, \quad K_\mu = \frac{p-1-q}{p}^{\frac{p}{p-1-q}} \left[\frac{p\,\lambda\,\mu}{(p-1)(q+1)}\right]^{\frac{1}{p-q-1}} \tag{41}$$

and substituting in expression (38)

$$\mathbb{L}(\eta(r:\mu)) = [(1-\mu)C_1 - (N-1)C_2\mu]^{\frac{p-q}{p-1-q}}$$

for some positive constants C_1 and C_2 depending only on p,q and λ.Then in the case, u_0 can be found through any root μ_0 of the algebraic equation

$$(1-\mu)C_1 - (N-1)C_2\mu^{\frac{p-1}{p-1-q}} = 0. \tag{42}$$

Instead of trying to solve (42), inspired by (41), it is easier to seek the solution of the Cauchy problem (31),(32) as a function of the form $u(r) = Cr^{p/(p-1-q)}$ for some C to be determined. Some straightforward calculations allow us to conclude the following result

<u>Lemma 1.6.</u> *Assume* $p > 1$ *and* $f(s) = \lambda|s|^{q-1}$ *with* $\lambda > 0$ *and* $0 < q < p - 1$. *Consider the function*

$$u(r) = C\,r^{\frac{p}{p-q-1}}, \quad C > 0 \tag{43}$$

and define

$$K_{N,\lambda} = \left[\frac{\lambda(p-1-q)^p}{p^{(p-1)}(qp + N(p-1-q))} \right]^{\frac{1}{p-1-q}} \tag{43}$$

Then

$$\mathbb{L}(u(r)) = \left[\lambda C^q - C^{p-1} \frac{p^{(p-1)}(pq+N(p-1-q))}{(p-1-q)^p} \right] r^{\frac{pq}{p-1-q}} \tag{44}$$

In particular, if $C = K_{N,\lambda}$ *then* $\mathbb{L}(u) = 0$ *and if* $C < K_{N,\lambda}$ *(resp.* $C > K_{N,\lambda}$*) then* $\mathbb{L}(u) > 0$ *(resp.* $\mathbb{L}(u) < 0$*).* □

Comparing (42) and (44) we see that a root μ_0 of equation (42) is

$$\mu_0 = \frac{pq + p - 1 - q}{pq + N(p-1-q)} .$$

The hypothesis (34) in Theorem 1.5 is, in some sense, optimal as the following result shows:

Proposition 1.7. *Let* $u \geqslant 0$ *be a nontrivial solution of the Cauchy problem (31),(32), defined at least on* $[0, \psi_1(\infty))$. *Then necessarily,*

$$\psi_1(u(r)) < r \quad \textit{for every } r \in (0, \psi_1(\infty)). \tag{45}$$

In particular, if

$$\int_{0^+} \frac{ds}{F(s)^{1/p}} = +\infty \tag{46}$$

the only nonnegative solution of (31),(32) is $u \equiv 0$.

Proof. Let $u(r) \geqslant 0$ be a solution of (31),(32) defined on $[0,R]$. Define $\varepsilon = \min\{ r : u(r) > 0\}$. Clearly $\varepsilon < R$, and since $u \in C^1([0,R])$ we have $u(\varepsilon) = u'(\varepsilon) = 0$. From the equation we deduce that $u'(r) > 0$ if $r > \varepsilon$, and so u is a bijection from $[\varepsilon,R]$ on $[0,u(R)]$. Then

$$\psi_1(u(R)) = \int_0^{u(R)} \frac{ds}{A^{-1}(F(s))} = \int_\varepsilon^R \frac{u'(r)dr}{A^{-1}(F(u(r)))}$$

where $A^{-1}(s) = [p\, s/(p-1)]^{1/p}$. Multiplying (31) by u' we obtain

$$f(u(r))u'(r) = F(u(r))' = A(u'(r))' + \frac{(N-1)}{r}(u'(r))^p.$$

So integrating between ε and R we have

$$F(u(r)) = A(u'(r)) + \int_\varepsilon^r \frac{(N-1)}{s} (u'(s))^p \, ds > A(u'(r)).$$

Then

$$\psi(u(R)) = \int_\varepsilon^R \frac{u'(r)dr}{A^{-1}(F(u(r)))} < \int_\varepsilon^R dr = R - \varepsilon$$

which proves (45). Finally, if (46) holds then $\psi_1(s) \equiv +\infty$ if $s > 0$, and (45) is impossible except if $u \equiv 0$. □

With respect to the symmetric boundary problem (7),(8) we can state a first result about to the null set $N(u)$ which will be much more improved in the next subsection. (Note that, for $N \geqslant 2$, the homogeneity argument used in Theorem 1.4 does not work, although solutions of (7),(8) vanishing in some small ball $B_\tau(0)$ are been given in Theorem 1.5*).

Corollary 1.8. *Let* $p > 1$, f *satisfying* (3). *Assume* (34) *and* k *such that* $u_\tau(R) = k$, *where* u_τ *is given in Theorem 1.5**, *(this certainly holds if* $\psi_{1/N}(k) = R$) *Then, if* u *is the solution of* (7), (8), *at least* $N(u) \supset B_\tau(0)$. *If, on the contrary,* (48) *holds, then* $u > 0$ *on* $(-R,R)$. □

Remark 1.2. The four last results hold under more general circumstances: with respect to the diffusion term, it is possible to consider the more general equation which is invariant by symmetries

$$-\operatorname{div}\left(a\frac{(|\nabla u|)}{|\nabla u|}\nabla u\right) + f(u) = 0$$

where $a \in C^0([0,\infty)) \cap C^1((0,\infty))$, $a(0) = 0$ and $a'(r) > 0$ if $r > 0$. Note that equation (7) corresponds to $a(r) = r^{p-1}$. For other important special cases see Talenti [3]. Now the operator $\mathbb{L}$ is defined by

$$\mathbb{L}(u) := -\frac{1}{r^{N-1}} \frac{d}{dr}\left(r^{N-1} a\left(\frac{du}{dr}\right)\right) + f(u)$$

and the results can be easily extended to this situation. With respect to the term f we point out that the hypothesis (3) will be weakened in different ways in Section 2.1. □

Remark 1.3. The optimal regularity of solutions can be known through the explicit solution u given in Theorem 1.4. So, in the homogeneous case, if $q < p-1$, $u \in C^{1,\alpha}$ with $\alpha = (1+q)/(p-1-q)$, and in fact $u \in C^{2,\beta}$ with $\beta = [2(1+q)-p]/(p-1-q)$ if $p \leqslant 2(1+q)$. □

1.1b. Interior estimates. Local super and subsolutions.

In this section we shall study the formation of the free boundary for the general boundary problem (1),(2). As in the radially symmetric case we shall need the auxiliary functions F and ψ_μ defined in (17) and (33), respectively. For simplicity in the statements we shall assume that

$$f \text{ is an \underline{odd} nondecreasing continuous function.} \tag{47}$$

(for f not necessarily odd see Remark 1.6).

Theorem 1.9. *Let* $p > 1$, *and assume* f *satisfying* (47) *as well as*

$$\int_{0^+} \frac{ds}{F(s)^{1/p}} < +\infty \tag{48}$$

Let g *and* h *satisfy* (4),(5),(*resp.* (4*) *and* (5*)*for* $p = 2$) *and let* $u \in W^{1,p}(\Omega)$ (*resp.* $u \in W^{1,1}(\Omega)$) *be the solution of* (1),(2) (*resp.* (1*), (2)).*Then the null set* $N(u)$ *contains at least, the set of* $x \in N(g) \cup N(h|_{\partial\Omega})$ *such that*

$$d \geqslant \varepsilon + \psi_{1/N}(M(\varepsilon)) \quad , \textit{ for some } \quad \varepsilon \geqslant 0, \tag{49}$$

where $d = d(x, S(g) \cup S(h|_{\partial\Omega})$ *and* $M(\varepsilon)$ *is given by the* L^∞*-estimate*

$$||u||_{L^\infty(D_\varepsilon)} \leqslant M(\varepsilon), \tag{50}$$

being $D_\varepsilon = \overline{\Omega \cap B_{d-\varepsilon}(x)}$.

Before giving the proof of this theorem we shall interpret its statement. The existence of the free boundary $\mathcal{F}(u)$ is assured under the diffusion-absorption balance given by (48) as well as a balance between the "size" of the set $N(g) \cup N(h|_{\partial\Omega})$ and of the solution. This is given by (49). Note that if u is not globally bounded then the L^∞-estimate (50) is such that $M(\varepsilon) \uparrow +\infty$ when $\varepsilon \downarrow 0$ (see theorems 4.8 and 4.18). Nevertheless, if $u \in L^\infty(\Omega)$ (for instance when g and h satisfy (4**) and (5**)),then $M(\varepsilon)$ may be taken independent of ε ($M(\varepsilon) = M$ given in (6**)) and hence the conclusion is

$$N(u) \supset \{x \in N(g) \cup N(h|_{\partial\Omega}) : d(x, S(g)) \cup S(h|_{\partial\Omega})) \geqslant L\}. \tag{51}$$

where

$$L = \Psi_{1/N}(M). \tag{52}$$

In this last case , $\mathcal{F}(u)$ is assured if, for instance, the radius ρ of the largest ball contained in the set $N(g) \cup N(h|_{\partial\Omega})$ satisfies $\rho > L$. The Figure 1 shows the estimate on the location of $\mathcal{F}(u)$ obtained.

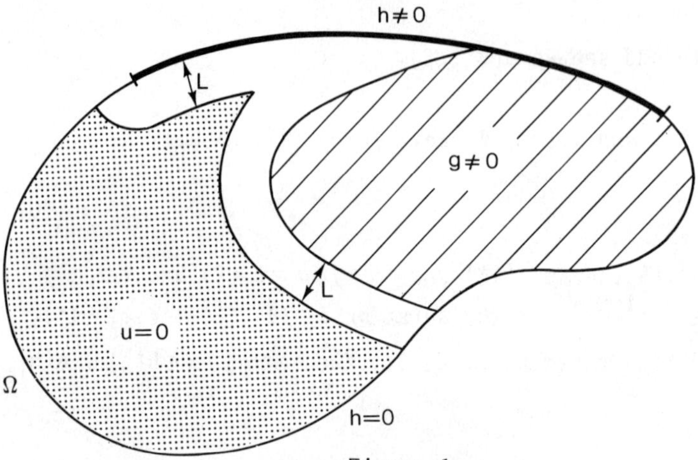

Figure 1.

Proof of Theorem 1.9. Let $x_0 \in N(g) \cup N(h|_{\partial\Omega})$ and let $R = d - \varepsilon$, ε given in (49). On the ball $B_R(x_0)$, let $\bar{u}(x) = \bar{u}(x:x_0) \in W^{1,p}(B_R(x_0)) \cap L^\infty(B_R(x_0))$ satisfy

$$- \Delta_p u + f(u) \geq 0 \qquad \text{in} \qquad B_R(x_0) \tag{53}$$

$$u \geq M(\varepsilon) \qquad \text{on} \qquad \partial B_R(x_0) \tag{54}$$

where $M(\varepsilon)$ is the constant appearing in (50). It is clear that if u is any solution of (1),(2) then we have

$$u(x) \leq \bar{u}(x : x_0) \quad \text{a.e.} \quad x \in \Omega \cap B_R(x_0) \tag{55}$$

Indeed, let $\tilde{\Omega} = \Omega \cap B_R(x_0)$. From the choice of x_0 and R we know that

$$0 = - \Delta_p u + f(u) \leq - \Delta_p \bar{u} + f(\bar{u}) \quad \text{in} \quad \tilde{\Omega} .$$

On the other hand, on the boundary $\partial\tilde{\Omega}$ we have the inequality

$$u \leqslant \overline{u} \qquad \text{on} \qquad \partial\tilde{\Omega} , \tag{56}$$

thus, (55) follows from the application of the comparison principle on $\tilde{\Omega}$. In order to check (56), note that $u = 0$ on $\partial\tilde{\Omega} \cap \partial\Omega$. The rest of $\partial\tilde{\Omega}$ is contained in $\partial B_R(x_0)$ and so, by the choice of $M(\varepsilon)$, the expression (56) holds by the construction of $\overline{u}$ (recall (54)). Now we shall prove that, if condition (49) holds, then a function $\overline{u}$ can be taken, such that it satisfies (53) and (54) as well as $\overline{u} \in C(B_R(x_0))$ and $\overline{u}(x;x_0) = 0$. To do this, we define

$$\overline{u}(x : x_0) = \eta(|x - x_0| : 1/N) \tag{57}$$

where $\eta(r : \mu) = \psi_\mu^{-1}(r)$ i.e. η is defined by (35). From the Theorem 1.5 and the symmetry of $\overline{u}$, such a function satisfies (53). Moreover, (54) leads to

$$\eta(R : 1/N) \geqslant M(\varepsilon)$$

and by the monotonicity of $\psi_\mu(r)$ with respect to r this condition holds if $R \geqslant \psi_{1/N}(M(\varepsilon))$. Now, condition (56) is obvious. Analogously, if $\underline{u}(x : x_0)$ is a function in $W^{1,p}(B_R(x_0)) \cap L^\infty(B_R(x_0))$

$$- \Delta_p \underline{u} + f(\underline{u}) \leqslant 0 \qquad \text{in} \qquad B_R(x_0)$$

$$\underline{u} \leqslant - M \qquad \text{on} \qquad \partial B_R(x_0)$$

we can prove that $\underline{u}(x : x_0) \leqslant u(x)$ a.e. $x \in \Omega \cap B_R(x_0)$. Since f has been assumed odd, such a function $\underline{u}(x : x_0)$, also satisfying $\underline{u} \in C(B_R(x_0))$ and $\underline{u}(x_0 : x_0) = 0$, can be found from Theorem 1.5, by taking

$$\underline{u}(x : x_0) = - \eta(|x - x_0| : 1/N)$$

and R such that

$$- \eta(R : 1/N) \leqslant - M(\varepsilon)$$

or, equivalently, $R \geqslant \psi_{1/N}(M(\varepsilon))$. Finally, by Theorem 1.1 $u \in C^0(\Omega)$ and, then $0 = \underline{u}(x_0 : x_0) \leqslant u(x_0) \leqslant u(x_0 : x_0) = 0$, which ends the proof. The proof for solutions $u \in W^{1,1}(\Omega)$ is the same. It suffices to remark that condition (56) follows from Theorem 1.2 $(u \in C^{1,\alpha}(\Omega))$ and hypothesis (50). □

Corollary 1.10. *Let* $p > 1$, $\lambda > 0$ *and* $q > 0$. *Let* g *and* h *satisfy* (4) *and* (5) *respectively. Then, if* $q < p - 1$, *the null set of the solution of the problem*

$$- \Delta_p u + \lambda |u|^{q-1} u = g \qquad in \qquad \Omega \tag{58}$$

$$u = h \qquad on \qquad \partial\Omega \tag{59}$$

satisfies the estimate

$$N(u) \supset \{x \in N(g) \cup N(h|_{\partial\Omega}) : d(x;S(g) \cup S(h|_{\partial\Omega}) \geq \varepsilon + L(\varepsilon) \text{ for some } \varepsilon \geq 0\},$$

where

$$L(\varepsilon) = [\frac{M(\varepsilon)}{K_{N,\lambda}}]^{\frac{p-1-q}{p}} , \quad K_{N,\lambda} = [\frac{\lambda(p-1-q)^p}{p^{(p-1)}(qp+N(p-1-q)}]^{\frac{1}{p-1-q}} \tag{60}$$

and $M(\varepsilon)$ *satisfies* (50).□

Remark 1.5. As it has been pointed out in the Introduction, the problem (58),(59) appears very often in the study of single isothermical q-order reactions (there is $p=2$, $g \equiv 0$ on Ω, and $h=1$, i.e. $N(g) \cup N(h|_{\partial\Omega}) = \Omega$ and $S(g) \cup S(h|_{\partial\Omega}) = \partial\Omega$) The number λ is an important parameter and, as we can see in (60), the measure and location of the null set $N(u)$, there called "dead core", depends on λ. In particular, we note that the null set $N(u)$ "tends" to the whole domain Ω when $\lambda \uparrow +\infty$. Some delicate estimates of $N(u)$ in terms of λ will be given in Section 1.4.□

Remark 1.6. The simplification made in Theorem 1.9 by assuming f as an odd function, can be easily removed. Indeed, the only change to be made is to construct the local subsolution $\underline{u}(x:x_0)$ in a different way. To do that, we first extend the function ψ_μ to the whole $\mathbb{R}$ in a natural way, i.e., if $r < 0$ we define

$$\psi_\mu(\tau) = (\frac{p-1}{p\mu})^{1/p} \int_0^\tau \frac{ds}{F(s)^{1/p}} = -(\frac{(p-1)}{p\mu})^{1/p} \int_\tau^0 \frac{ds}{F(s)^{1/p}}$$

Now, it is clear that $\psi_\mu((-\infty,0)) \subset [-\infty,0)$ and then, if we make the assumption

$$\max \{ \int_{0-} \frac{ds}{F(s)^{1/p}} , \int_{0+} \frac{ds}{F(s)^{1/p}} \} < +\infty \tag{61}$$

instead of (48) then we can define $\eta(r:\mu)$ for $r < 0$ as the inverse of ψ_μ. Finally, it suffices to take

$$\underline{u}(x : x_0) = \eta(-|x - x_0| : 1/N).$$

We leave the details for the reader.□

Remark 1.7. A curious phenomenon appears if the balance between the diffusion and the absorption terms is such that not only assumption (48) holds, but even more, if

$$\int_0^{+\infty} \frac{ds}{F(s)^{1/p}} < +\infty .$$

In this case, for instance, the estimate (51) can be replaced by the "uniform estimate"

$$N(u) \supset \{x \in N(g) \cup N(h|_{\partial\Omega}) : d(x : S(g) \cup S(h|_{\partial\Omega})) \geq \psi_1(+\infty)\} ,$$

which expresses that the null set of the solutions corresponding to different data g and h, with the same set $N(g) \cup N(h|_{\partial\Omega})$, contains at least the subset $\{x \in N(g) \cup N(h|_{\partial\Omega}) : d(x : S(g) \cup S(h|_{\partial\Omega})) \geq \psi_1(+\infty)\}$, i.e, with independence of which are the values of g and h when their respective supports are fixed.□

The method used in the proof of Theorem 1.9 has a local character and this allows us to apply it even in the case in which there is no global comparison principle on the whole domain Ω. This situation occurs very often when the function f depends also on x:

Proposition 1.11. *Let* $\theta \in L^1_{loc}(\Omega)$, $p > 1$ *and* f, g *and* h *as in Theorem* 1.9. *Given* $\lambda > 0$ *let* Ω_λ *denote the set*

$$\Omega_\lambda = \{x \in \Omega : \theta(x) \geq \lambda\} .$$

Let $u \in W^{1,p}(\Omega)$ *be a weak solution of the problem*

$$\begin{aligned} -\Delta_p u + \theta(x)f(u) &= g && \text{in } \Omega \\ u &= h && \text{on } \partial\Omega \end{aligned} \tag{62}$$

Then, if (48) *holds, the null set of* u *satisfies the estimate*

$$N(u) \supset \{x \in (N(g) \cup N(h|_{\partial\Omega})) \cap \bar{\Omega}_\lambda : d(x,[S(g|_{\partial\lambda}) \cup S(h_{\partial\Omega\cap\partial\Omega_\lambda}) \cup (\partial\Omega_\lambda - \partial\Omega)]) \geq \varepsilon + L(\varepsilon)$$

$$\textit{for some } \varepsilon \geq 0\},$$

where $L(\varepsilon) = \Psi_{\lambda/N}(M(\varepsilon))$ *and* $M(\varepsilon)$ *is given by* (50).

Proof. It is the same as in Theorem 1.9 but choosing in this case x_0 belonging to $(N(g) \cup N(h|_{\partial\Omega})) \cap \bar{\Omega}_\lambda$, and $R > 0$ given by

$$R = d(x_0, (S(g|_{\partial\Omega}) \cup S(h|_{\partial\Omega\cap\partial\Omega}) \cup (\partial\Omega_\lambda - \partial\Omega))) - \varepsilon.$$

Note that any solution of (62) satisfies

$$-\Delta_p u + \lambda f(u) \leq 0$$

on the set $N(g) \cap \Omega_\lambda$. We also use in this occasion the fact that

$$-M(\varepsilon) \leq u|_{\partial\Omega_\lambda - \partial\Omega} \leq M(\varepsilon). \quad \square$$

Remark 1.8. The above result is of particular interest in some problems leading to equations such as (62), for instance, the following:

i) Biological populations: In Gurney-Nisbet [1], the equation (62) for $p = 2$ and $f(u) = (u)^{1/2}$ if $u \geq 0$, is introduced to model the asymptotic state of a biological population. The sign of θ describes the hostile or favourable effect of the environment on the population, $\theta \geq 0$ or $\theta \leq 0$, respectively. A systematic treatment of this equation is given in Schatzman [1] (see also other references in this work), under the assumptions: $\Omega = \mathbb{R}^N$, $g \geq 0$ with compact support and $\theta \in C^{0,\alpha}(\mathbb{R}^N)$ such that

$$\lim_{|x|\to\infty} \theta(x) > 0 \quad , \quad \text{and} \quad \{x \in \mathbb{R}^N : \theta(x) < 0\} \neq \emptyset .$$

There, the compactness of a solution of (62) is proved by constructing a global supersolution with compact support (see also the subsection 1.1d). This conclusion can also be obtained as an easy consequence of Proposition 1.12. Other results on the free boundary for this equation can be found in Section 1.3.

ii) <u>nonlinear elliptic systems</u>. Weakly coupled nonlinear elliptic systems such as, for instance,

$$- \Delta u + n(v)f(u) = g(x)$$

$$- \Delta v + m(x,u,v) = 0$$

lead to the equation (62) for $p = 2$ and $\theta(x) = n(v(x))$. Such systems, with f satisfying (48), play an important part in the study of non-isothermal simple chemical reactions (see e.g. Aris [1] and Diaz-Hernandez [1]) as well as in some predator-prey systems (see Hernandez [1]). The Proposition 1.12 gives some estimates on the location of the null set of u. Some "a priori" information on the set $\Omega_\lambda = \{x \in \Omega : n(v) \geq \lambda\}$ can be obtained, for instance, from the peculiar form of n (case of chemical reactions) or by the maximum principle for v (case of predator-prey systems). (See details in the mentioned works).□

Other interesting generalizations of Theorem 1.9 are related to some other diffusion terms. In this sense, it is very easy to extend that result to the solutions of the more general (invariant by symmetries) equation

$$- \operatorname{div} \left(\frac{a(|\nabla u|)}{|\nabla u|} \nabla u \right) + f(u) = g. \tag{63}$$

Indeed, the comparison principle and the local boundedness of the weak solutions are still true, now in the framework of the Orlicz-Sobolev spaces (see Chapter 4) and the local super and subsolutions can also be constructed (see Remark 1.2).

We can also replace the operator Δ_p by a general second order elliptic linear operator L as the one given by

$$Lu = \sum_{i,j=1}^{N} D_j(a_{ij}(x)D_i u) + \sum_{j=1}^{N} D_j(b_j(x)u) + c(x)u \tag{64}$$

where $D_j = \partial/\partial x_j$, and the functions a_{ij}, b_j and c are such that

$$a_{ij}, b_j \in W^{1,\infty}(\Omega), \; c \in L^\infty(\Omega) \text{ and } c + D_j b_j \leq 0 \text{ on } \Omega, \tag{65}$$

$$\left.\begin{array}{l} \text{and there exists } \Lambda \text{ and } \lambda \in L^\infty_{loc}(\Omega), \; \lambda(x) > 0 \text{ satisfying} \\ \Lambda(x)|\xi|^2 \geq \sum_{ij} a_{ij}(x)\, \xi_i \xi_j \geq \lambda(x)|\xi|^2 \text{ for every } \xi \in \mathbb{R}^N, \xi \neq 0 \end{array}\right\} \tag{66}$$

We recall that the existence and uniqueness of weak solutions $u \in W^{1,2}(\Omega) \cap L^\infty(\Omega)$ of the semilinear problem

$$- Lu + f(u) = g \quad \text{in} \quad \Omega \tag{67}$$

$$u = h \quad \text{on} \quad \partial\Omega \tag{68}$$

can be easily obtained under the above conditions on L . . We refer reader,again, to Chapter 4 where, in fact, a complete alternative to Theorem 1.1 can be found through the comments made there. With respect to the free boundary $\mathcal{F}(u)$ we have:

<u>Theorem 1.13.</u> *Suppose* (65),(66), $c(x) \leq 0$ *and assume that* f *is an odd nondecreasing function such that*

$$\int_{0^+} \frac{ds}{F(s)^{1/2}} < +\infty \tag{69}$$

where F *is the primitive of* f , $F(0) = 0$. *Let* g *and* h *satisfy* (4) *and* (5) *respectively for* $p = 2$. *Then the null set* N(u) *of any weak solution of* (67),(68) *satisfies*

$$N(u) \supset \{x \in N(g) \cup N(h|_{\partial\Omega}) : d(x, S(g) \cup S(h|_{\partial\Omega})) \geq \varepsilon + L(\varepsilon) \text{ for some } \varepsilon \geq 0\}$$

where $L(\varepsilon) = \Psi_\mu(M(\varepsilon))$ *for some suitable* $\mu > 0$ *and* $M(\varepsilon)$ *given in* (50).

<u>Proof</u>. As in Theorem 1.9 the main difficulty is the construction of the local supersolution $\bar{u}(x:x_0)$ defined on $B_R(x_0)$ for each $x_0 \in N(g) \cup N(h|_{\partial\Omega})$. This will be done by means of a radially symmetric function $\bar{u}(x:x_0) = \eta(r)$, $r = |x - x_0|$, such that $\eta \geq 0$, $\eta'' \geq 0$. Setting $x_0 = 0$, for simplicity $r = |x|$ and we have

$$L\eta(r) = \eta''(r) \sum_{ij} a_{ij}(x) \frac{x_i x_j}{r^2} + \frac{\eta'(r)}{r} [\sum_{i,j} a_{ij}(x) \frac{x_i x_j}{r^2} - \{ \sum_i a_{ii}(x) + \sum_{ij} D_j a_{ij}(x) x_i +$$

$$+ \sum_j b_j(x) x_j \}] + \eta(r) [\sum_j D_j b_j(x) + c(x)]$$

Using the assumptions (65),(66), and since $c \leq 0$, we have

$$- L\eta(r) \geq - \Lambda^0 \eta''(r) - K_1 \frac{\eta'(r)}{r} \quad \text{in} \quad B_R(x_0) ,$$

for some $K_1 > 0$,(note that , by (66) , $\sum a_{ij}(x) x_i x_j r^{-2} \leq \Lambda^0$ where $\Lambda^0 = \text{ess sup } \Lambda(x)$ on $B_R(x_0)$). Therefore, it suffices to take $\eta(r)$

as a nontrivial solution of the Cauchy problem

$$- \eta''(r) - K_2 \frac{\eta'(r)}{r} + \frac{1}{\Lambda^0} f(\eta(r)) = 0$$

$$\eta(0) = \eta'(0) = 0$$

where $K_2 = K_1/\Lambda^0$. The existence of such a solution is assured by assumption (69). In fact, as in Theorem 1.5, η is given by the inverse of ψ_μ, defined in (33), for some $\mu > 0$. We leave details to the reader. □

Note that in the above results L is not necessarily uniformly elliptic ($\lambda(x) \geq \lambda_0 > 0$). The hypothesis (65) can also be avoided by the consideration of functions f not necessarily monotone (see Section 2.1). With respect to general nonlinear diffusion terms we refer the reader to subsection 1.2c for the study of the free boundary

We shall end this subsection by referring to some contexts in which the free boundary is slightly different. For instance, sometimes $f(0) \neq 0$, or, in other cases, it is known that the external perturbation $g(x)$ is identically a constant $k > 0$ on a positively measured subset of Ω. This happens in the study of non-Newtonian fluids moving in a flat channel under the action of a constant pressure gradient (Hartmann flow). In this case, the velocity u satisfies

$$- \frac{d}{dx} \left(\left| \frac{du}{dx} \right|^{p-2} \frac{du}{dx} \right) + \lambda u = P$$

where λ and P are positive physical constants (see Martinson-Pavlov [1]). As in the case of the Couette flow ($P \equiv 0$), mentioned in the Introduction, the physics of the problem shows that for dilatant fluids ($p > 2$) quasi-solid zones may appear where the velocity is constant but, now, non zero, $u = P/\lambda$. More generally, we have

Theorem 1.14. *Let* g *and* h *satisfy* (4) *and* (5), *respectively, and assume that there exists* $k \in \mathbb{R} - \{0\}$ *such that*

$N(g - k)$ *has a positive measure*

Let $p > 1$ *and* f *be a nondecreasing continuous function satisfying* $f(a) = k$, *for some* $a \in \mathbb{R}$, *as well as*

$$\int_{0^+} \frac{ds}{\left(\int_a^{s+a} f(t)dt - ks\right)^{1/p}} < +\infty \tag{70}$$

Then the solution of (1) , (2) *satisfies*

$$u(x) = a \quad a.e. \ in \ \{x \in N(g-k) \cup N(h-k|_{\partial\Omega}) : d(x, S(g-k) \cup S(h-k|_{\partial\Omega})) \geq \varepsilon + L(\varepsilon) \ for \ some \ \varepsilon \geq 0\},$$

where $L(\varepsilon) = \Psi_{1/N}(M(\varepsilon) + |a|)$ *with* $M(\varepsilon)$ *given in* (50).

Proof. It suffices to apply Theorem 1.9 to the function $w = u - a$. Note that w satisfies

$$-\Delta_p w + \tilde{f}(w) = \tilde{g} \quad \text{in} \quad \Omega$$

and $w = \tilde{h}$ on $\partial\Omega$, where $\tilde{g} = g - k$, $\tilde{h} = h - k$ and $\tilde{f}(s) = f(s+a) - k$ The function $\tilde{f}$ is still a nondecreasing continuous function satisfying $\tilde{f}(0) = 0$ and hypothesis (48) (due to (70)).□

It is easy to see that in the semilinear case ($p = 2$), condition (70) implies that the slope of the function f at the point $s = a$ must be infinity. This happens in some formulations of the Stefan problem (see an analogous point of view in Bertsch-DeMottoni-Peletier [1]). In contrast to this, if $p > 2$ condition (70) holds for any Lipschitz continuous function f and no special behaviour of f in $r = a$ is needed. In this case, Theorem 1.14 holds because the operator $\Delta_p u$ degenerates near the set $\{x \in \Omega: \nabla u = 0\}$.

1.1c. Boundary estimates. Non diffusion of the support.

The method of local super and subsolutions used in the proof of Theorem 1.9 can also be applied in order to study the vanishing of the solution at points of the boundary of the support of g and $h|_{\partial\Omega}$. It seems natural that the effect of external influences may make difficult the vanishing of the solution at these points. Nevertheless, we shall see that this effect appears under suitable additional assumptions on g or h, respectively.

In contrast with the results of the above subsection, now we shall need the local boundedness of the solution in Ω and not only in int $N(g)$. For this reason, we shall assume the stronger hypotheses (4**) and (5**) on the data. They can be weakened if some additional growing conditions on f are known (see Chapter 4).

We first consider the vanishing of the solution u in a point x_0, of $\partial S(g) \cap \partial N(g)$.

Theorem 1.15. *Assume* $p > 1$ *and let* f *be an odd nondecreasing continuous function satisfying the integral condition* (48). *Let* g *and* h *satisfy* (4**) *and* (5**), *respectively. Consider* $x_0 \in \partial S(g) \cap \partial N(g)$ *and assume that there exists* $R > 0, \mu_1$ *and* $\mu_2 > 0$ *such that*

$$\mathbb{L}(-\eta(|x - x_0|:\mu_1)) \leqslant g(x) \leqslant \mathbb{L}(\eta(|x - x_0|:\mu_2)) \text{ a.e. } x \in \Omega \cap B_R(x_0) \quad (71)$$

where $\eta(r:\mu) = \psi_\mu^{-1}(r), \psi_\mu$ *given in* (33) *and* $\mathbb{L}$ *is the differential operator defined in* (31). *Moreover, we assume one of the following conditions*

$$\partial(\Omega \cap B_R(x_0)) \subseteq N(h|_{\partial\Omega}) \quad (72)$$

or

$$d(x_0, \partial S(h|_{\partial\Omega})) \geqslant R \geqslant \max\{\psi_{\mu_1}(M), \psi_{\mu_2}(M)\} \quad (73)$$

where $\|u\|_\infty \leqslant M$ *(for instance,* M *given in* (6**)) . *Then, for any solution* u *of* (1),(2) *we have*

$$-\eta(|x - x_0|:\mu_1) \leqslant u(x) \leqslant \eta(|x - x_0|:\mu_2) \text{ a.e. } x \in \overline{\Omega} \cap B_R(x_0).$$

In particular $u(x_0) = 0$.

Before proving the above result, we shall try to clarify the meaning of the statement. First we note that (72) holds if $\partial(\Omega \cap B_R(x_0))$ is the whole (or a part of) $\partial\Omega$, where $h = 0$. This is the case if, for instance, $B_R(x_0) \supset \Omega$ and $h = 0$ on $\partial\Omega$. If condition (72) fails, then the region where inequalities (71) hold must be sufficiently large, depending on the bounds M on the solution u. In the particular case of $f(s) = \lambda|s|^{q-1}$, with $\lambda > 0$ and $q > 0$, the hypothesis (48) is equivalent to $q < p - 1$. Using Lemma 1.5, instead of computing the functions $\eta(r:\mu)$, we find that condition (71) can be expressed as

$$-K_1|x - x_0|^{\frac{pq}{p-1-q}} \leqslant g(x) \leqslant K_2|x - x_0|^{\frac{pq}{p-1-q}} \text{ a.e. } x \in \Omega \cap B_R(x_0) \quad (74)$$

and (73) means that

$$R = d(x_0, \partial S(h|_{\partial\Omega})) \geqslant M^{\frac{p}{p-1-q}} \max\{C_1, C_2\} \quad (75)$$

for some $C_1, C_2 > 0$ and K_1, K_2 satisfying

$$K_i = [\lambda C_i^q - C_i^{p-1} \frac{p^{(p-1)}(pq + N(p-1-q))}{(p-1-q)^p}] \quad ,i = 1,2.$$

In this case, any solution u of (58),(59) satisfies

$$- C_1|x-x_0|^{\frac{p}{p-q-1}} \leqslant u(x) \leqslant C_2|x - x_0|^{\frac{p}{p-q-1}} \quad \text{a.e.} \quad x \in \bar{\Omega} \cap B_R(x_0). \tag{76}$$

In subsection 1.3, we shall show that the inequality (76) is optimal.

Proof of Theorem 1.15. Let $x_0 \in \partial N(g)$. From the considerations of Theorem 1.5a we know that the function $\bar{u}(|x:x_0|) = \eta(|x - x_0|:\mu_2)$ satisfies

$$- \Delta_p\bar{u} + f(\bar{u}) = \mathbb{L}(\bar{u}) \geqslant g(x) \quad \text{a.e.} \quad x \in B_R(x_0) \cap \Omega$$

On the other hand, if $\tilde{\Omega} = \Omega \cap B_R(x_0)$, on the boundary $\partial\tilde{\Omega}$ we have

$$u \leqslant \bar{u} \quad \text{on} \quad \partial\tilde{\Omega} .$$

Indeed, if (72) i.e. if $\partial(\Omega \cap B_R(x_0)) \subseteq N(g|_{\partial\Omega})$, then it is trivial because on $\partial\tilde{\Omega}$ $u = 0$ and , by definition, $\bar{u} \geqslant 0$. When (72) fails, we can write $\partial\tilde{\Omega} = (\partial\tilde{\Omega} \cap \partial\Omega) \cup (\partial\tilde{\Omega} - \partial\Omega)$. As $R = d(x_0, S(h|_{\partial\Omega}))$, it is clear that

$$0 = u \leqslant \bar{u} \quad \text{on} \quad \partial\tilde{\Omega} \cap \partial\Omega ,$$

On the other hand, if $x \in \partial\tilde{\Omega} - \partial\Omega$ then $|x - x_0| = R$ and so, using the boundedness of u and hypothesis (73) we have

$$u \leqslant M \leqslant \eta(R : \mu_2) = \bar{u}$$

(recall that ψ_μ is nondecreasing and that $\eta(r : \mu_2) = \psi_{\mu_2}^{-1}(r))$. In any case, by comparison of solutions on $\tilde{\Omega}$ we conclude that $u(x) \leqslant \bar{u}(x : x_0)$ a.e. $x \in \tilde{\Omega}$. Analogously, defining $\underline{u}(x : x_0) = -\eta(|x - x_0|: \mu_1)$, the inequality $\underline{u}(x : x_0) \leqslant u(x)$ a.e. $x \in \tilde{\Omega}$ is proved, which ends the proof. □

Remark 1.9. It is clear that assumptions (71) and (73) may be simplified if $g \geq 0$ and $h \geq 0$ in Ω and $\partial\Omega$, respectively. Indeed, in this case any solution u of (1),(2) is such that $u \geq 0$ in $\overline{\Omega}$ and (71) can be replaced by

$$0 \leq f(x) \leq \mathbb{L}(\eta(|x - x_0|:\mu_2)) \quad \text{a.e.} \quad x \in \Omega \cap B_R(x_0).$$

Note also that, from the results of Theorem 1.5 we know that necessarily, $\mu_2 \leq 1$, and that, in fact, the condition $\mathbb{L}(\eta(r;\mu_2)) \geq 0$ is assured if, for instance, $\mu_2 \leq 1/N$. □

The above results can be applied to obtain global consequences. For the sake of simplicity we shall only consider a particular case in which we shall get the curious fact that, under suitable hypotheses, the support of the solution coincides with that of the function g if, for instance, $h = 0$.

Theorem 1.16. *Assume* $p > 1$ *and let* f *be an odd nondecreasing continuous function satisfying* (48) . *Let* g *and* h *satisfy* (4**) *and* (5**) *respectively and assume that there exists* $R > 0$, μ_1 *and* $\mu_2 \in [0,1/N]$ *such that*

$$\left.\begin{aligned} &-(1-\mu_1 N)f(\eta(d(x,\partial S(g)):\mu_1)) \leq g(x) \leq (1-\mu_2 N)f(\eta(d(x,\partial S(g):\mu_2)) \\ &\text{a.e. } x \in S(g) \quad , \ d(x, \partial S(g)) \leq R. \end{aligned}\right\} \tag{77}$$

In addition, we assume that

$$R = d(S(h|_{\partial\Omega}) \ , \ S(g)) \geq R \geq 2 \max\{\psi_{\mu_1}(M) \ , \ \psi_{\mu_2}(M)\} \tag{78}$$

where $M \geq ||u||_{L^\infty}$ *(for instance* M *given by* (6**)*). Then, if* u *is any solution of* (1),(2), *we have*

$$u(x) = 0 \quad \textit{if} \quad x \in \overline{N(g)} \textit{ and } d(x,S(h|_{\partial\Omega})) \geq R.$$

In particular, if $h = 0$ *on* $\partial\Omega$, *then*

$$\text{supp } u = \text{supp } g.$$

Proof. First, we claim that $u = 0$ on $\partial S(g)$. Indeed, let $x_0 \in \partial S(g)$. If $x_0 \in \partial\Omega \cap \partial S(g)$ the conclusion is obvious by (78), so let $x_0 \in \partial S(g) - \partial\Omega$, and consider the ball $B_R(x_0)$. It is clear that, for every $x \in \Omega \cap B_R(x_0)$

$d(x,\partial S(g)) \leqslant |x - x_0|$, because $x_0 \in \partial S(g)$. Then, by (77), the monotonicity of the functions f and η, and the relation (35) , we have that

$$g(x) \leqslant (1-\mu_2 N)f(\eta(|x-x_0|,\mu_2)) \leqslant L(\eta(|x-x_0|,\mu_2)) \text{ a.e. } x \in \Omega \cap B_R(x_0)$$

(recall that $\mathbb{L}(\eta(r\ ;\ \mu_2)) \geqslant 0$ because $\mu_2 \leqslant \frac{1}{N}$ and that $g(x) = 0$ in $B_R(x_0) \cap N(g))$. On the other hand, from (78) we know that

$$d(x_0, S(h|_{\partial\Omega})) \geqslant \max\{\ \psi_{\mu_1}(M)\ ,\ \psi_{\mu_2}(M)\}\ .$$

Hence, hypothesis (73) is satisfied and $u(x_0) = 0$ by Theorem 1.15. Thus $u = 0$ on $\partial S(g)$. Now consider the region $\tilde{\Omega} = \{x \in N(g) : d(x,S(g)) \leqslant R\}$ with R given by (78) (see Figure 2). On this set the solution of (1),(2)

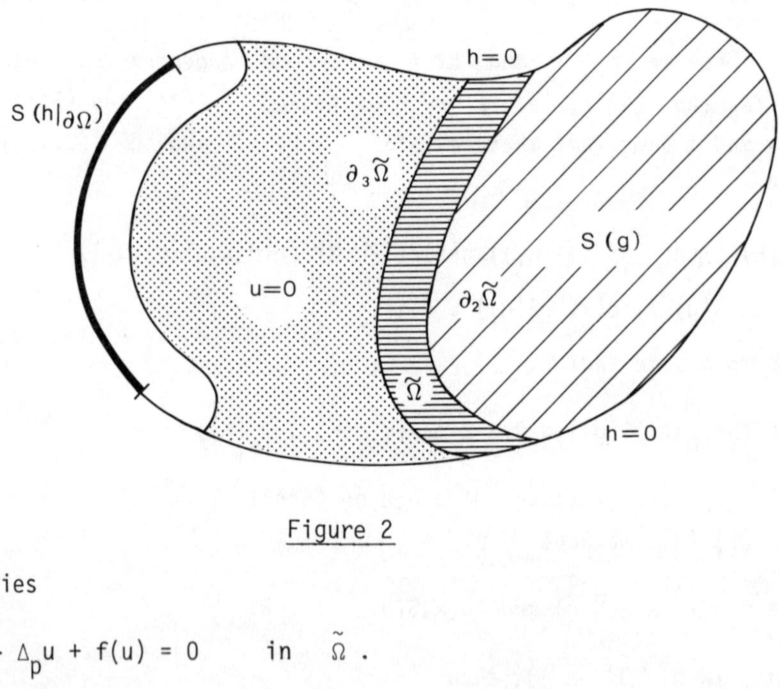

Figure 2

satisfies

$$-\Delta_p u + f(u) = 0 \qquad \text{in} \quad \tilde{\Omega}\ .$$

On the other hand

$$u = 0 \qquad\qquad \text{on} \quad \partial\tilde{\Omega}.$$

Indeed, let $\partial\tilde{\Omega} = \partial_1\tilde{\Omega} \cup \partial_2\tilde{\Omega} \cup \partial_3\tilde{\Omega}$ where $\partial_1\tilde{\Omega} = \partial\tilde{\Omega} \cap \partial\Omega$, $\partial_2\tilde{\Omega} = \partial S(g) \cap \tilde{\tilde{\Omega}}$ and $\partial_3\tilde{\Omega} = \{x \in \overline{N(g)} : x \in \partial\tilde{\Omega}\ ,\ x \notin \partial\Omega$ and $d(x,\partial S(g)) = R\}$. It is clear that $u = 0$ on $\partial_1\tilde{\Omega} \cup \partial_2\tilde{\Omega}$. Moreover the function ψ_μ is decreasing in μ and so we have that

$$\max\{\psi_{\mu_1}(M), \psi_{\mu_2}(M)\} > \psi_{1/N}(M)$$

where $M \geq ||u||_{L^\infty}$. Then, from (78) we know that

$$d(\partial_3\tilde{\Omega}, S(h|_{\partial\Omega}) \cup S(g)) \geq L$$

and so $u = 0$ on $\partial_3\tilde{\Omega}$, by the Theorem 1.9. In consequence, from the uniqueness results in $\tilde{\Omega}$ we deduce that $u = 0$ in $\tilde{\Omega}$, which proves the theorem.□

Remark 1.10 If, for instance, $g \geq 0$, $h \geq 0$ and f is given by $f(s) = \lambda|s|^{q-1} s$ with , $\lambda > 0$ and $0 < q < p - 1$ then the condition (77) leads to the inequality

$$0 \leq g(x) \leq K \; d(x, \partial S(g))^{\frac{pq}{p-1-q}} \quad \text{a.e.} \quad x \in S(g) \text{ , } d(x, \partial S(g)) \leq R$$

for some suitable constant $K > 0$.□

We can also study the vanishing of the solution u at points $x_0 \in \partial\Omega$ of the boundary of the set $S(h|_{\partial\Omega})$.

Theorem 1.17. *Assume* $p > 1$ *and let* f *be an odd nondecreasing continuous function satisfying* (48). *Let* g *and* h *satisfying* (4**) *and* (5**) *respectively. Consider* $x_0 \in \partial\Omega \cap \overline{N(g)}$ *and assume that there exists* $R > 0$ *such that the boundary datum* h *verifies*

$$- \eta(|x - x_0|:1/N) \leq h(x) \leq \eta(|x - x_0|:1/N) \qquad x \in \partial\Omega \cap B_R(x_0) \tag{79}$$

where $\eta(r:\mu) = \psi_\mu^{-1}(r)$ *and* R *satisfies*

$$d(x_0, S(g)) \geq R \geq \psi_{1/N}(M).$$

where $M \geq ||u||_{L^\infty(\Omega)}$. *Then, for any solution* u *of* (1),(2) *we have*

$$- \eta(|x-x_0|:1/N) \leq u(x) \leq \eta(|x-x_0|:1/N) \quad \text{a.e.} \quad x \in \overline{\Omega} \cap B_R(x_0).$$

In particular, $u(x_0) = 0$.

Proof. It suffices to compare u with the auxiliary functions $\overline{u}(x : x_0) = \eta(|x - x_0|:1/N)$ and $\underline{u}(x;x_0) = -\eta(|x - x_0|:1/N)$ in the set $\tilde{\Omega} = \Omega \cap B_R(x_0)$. Note that $g = 0$ in $\tilde{\Omega}$ and that on $\partial\tilde{\Omega}$ we have $\underline{u} \leq u \leq \overline{u}$ by the hypothesis (79) and the choice of R. □

1.1d. Solutions with compact support. Global super and subsolutions.

For different purposes, it is sometimes interesting to exhibit global super and subsolutions, i.e., now defined in the whole domain Ω. There is a whole method which proves the existence of the solution from the existence of such super and subsolutions (see Chapter 4). In our case, it is not necessary because of the monotonicity of f (assumption (3)); nevertheless, it is useful for some variants of equation (1) such as the equation (62), in the case of biological populations (see Schatzman [1]). It is also useful, when Ω is unbounded and the equation is not coercive in order to find solutions in $W^{1,p}(\Omega)$ (see Remark 4.1).

Theorem 1.18. *Let Ω be a regular set (not necessarily bounded). Assume $p > 1$ and let f be an odd nondecreasing continuous function satisfying*

$$\int_{0+} \frac{ds}{F(s)^{1/p}} < +\infty ,$$

where F is the primitive of f. Let $g \in L^\infty(\Omega)$ and $h \in W^{1,p}(\Omega) \cap L^\infty(\Omega)$ with compact support, $S(g)$ and $S(h|_{\partial\Omega})$, respectively. Then there exists a unique $u \in W^{1,p}(\Omega) \cap L^\infty(\Omega)$, solution of (1),(2) having compact support. Moreover, for every $x_0 \in S(g) \cup S(h|_{\partial\Omega})$ the function $\overline{u}(x) = \pm G(|x - x_0|)$ is a global super (sub) solution of (1),(2) with G defined by

$$G(r) = \begin{cases} K_1 - K_2 r^{\frac{p}{p-1}} & \text{if} \quad 0 \leq r \leq R_0 \\ \eta(R_1 - r) & \text{if} \quad R_0 \leq r \leq R_1 \\ 0 & \text{if} \quad R_1 \leq r, \end{cases} \tag{80}$$

where $B_{R_0}(x_0) \supset (S(g) \cup S(h|_{\partial\Omega}))$, K_1, K_2 and R_1 are some suitable positive constants and $\eta(r) = \eta(r:1) = \psi_1^{-1}(r)$, ψ_μ defined by (33).

Proof. First of all, we shall prove that we can choose K_1, K_2 and R_1 such that the function $\overline{u}(x) = G(|x - x_0|)$ is a supersolution of (1),(2). Without loss of generality we can assume $x_0 = 0$. We also recall that the function $\eta(r) = \psi_1^{-1}(r)$ is a C^2 function defined in the interval $[0, \psi_1(+\infty))$ and such that $\eta(0) = \eta'(0) = 0$, $\eta(r) > 0$ and $\eta'(r) > 0$ if $r > 0$ and

$$-((\eta'(r))^{p-1})' + f(\eta(r))) = 0 \quad \text{in} \quad (0, \psi_1(\infty)) \tag{81}$$

(see, for instance, Lemma 1.2) . In order to have $\overline{u} \in W^{1,p}(\Omega)$, and $\overline{u}$ with compact support, it is enough to choose K_1, K_2 and R_1 such that $\overline{u} \in C^1(\Omega)$, or equivalently, that $G \in C^1([0,\infty])$ and $G'(0) = 0$. This leads to the conditions

$$K_1 - K_2 R_0^{\frac{p}{p-1}} = \eta(R_1 - R_0) \tag{82}$$

$$K_2 \frac{p}{p-1} R_0^{\frac{p}{p-1}} = \eta'(R_1 - R_0) \tag{83}$$

Now we shall impose the condition that

$$-\Delta_p \overline{u} + f(u) = \tilde{g} \quad \text{in} \quad \Omega \tag{84}$$

for some $\tilde{g} \geqslant g$. First, we consider the region $\Omega_0 = \Omega \cup B_{R_0}(0)$, where R_0 is such that $S(g) \cup S(h|_{\partial\Omega}) \subset \Omega_0$. As $\overline{u} \geqslant 0$, we have that $f(\overline{u}) \geqslant 0$. So the condition (84) holds if, for instance,

$$-\Delta_p \overline{u} = -\frac{d}{dr}\left(\left(\frac{dG}{dr}(r)\right)^{p-1}\right) + \frac{N-1}{r}\left(\frac{dG}{dr}(r)\right)^{p-1} \geqslant \|g\|_{L^\infty}, \; r < R \;,$$

that is, if

$$N\left(K_1 \frac{p}{p-1}\right)^{p-1} \geqslant \|g\|_{L^\infty(\Omega)} \tag{85}$$

In the region $\Omega_1 = \{x \in \Omega : R_0 < |x| < R_1\}$ we know that $g = 0$. Then, using (81), we have that

$$-\Delta_p \overline{u} + f(\overline{u}) = -(|\eta'|^{p-2}\eta')' + f(\eta) + \frac{N-1}{r}(\eta')^{p-1}.$$

In conclusion, as $\eta' \geqslant 0$ the condition (84) holds in the whole domain where

$$\tilde{g}(x) = \begin{cases} \|g\|_{L^\infty} & \text{if } |x| \leq R_0 , \quad x \in \Omega \\ \frac{N-1}{|x|}(\eta'(R_1-|x|))^{p-1} & \text{if } R_0 \leq |x| \leq R_1 , \quad x \in \Omega \\ 0 & \text{if } R_1 \leq |x| , \quad x \in \Omega \end{cases}$$

and assuming that K_1 satisfies (85). In order to have

$$h|_{\partial\Omega} \leq \bar{u}|_{\partial\Omega} \tag{86}$$

we note that since G is non increasing, it suffices to require that $G(R_0) \geq \|u\|_{L^\infty(\Omega)} \geq u|_{\Omega_0}$ (recall that $h = 0$ on $\partial\Omega \cap (\mathbb{R}^N - B_{R_0}(x_0))$). Hence, it is enough to have

$$M \leq \eta(R_1 - R_0) , \tag{87}$$

where $M \geq \|u\|_{L^\infty(\Omega)}$. Finally, to show how the constants K_1, K_2 and R_1 can be chosen verifying (82),(83),(85) and (87), we can proceed as follows: We shall first choose R_1, K_1 and K_2 in terms of R_0 , which will be later fixed only in terms of g, h and $\|u\|_\infty$. From (87), and recalling that the function $\psi_1 = \eta^{-1}$ is nondecreasing, we take

$$R_1 = R_0 + \psi_1(M). \tag{88}$$

This choice allows us to take

$$K_2 = \eta'(M) \frac{p-1}{p} R_0^{-\frac{1}{p-1}}$$

so that (83) is verified. In consequence, (82) holds if we choose

$$K_1 = M + \eta'(\psi_1(M))^{\frac{p-1}{p}} R_0 . \tag{89}$$

All the conditions are now reduced to (85) and $B_{R_0}(0) \supset S(g) \cup S(h|_{\partial\Omega})$, which is satisfied if R_0 is large enough that

$$R_0 \geq \max \{\frac{(\|g\|_\infty/N)^{1/(p-1)}(p-1)-pM}{(p-1)\eta'(\psi_1(M))} , \operatorname{diam}(S(g) \cup S(h|_{\partial\Omega})\} , \tag{90}$$

where M is any positive bound of $\|u\|_{L^\infty}$ such as, for instance, the one given in (6**). To end the proof of this theorem we shall show the existence of a solution of (1),(2) (the uniqueness is again a trivial consequence of the comparison principle). Let R_0 and R_1 be given by (90) and (88), respectively, and define $\tilde{\Omega} = \{x \in \Omega \ : |x| \leq R_1 + 1\}$. Let $\tilde{u} \in W^{1,p}(\Omega) \cap L^{\infty}(\Omega)$ be the unique solution of the problem

$$- \Delta_p u + f(u) = g \quad \text{in} \quad \Omega$$

$$u = h \quad \text{on} \quad \partial\tilde{\Omega} \cap \partial\Omega$$

$$u = 0 \quad \text{on} \quad \partial\tilde{\Omega} - \partial\Omega .$$

Obviously the functions $\overline{u}(x) = G(|x|)$ and $\underline{u}(x) = -G(|x|)$, with G defined in (80) and the constants taken as before, are still super and subsolution of this problem. Then, $\underline{u}(x) \leq u(x) \leq \overline{u}(x)$ a.e. $x \in \tilde{\Omega}$. In particular, $\tilde{u}(x) = 0$ on the set $\{x \in \tilde{\Omega} \ : R_1 \leq |x| \leq R_1 + 1\}$ and so the extended function u, defined by $u = \tilde{u}$ in $\tilde{\Omega}$ and $u = 0$ in $\Omega - \tilde{\Omega}$, is such that $u \in W^{1,p}(\Omega) \cap L^{\infty}(\Omega)$ and satisfies (1),(2).□

Remark 1.11 The compactness of the support of the solution of (1),(2) can also be obtained by means of local super and subsolutions, and even by obtaining a sharper estimate of the location of the support of u, better than that derived in the above theorem: $S(u) \subset B_{R_1}(x_0) \cap \Omega$, R_1 given by (88).□

Remark 1.12 With respect to the case of unbounded data, for instance , $g \in L^1(\Omega)$ and $h \in W^{1,1}(\Omega)$ with $\Delta h \in L^1(\Omega)$, the construction of global super and subsolutions is much more delicate because, in general, the solution is not bounded in $S(g) \cup S(h|_{\partial\Omega})$. Nevertheless, if g and $h|_{\partial\Omega}$ have compact supports, the existence of a unique L^1-solution of (1*) (2) with compact support can be obtained in the following way: first, the compactness of the support of any solution is established by using the functions $\pm G(|x-x_0|)$ as super-subsolutions, now in the set $\Omega - B_{R_0}(x_0)$, where R_0 is such that $S(g) \cup S(h|_{\partial\Omega}) \subset B_{R_0-1}(x_0)$, and by replacing the conditions (82) and (83) by

$$M \geq \eta(R_1 - R_0) .$$

Here, M is any bound of u in the set $N(g) \cup N(h|_{\partial\Omega})$ such as, for instance, that given in (6*) (this also holds for unbounded domains). Finally, the existence of such a solution is proved as in Theorem 1.18. We remark that this can also be applied to L^1-solutions of (1),(2).□

Using some elementary differential geometry it is possible to construct global super and subsolutions much better adapted to the peculiarity of the data than the ones exhibited in the above theorems. To explain this in a simple way, we shall only consider equation (1*) and assume that

$$S(g) \quad \text{is a bounded and convex set of } \mathbb{R}^N. \tag{91}$$

Given $R > 0$, let V_R be a tubular neighborhood of the boundary $\partial S(g)$, defined by the usual parametric representation

$$x = x(\omega,t) = \omega + t\,\vec{n}(\omega) \quad , \quad \omega \in \partial S(f) \quad , \quad t \in (-R,R)$$

where $\vec{n}(\omega)$ is the outward normal unit vector on $\partial S(g)$ at ω. Let $V_R^+ = \{ x = x(\omega,t) \in V_R \text{ with } t > 0 \}$

Theorem 1.19 *Let f be an odd nondecreasing continuous real function satisfying* (62). *Let* $g \in L^\infty(\Omega)$ *such that it verifies* (91) *and let* $u \in C^{1,\alpha}(\overline{\Omega}) \cap W^{2,\infty}(\Omega)$ *be the unique solution of the problem*

$$\begin{aligned} -\Delta u + f(u) &= g \quad \text{in} \quad \Omega \\ u &= 0 \quad \text{on} \quad \partial\Omega \end{aligned}$$

Consider the function $\overline{u}(x) \in C^{1,\alpha}(\Omega - S(g))$ *defined by*

$$\overline{u}(x) = \begin{cases} \eta(R_1 - t) & \text{if} \quad x \in V_{R_1}^+, \quad x = \omega + t\vec{n}(\omega) \quad , \omega \in \partial S(g) \ , 0 \leq t \leq R_1 \\ 0 & \text{if} \quad x \in \Omega - S(g) \quad , \quad d(x, \partial S(g)) > R_1 \end{cases}$$

where R_1 *is suitable positive constant and* $\eta = \psi_1^{-1}$, ψ_1 *given in* (33). *Then* $\overline{u}$ *is a supersolution in the set* $\Omega - S(g)$, *i.e.*, $u \leq \overline{u}$ *in the set* $\Omega - S(g)$.

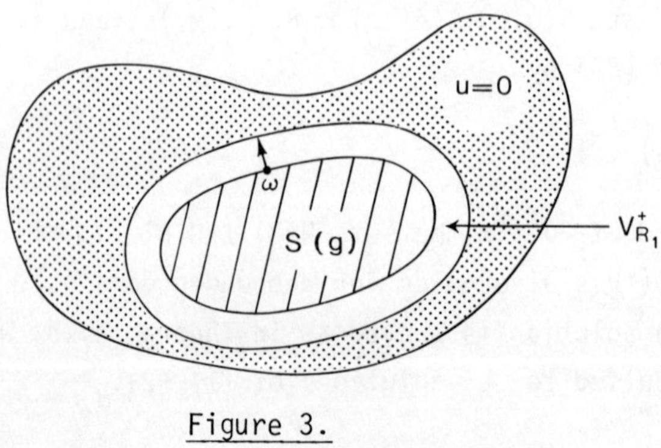

Figure 3.

radial and non-increasing. Now, if we multiply (62) by $-f(t)$ and integrate between 0 and M, where $M > \max_{\overline{\Omega}} u$, we obtain

$$4\pi \int_0^M -f(t)\mu(t)dt \leq \int_0^M \left(\int_{\{u>t\}} -f(u)dx \right) \left(-\frac{d\mu}{dt}\right) f(t)dt$$

But we have

$$-\int_0^M f(t)\mu(t)dt = -\int_0^M F(t)(-d\mu(t)) = -\int_\Omega F(u)dx$$

where $F(r) = \int_0^r f(s)ds$, and

$$\int_{\{u>t\}} -f(u)dx = -\int_t^M f(s)(-d\mu(s)).$$

Then

$$\int_0^M \left(\int_t^M -f(s)(-d\mu(s))\right)\left(-\frac{d\mu}{dt}\right)(-f(t))dt \leq \frac{1}{2}\left(\int_0^M (-f(s))(-d\mu(s))\right)^2 \leq \frac{1}{2}\left(\int_\Omega f(u)dx\right)^2.$$

In conclusion, we have proved

$$8\pi \int_\Omega (-F(u))dx \leq \left(\int_\Omega -f(u)dx\right)^2 \tag{63}$$

and the equality holds only if u is radial and decreasing. Then using the Pohozaev identity (see Lemma 2.30)

$$\int_\Omega -f(u)dx = \frac{1}{4}\int_{\partial\Omega} \left(\frac{\partial u}{\partial n}\right)^2 ds ,$$

where n is the outward unit normal to $\partial\Omega$, we have

$$\int_\Omega F(u) \geq \frac{1}{8\pi}\left(\int_{\partial\Omega} \left|\frac{\partial u}{\partial n}\right| ds\right) = \frac{1}{8\pi}\left(\int_{\partial\Omega} \left(-\frac{\partial u}{\partial n}\right)ds\right)^2 \geq \frac{1}{8\pi}\left(\int_\Omega -f(u)dx\right)^2.$$

Comparing with (63) we conclude the proof easily. □

<u>Remark 1.20</u>. The above proof is an adaptation of a similar result of Crooke-Sperb [1]. The case of higher dimensions seems to be open. □

<u>Remark 1.21.</u> Let u satisfy

$$-\Delta_p u + f(u) = g \quad \text{in} \quad \Omega$$

$$u = 0 \quad \text{on} \quad \partial\Omega .$$

If g is a radially symmetric function and, for instance, f is strictly increasing, it is clear that the unique solution will also be radially symmetric and so also its null set $N(u)$. Nevertheless, we remark that, in general, the mere symmetry of the boundary of the support $S(g)$ of g does not imply the symmetry of the support $S(u)$ of the solution u. Indeed it is not difficult to construct a non radially-symmetric function g satisfying the growing condition (77) of Section 1.1 in some strict part Γ of the boundary of its support. In this case, the solution u must vanish on Γ but not on the rest of $\partial S(g)$ if g is suitably chosen (see Figure 6).

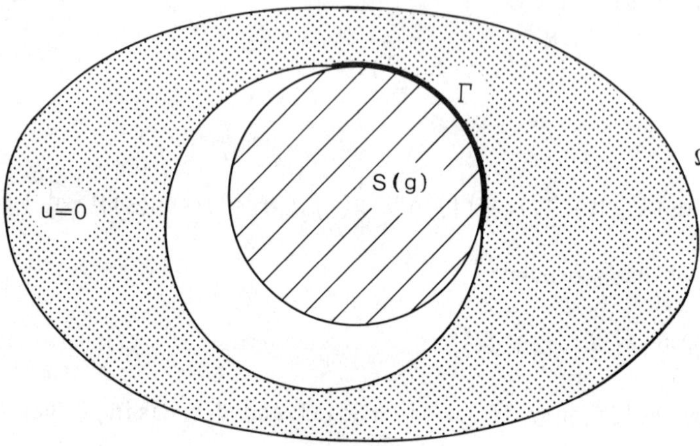

Figure 6

1.3c. The free boundary for equations with a general nonlinear diffusion term.

We have already seen several existence criteria of the null set of the solution of second order quasilinear equations of the form

$$- \Delta_p u + f(u) = g \tag{64}$$

More general absorption terms were discussed in Proposition 1.11 (see also Section 2.1). Another generalization was given in Theorem 1.13 for the case of a general linear diffusion (see also the comments made about the nonlinear equation (63) of Section 1.1). The main goal of this subsection is to study the formation of the free boundary $\mathcal{F}(u)$, for any $u \in W^{1,p}(\Omega) \cap L^\infty(\Omega)$ solution of the equation

$$- \operatorname{div} A(u,\nabla u) + f(u) = g \quad \text{in} \quad \Omega \tag{65}$$

$$u = h \qquad \text{on} \quad \partial\Omega \tag{66}$$

The main idea will be to obtain an adequate relation between u and v, where v is the solution of (20),(21). In this respect we remark that the structural assumption (5) made on the term $A(u,\nabla u)$ in Theorem 1.26 gives only a negative result (see Theorem 1.28). Roughly speaking, if the absorption term f is fixed, the assumption (5) says that the diffusion term of (65) is faster than that of the equation (64) and so, it is possible to show that the null set of the solution of (65)(66) is the empty set, even though the null set of v, solution of (64)(66), is of positive measure. In consequence, condition (5) must be substituted by another of a different nature, expressing that now the diffusion term of (65) is slower than that of the equation (64). A natural condition in this direction is

$$|A(r,\xi)| \leqslant C|\xi|^{p-1} \quad \text{for every} \quad (r,\xi) \in \mathbb{R} \times \mathbb{R}^N, \text{ for some } \; C > 0 \tag{67}$$

Nevertheless, one cannot expect, in general, to obtain the corresponding inequalities opposite to those given in Theorem 1.27 $f(v) \lesssim f(u^*)$ (where u^* is the symmetric rearrangement of any solution of (65) vanishing on $\partial\Omega$ and v is the corresponding radially symmetric solution of the associated equation (64)). The main reason for this is that the isoperimetric inequality makes the method followed in the proof of Theorem 1.27 irreversible. In spite of this difficulty, the relation $f(v) \lesssim f(u^*)$ holds if u is radially symmetric:

Proposition 1.37. *Let Ω be a ball of $\mathbb{R}^N$. Let $A(r,\xi)$ satisfy (67) and f be a continuous non decreasing real function. Given $K \geqslant 0$ let u and v be in $W_0^{1,p}(\Omega)$ and satisfy*

$$- \operatorname{div} A(u,\nabla u) + f(u) = K \qquad \textit{in} \quad \Omega \tag{68}$$

and

$$- C \Delta_p v + f(v) = K \qquad \textit{in} \quad \Omega \tag{69}$$

respectively. Let us assume that $u = u^*$. *Then* $f(v) \lesssim f(u^*)$

As in previous sections we postpone the proof of this result in order to explain (roughly) how it can be applied to the study of the free boundary of (65)(66).

Theorem 1.38. *Let* Ω *be an open regular set of* $\mathbb{R}^N$, $p > 1$ *and consider* $g \in L^{p'}(\Omega)$ *and* $h \in W^{1,p}(\Omega)$. *Let* $u \in W^{1,p}(\Omega) \cap L^\infty(\Omega)$ *be any nonnegative solution of the problem* (65)(66). *Assume the growing condition* (67), *as well as*

$$\text{"the comparison principle holds for the problem (65)(66)",} \tag{70}$$

$$\left.\begin{array}{l}\text{"any solution of (68) on a ball and vanishing on the boundary}\\ \text{coincides with its symmetric rearrangement"}\end{array}\right\} \tag{71}$$

Finally suppose that

$$\int_{0^+} \frac{ds}{F(s)^{1/p}} < +\infty \tag{72}$$

where F *is the primitive of* f *with* $F(0)=0$. *Then the null set* $N(u)$ *of* u *satisfies*

$$N(u) \supset \{x \in N(g) \cup N(h|_{\partial\Omega}) : d(x, S(g) \cup S(h|_{\partial\Omega})) > L\}.$$

where $L = \psi_{1/N}(M)$, M *is any bound of* $\|u\|_\infty$ *and* ψ_μ *is defined in* (33) *of Section 1.1.*

Proof. As in the proof of Theorem 1.9, by (70) it suffices to show that given $x_0 \in N(g) \cup N(h|_{\partial\Omega})$ and $R = d(x_0, S(g) \cup S(h|_{\partial\Omega}))$, if $\bar{u}(x) = \bar{u}(x:x_0)$ satisfies

$$\begin{aligned} -\operatorname{div} A(u, \nabla u) + f(u) &= 0 && \text{in} \quad B_R(x_0) \\ u &= M && \text{on} \quad \partial B_R(x_0). \end{aligned}$$

then $\bar{u}(x_0) = 0$ if $R > L$. In order to prove this, note that without loss of generality we can assume $C = 1$. Consider $v \in W^{1,p}(B_R(x_0))$ satisfying

$$\begin{aligned} -\Delta_p v + f(v) &= 0 && \text{in} \quad B_R(x_0) \\ v &= M && \text{on} \quad \partial B_R(x_0). \end{aligned}$$

Then, by Proposition 1.37, we have $f(M-v) \lesssim f(M-\bar{u})$. But (72) implies that $f^{-1}(0) = 0$. Then, arguing as in Theorem 1.28 we obtain that meas $N(v) \leq$ meas $N(\bar{u})$. Moreover, if $R \geq \psi_{1/N}(M)$, by Theorem 1.9 we know that $v(x_0) = 0$ and that meas $N(v) > 0$. This proves that meas$N(\bar{u})>0$. Finally, by (71) $\bar{u}$ is a non-increasing function of $|x|$ and we deduce that necessarily $\bar{u}(x) = 0$ a.e. $x \in B_\varepsilon(x_0)$ for some small enough $\varepsilon > 0$. □

Remark 1.22. A comparable result, but without the assumptions (70),(71), will be obtained by means of an energy method in Section 3.1. □

Proof of Proposition 1.37. Let u satisfy (68). Since the right hand side of (68) is a constant, the proof of Lemma 1.29 gives us that

$$-\frac{d}{dt}\int_{\{u>t\}} A(u,\nabla u)\cdot\nabla u \, dx = \int_0^{\mu(t)} [K-f(\tilde{u}(s))]ds ,$$

where μ and u are the distribution function and the decreasing re-arrangement of u. Moreover, since Ω is a ball and u is symmetric (and so $|\nabla u|$ is constant on the level set $\{u = t\}$), it is not difficult to show (see e.g. Talenti[3] p. 362) that u satisfies (42) with the equality sign, i.e.

$$\left(-\frac{d\mu}{dt}\right)^{1/p'}\left(-\frac{d}{dt}\int_{\{u>t\}}|\nabla u|^p\,dx\right)^{1/p} = N\,\omega_N^{1/N}\,\mu(t)^{(N-1)/N} \quad \text{for a.e. } t > 0.$$

On the other hand, using the Schwartz inequality on the set $\{x\in\Omega : t<u(x)\leq t+h\}$, the assumption (67), and passing to the limit as $h \downarrow 0$, we obtain

$$-\frac{d}{dt}\int_{\{u>t\}} A(u,\nabla u)\cdot\nabla u \, dx \leq -\frac{d}{dt}\int_{\{u>t\}}|\nabla u|^p dx \qquad \text{a.e.} \quad t > 0,$$

(again, it suffices to consider $C = 1$ in (68)). In conclusion, as in Lemma 1.31, we obtain now that

$$-\frac{d\tilde{u}}{ds}(s) \geq \left(\frac{1}{N\,\omega_N^{1/N} s^{(N-1)/N}}\right)^{p/(p-1)}\left[\int_0^s (K-f(\tilde{u}(\theta))d\theta\right]^{1/(p-1)} \quad \text{a.e. } s\in(0,|\Omega|) \tag{73}$$

Finally, the conclusion holds as in Theorem 1.26 by noting that if $v\in W_0^{1,p}(\Omega)$ satisfies (69), then its decreasing rearrangement satisfies (73) with the equality sign (Lemma 1.32). □

1.4. FURTHER RESULTS ON THE FREE BOUNDARY FOR SEMILINEAR EQUATIONS.

The analysis made in the previous section can be completed by considering some other questions such as the regularity and measure properties of the free boundary. This kind of property seems to be hard to establish in the general setting we are considering and this is why some simplification is needed. In this section we shall only consider some semilinear problems such as for instance,

$$- \Delta u + f(u) = 0 \quad \text{in} \quad \Omega \tag{1}$$

$$u = 1 \quad \text{on} \quad \partial\Omega \quad , \tag{2}$$

where Ω is an open bounded set with regular $\partial\Omega$ and f is a continuous nondecreasing function such that $f(0) = 0$.

Due to the semilinear character of (1), precise information on the behaviour of solutions near the free boundary will be obtained as a consequence of the main result of Subsection 1.4a, relative to a Harnack type inequality. This inequality is also proved for solutions of the equation with an x-dependent absorption term. The locally finite character of the perimeter of the support $S(u)$ and the zero Lebesgue measure of the free boundary $\mathcal{F}(u)$, are some of the consequences of the Hausdorff measure estimates given in Subsection 1.4b. The stability of the free boundary with respect to a parameter λ in the equation, is considered in Subsection 1.4c. Finally, some remarks on the convexity and other geometrical properties of the free boundary are made in Subsection 1.4d.

The results of this section are due to Phillips[2], Friedman -Phillips [1] , Spruck [1] and Alt-Phillips [1].

1.4a. On the behaviour of solutions near the free boundary.

The results of this subsection concern mainly nonnegative solutions of the homogeneous semilinear problem

$$- \Delta u + u^q = 0 \quad \text{in} \quad \Omega \tag{3}$$

$$u = h \quad \text{on} \quad \partial\Omega \tag{4}$$

where Ω is an open bounded set with $\partial\Omega$ locally Lipschitz, $0 < q < 1$, and

$h \in H^1(\Omega)$,$h \geq 0$. We recall that the existence and uniqueness of the solution $u \in H^1(\Omega)$ can be obtained, for instance, by minimizing the functional

$$J(v) = \int_\Omega \left(\frac{1}{2} |\nabla v|^2 + \frac{1}{q+1} |v|^{q+1} \right) dx \tag{5}$$

on the convex set $K = \{v \in H^1(\Omega) : v-h \in H^1_0(\Omega)\}$. Due to the assumption $0 < q < 1$, the free boundary $F(u)$ may exist depending on the "sizes" of Ω and h (see Theorem 1.9). By standard arguments of regularity,$u \in C^{2,\delta}(D)$ for $\overline{D} \subset \Omega$ and some $0 < \delta < 1$. Moreover, if D is such that $u(x) \geq \tau$ in D for some $\tau > 0$, then $u \in C^\infty(D)$. In this way, a loss of regularity is expected to occur at the free boundary $F(u)$ (see examples in Section 1.1a) The study of the behaviour of u near $F(u)$ is then of a great interest, and will be carried out locally. The main ingredient in such a study is the following Harnack type inequality, controlling (in a weak sense) the behaviour of u near a point $x_0 \in F(u)$ by means of the average of u over $\partial B_r(x_0)$

$$⨍_{B_r} u = \frac{1}{\alpha_N r^{(N-1)}} \int_{\partial B_r(x_0)} u \, ds , \tag{6}$$

where $B_r(x_0)$ is the closed ball with radius r, and center x , and α_N the surface area of the unit ball in $\mathbb{R}^N$.

<u>Lemma 1.39</u>. *Let* $0 < q < 1$. *Then, there exist constants* C_0, $\tau > 0$, *depending only on* N *and* q *such that if* $u \geq 0$ *satisfies the equation* (3) *in* $B_r = B_r(0)$, *and*

$$⨍_{\partial B_r} u \geq C_0 \, r^{\frac{2}{1-q}} \tag{7}$$

then

$$u(x) \geq \tau ⨍_{\partial B_r} u \qquad \textit{for} \qquad x \in B_{r/4}. \tag{8}$$

<u>Remark 1.23</u> A crucial point in the proof of the above result is the <u>homogeneity</u> of the equation (3). Due to this property we can rescale the domain and solutions: if $u \in H^1(B_r)$ satisfies (3) in B_r, then for every $s > 0$, the function u^s, defined by

$$u^s(x) = \frac{u(sx)}{s^{2/(1-q)}} , \qquad (9)$$

is such that $u^s \in H^1(B_{r/s})$ and satisfies (3) on $B_{r/s}$ (note that

$$J_r(v) = s^{(N+\frac{4}{1-q}-2)} J_{r/s}(v^s)$$

for every $v \in H^1(B_r)$, if J_r denotes the functional (5) on B_r instead of on Ω). □

Proof of Lemma 1.39. It is enough to consider the case $r=1$ and the local solution of (3) on B_1 defined by

$$u^r(x) = \frac{u(rx)}{r^\beta} , \quad \beta = 2/(1-q)$$

Note that

$$\frac{1}{r^\beta} ⨍_{\partial B_r} u = ⨍_{\partial B_1} u^r. \qquad (10)$$

We first determine a constant $C_1 > 0$ so that, if $⨍_{\partial B_1} u \geq C_1$, then $u(0) \geq \frac{1}{2} ⨍_{\partial B_1} u$. Since $\Delta u = f(u)$, from the representation formulas (see Gilbarg-Trudinger [1])

$$⨍_{\partial B_1} u - u(0) = C \int_{B_1} \Delta u \, dx = \int_0^1 (⨍_{\partial B_r} u^q) r^{N-1} G(r)dr \leq \bar{C}(⨍_{\partial B_r} u)^q$$

where $G(r) = -\log r$ if $N=2$ and $G(r) = r^{2-N}-1$ if $N \geq 3$, C and $\bar{C}$ are constants and where the last inequality holds because $0 < q < 1$. Hence, if

$$⨍_{\partial B_1} u \geq (2\bar{C})^{\frac{1}{1-q}} \qquad (11)$$

then $u(0) \geq \frac{1}{2} ⨍_{\partial B_1} u$. So, it suffices to take $C_1 = (2\bar{C})^{1/(1-q)}$.

Now, let $x \in B_{1/4}(0)$. Since u is subharmonic

$$u(0) \leq C(N) ⨍_{\partial B_{1/2}(x)} u \qquad (12)$$

Using (11) to estimate u(0) and assuming further that

$$⨍_{\partial B_1(0)} u \geq 2^{1-\beta} C(N) C_1$$

we get, from (12), that

$$\fint_{\partial B_{1/2(x)}} u \geq 2^{-\beta} C_1 .$$

Hence, by (11) we have

$$u(x) \geq \frac{1}{2} \fint_{\partial B_{1/2}(x)} u .$$

Finally, estimating the right-hand side by (11),(12) we obtain

$$u(x) \geq (4\,C(N))^{-1} \fint_{\partial B_1(0)} u \qquad \text{for} \qquad x \in B_{1/4(0)}$$

provided

$$\fint_{\partial B_1(0)} u \geq C_1 \cdot \max(1, 2^{1-\beta} C(N)) =: C_0 \ . \ \square$$

Before giving some applications of this result, it is interesting to see a similar Harnack inequality, true for semilinear equations satisfying the homogeneity property in a weak sense. This will be applied to the case of a x-dependent absorption term in the equation:

$$-\Delta u + \theta(x) u^q = 0 \quad , \quad u \geq 0, \tag{13}$$

where $0 < q < 1$.(For the existence of $\mathcal{F}(u)$ see Proposition 1.11).

<u>Lemma 1.40</u>. *Let* $0 < q < 1$, $0 \leq \theta \leq M$ *and* $u \geq 0$ *satisfying* (13) *in the ball* B_1. *Then*

$$\left(\fint_{\partial B_r} u\, ds\right)^{1-q} \geq \left(\fint_{\partial B_1} u\, ds\right)^{1-q} - (1-q)\frac{M}{N}(1-r^2). \tag{14}$$

In particular, if $\fint_{\partial B_r} u\, ds \geq c_0 r^{2/(1-q)}$, $c_0 = [(1-q)M/N]^{1/(1-q)}$, u *is positive in a neighbourhood of the origin.*

<u>Proof</u>. Let $m(r) = \fint_{\partial B_r} u\, ds$. Then

$$\fint_{\partial B_r} \Delta u\, ds = \fint_{\partial B_r} \left(\frac{\partial^2 u}{\partial r^2} + \frac{N-1}{r}\frac{\partial u}{\partial r}\right) ds = \fint_{\partial B_r} \theta(x) u^q\, ds \leq$$

$$\leq M \fint_{\partial B_r} u^q\, ds \leq M\, m^q \tag{15}$$

by Jensen's inequality. Also,

$$m'(r) = \fint_{\partial B_r} \frac{\partial u}{\partial r}\, ds = \frac{1}{\omega_N r^N} \int_{B_r} \Delta u \geq 0$$

Integrating (15)

$$r^{N-1}\, m'(r) \leq M \int_0^r s^{N-1} m(s)^q\, ds \leq \frac{M}{N}\, r^N\, m(r)^q.$$

Hence

$$m'(r) \leq \frac{M}{N}\, r\, m(r)^q \tag{16}$$

Integrating (16) from r to 1 gives

$$\frac{m(1)^{1-q}}{1-q} - \frac{m(r)^{1-q}}{1-q} \leq \frac{M}{N}\,(1-r^2)\ ,$$

which shows (14). Finally, we note that, if $u^\lambda(x) = \lambda^{\frac{2}{1-q}}\, u(\lambda x)$ and u satisfies (13) on B_r, taking $\lambda = r$ we have

$$-\,\Delta u^\lambda + \theta(\lambda x)(u^\lambda)^q = 0 \quad \text{in} \quad B_1.$$

Then the second conclusion of the statement is an easy consequence of (13) and (10).□

A first consequence of the above lemmas is the following:

Corollary 1.41. *Let* u *be a nonnegative solution of equation* (3) (*resp.* (13)) *on the ball* $B_r(0)$ *and suppose that* $0 \in F(u)$. *Then*

$$\fint_{\partial B_r(0)} u\, ds \leq c_0\, r^{\frac{2}{1-q}}\ ,$$

(c_0 *given in Lemma 1.40*). *Moreover there exists* $C = C(N,q)$(*resp* $C(N,q,\theta)$) *such that*

$$0 \leq u(x) \leq C|x|^{\frac{2}{1-q}} \qquad x \in B_{r/2}\ . \tag{17}$$

Proof. The first part follows obviously from Lemma 1.39 or 1.40, respectively. Now, let $x \in B_{r/2}(0)$ and set $R = |x|$. Let $w(x)$ be the harmonic function on $B_{2R}(0)$ satisfying $w = u$ on ∂B_{2R}. Then by Lemma 1.40,

$$C_0(2R)^{\frac{2}{1-q}} \geq \fint_{\partial B_{2R}} u\, ds = \fint_{\partial B_{2R}} w\, ds = w(0) \geq C_1 w(x) \geq C\, u(x)\ ,$$

where we have used the classic Harnack's inequality (for w) and the subharmonicity of u. Hence

$$u(x) \leqslant C|x|^{\frac{2}{1-q}}, \quad C = 2^{\frac{2}{1-q}} \frac{C_0}{C_1}. \square$$

Remark 1.24. The inequality (17) gives a direct proof of the continuity of the solution on the free boundary $F(u)$ (and then on D, $\overline{D} \subset \Omega$) as well as that $\nabla u(P) = 0$ for every $P \in F(u)$. Indeed, by (17) we have that

$$0 \leqslant u(x) \leqslant C|x - P|^{\frac{2}{1-q}} \quad \text{for } x \text{ such that } |x-P| \leqslant d(P,\partial\Omega)/2.$$

We also note that (17) coincides with the inequality obtained in Section 1.1 to estimate the location of the free boundary. $\square$

Next, we derive lower estimates of growth for u, in a sense of average, near $F(u)$.

Proposition 1.42. *Let* u *be a nonnegative solution of equation* (3) (*resp.* (13)) *on* $\Omega \supset B_r(x_0)$ *and suppose that* $0 \in F(0)$ *and that* $B_r(x_0) \subset \{x: u > 0\}$ (*resp* $B_r(x_0) \subset \{x: u(x) > 0\} \cap \{x: \theta(x) > \lambda\}$ *for some* $\lambda > 0$). *Then*

$$⨍_{\partial B_r(x_0)} u \, ds \geqslant Cr^{2/(1-q)} \tag{18}$$

for some $C > 0$.

Proof. Again, by scaling, it suffices to assume $r = 1$ (see (10)). Since u is subharmonic

$$\sup_{B_{1/2}(x_0)} u \leqslant C_2 ⨍_{\partial B_1(x_0)} u \, ds \tag{19}$$

We claim that

$$\sup_{B_{1/2}(x_0)} u > \left(\tfrac{1}{2}\right)^{2/(1-q)} K_{N,\lambda} \tag{20}$$

where $K_{N,\lambda}$ was defined in (43) of Section 1.1, and $\lambda = 1$ if u satisfies (3). Indeed, otherwise, if we define $v(x) = K_{N,\lambda}|x-x_0|^{2/(1-q)}$, we have that $u \leqslant v$ on $B_{1/2}(x_0)$. Moreover, by Lemma 1.6, v satisfies $-\Delta v + \lambda v^q = 0$ on $B_{1/2}(x_0)$. Then by the comparison principle $u \leqslant v$ in $B_{1/2}(x_0)$, a contradiction, since $u(x_0) > 0$. The combination of (19) and (20) proves the result. □

Remark 1.25. It is clear that (18) implies that

$$\sup_{x\in\partial B_r(x_0)} u \geq C\,|x - x_0|^{\frac{2}{1-q}} \tag{21}$$

This inequality shows a kind of nondegeneracy of u on the set $\tilde{S}(u) = \{x \in \Omega : u(x) > 0\}$, in the sense that u cannot be uniformly small in some neighborhood of any point of $\tilde{S}(u)$. A similar result will be proved later (see Theorem 1.44). □

Another important growth estimate is related to the correct balance between u and $|\nabla u|$ near the free boundary $\mathcal{F}(u)$. In this respect we first recall the estimate

$$|\nabla u(x)|^2 \leq \frac{2}{q+1}\,(u(x))^{q+1} \qquad \text{for} \qquad x \in \overline{\Omega} \tag{22}$$

given in Theorem 1.24 for the functions solutions of equation (3) and constant on $\partial\Omega$. In the next subsection we shall need estimates like (22) but when u is merely a local solution of (3). This is the content of the following result due to Phillips [2].

Lemma 1.43. *Let* $0 < q < 1$. *Given* $k > 0$ *there exists an* $\varepsilon = \varepsilon(k) > 0$ *such that, if* u *satisfies equation* (3) *in* $B_{2r}(x_0)$ *for some* $r > 0$ *and* $x_0 \in \mathcal{F}(u)$, *then*

$$\frac{|\nabla u|^2}{u^{(q+1)}} \leq \frac{2}{q+1} + k \qquad \text{on} \qquad \{u > 0\} \cap B_{\varepsilon r}(x_0)\ . \ \square \tag{23}$$

1.4b Lebesgue and Hausdorff measure of the free boundary. Application to domains of boundary having nonnegative mean curvature.

It seems natural to expect the free boundary $\mathcal{F}(u)$ with a zero N-dimensional Lebesgue measure. We shall prove this by a stronger result about the density of the boundary points of the strict support of u, $\tilde{S}(u)$, defined by

$$\tilde{S}(u) \equiv \{x \in \Omega : u(x) > 0\} \equiv \{u > 0\}. \tag{24}$$

Theorem 1.44. *Let* u *be the solution of* (3),(4). *Assume* $0 \in F(u)$ *and* $B_{4r}(0) \subset \Omega$. *Then there exist* $\varepsilon_0 > 0$, $C > 0$ *and* $y \in \partial B_\varepsilon(0)$ *such that*

$$u(y) \geq C\,\varepsilon^{\frac{2}{1-q}} \tag{25}$$

for every $\varepsilon < \varepsilon_0 r$. *Moreover, there exists* $\theta = \theta(N,q) \in (0,1/4)$ *such that* $B_{\theta\varepsilon}(y) \subset \tilde{S}(u)$. *In particular*

$$\frac{|\tilde{S}(u) \cap B_{2\varepsilon}(0)|}{|B_{2\varepsilon}(0)|} \geq (\frac{\theta}{2})^N \quad \text{if} \quad 2\varepsilon < \varepsilon_0 r. \tag{26}$$

Proof. Take $\varepsilon \leq r/2$, let $y_0 \in \tilde{S}(u) \cap B_\varepsilon(0)$ and define

$$w(x) = u(x)^{1-q} - \frac{K(1-q)}{2N}|x - y_0|^2 \quad \text{on} \quad \tilde{S}(u) \cap B_\varepsilon(y_0) \tag{27}$$

where K is a positive constant to be chosen. By the regularity results, $w \in C(\bar{S}(u) \cap B_\varepsilon(y_0)) \cap C^\infty(\tilde{S}(u) \cap B^0_\varepsilon(y_0))$, where $B^0_\varepsilon(y_0) = \text{int}\,(B_\varepsilon(y_0))$. Furthermore,

$$\Delta w = (1 - q)\,[- q\,\frac{|\nabla u|^2}{u^{q+1}} + 1 - K\,] \quad \text{on} \quad \tilde{S}(u) \cap B^0_\varepsilon(y_0).$$

We shall prove that $\Delta w \geq 0$. Let us set $K = (1-q)/(1+q)^2$. By Lemma 1.43, there exists $\varepsilon_0 > 0$ such that

$$\frac{|\nabla u|^2}{u^{q+1}} \leq \frac{2}{q+1} + K \quad \text{on} \quad \tilde{S}(u) \cap B^0_{\frac{\varepsilon_0 r}{2}}(0) .$$

Hence

$$\Delta w \geq (1-q)[- q\,(\frac{2}{q+1} + K) + 1 - K] = 0 \quad \text{on} \quad \tilde{S}(u) \cap B_\varepsilon(0), \text{ if } \ 2\varepsilon < \varepsilon_0 r.$$

Since $w(y_0) = u(y_0)^{1-q} > 0$, by the maximum principle, $w > 0$ at some point of $\partial(\tilde{S}(u) \cap B_\varepsilon(y_0))$. But $w < 0$ on $F(u)$, so there exists a $z \in \partial B_\varepsilon(y_0)$ such that $w(z) > 0$, i.e. $u(z) \geq C\,\varepsilon^{2/1-q}$ with $C^{(1-q)} = [(1-q)^2/2N(1+q)^2]$. Finally, if we let $y_0 \to 0$, then a sequence of points z converges to some $y \in \partial B_\varepsilon(0)$, and (25) holds by continuity.

To prove (26), let $\tilde{C} = \min\{C, C_0\}$, where C_0 is given in Lemma 1.39, and define $\theta = \frac{1}{4}\left(\frac{\tilde{C}}{C_0}\right)^{2/(1-q)}$. Then

$$u(y) \geq \tilde{C}(4\theta)^{-2/(1-q)}(4\,\theta\varepsilon)^{2/(1-q)} = C_0(4\theta\varepsilon)^{2/(1-q)}$$

Since u is subharmonic

$$⨍_{\partial B_{4\theta\varepsilon}(y)} u \geq C_0(4\theta\varepsilon)^{2/(1-q)} .$$

From Lemma 1.39 (or Lemma 1.40), it follows that $B_{\theta\varepsilon}(y) \subset \tilde{S}(u)$. Hence

$$\frac{|\tilde{S}(u) \cap B_{2\varepsilon}(0)|}{|B_{2\varepsilon}(0)|} \geq \frac{|B_{\theta\varepsilon}(y)|}{|B_{2\varepsilon}(0)|} = \left(\frac{\theta}{2}\right)^N > 0. \square$$

<u>Remark 1.25.</u> Note that, by Proposition 1.42, we know more than (25). Indeed, $\sup_{\partial B_{\theta\varepsilon}(x_0)} u \geq K(\theta\varepsilon)^{2/1-q}$ for some constant $K > 0$ and for every $\varepsilon < \varepsilon_0 r$. $\square$

<u>Corollary 1.45</u>. *Let* u *be the solution of* (3), (4). *Then* $|F(u)| = 0$.

<u>Proof</u>. From a well-known result of measure theory (see e.g. Munroe [1]), it is enough to show that no point of $F(u)$ can be a density point of $F(u)$. Now, by Theorem 1.44, if $x_0 \in F(u)$, then

$$\frac{|F(u) \cap B_r(x_0)|}{|B_r(x_0)|} \leq \frac{|N(u) \cap B_r(x_0)|}{|B_r(x_0)|} \leq \frac{|B_r(x_0) - B_{\varepsilon_0 r/2}(y)|}{|B_{\varepsilon_0 r/2}(y)|} \leq 1 - \left(\frac{\theta}{2}\right)^N < 1 ,$$

if $B_{4r}(x_0) \subset \Omega$ and y, θ and ε_0 are taken as in Theorem 1.44. $\square$

The above result can be improved by using some other measure different from the Lebesgue measure which allows us to distinguish among the zero Lebesgue sets. That is the case of the Hausdorff measure. We recall the exact notion:

<u>Definition 1.3.</u> *Given a real number* $a > 0$, *the* a*-dimensional Hausdorff outer measure of a set* $E \subset \mathbb{R}^N$ *is given by*

$$H^a(E) = c(a) \lim_{d \to 0} \left(\inf \sum_j \operatorname{diam}(E_j)^a\right), \tag{28}$$

where the infimum is taken over all countable coverings of E *by sets* E_j *with* $\operatorname{diam}(E_j) < d$ *and where*

$$c(a) = 2^{-a}\,\Gamma(\tfrac{1}{2})^a / \ (\tfrac{a}{2}+1)\ , \quad \Gamma \equiv \textit{the gamma function.}$$

If $a = 0$, $H^0(E)$ = *cardinality of* E *and if* $a < 0$, $H^a(E) = \infty$ *if* $E \neq \emptyset$. *Finally the Hausdorff dimension of* E *is defined by*

$$H\text{- dim } E = \inf\{a \in \mathbb{R} : H^a(E) = 0\}.$$

Systematic treatments of the Hausdorff measure are available in the books by Federer [1], Rogers [1] and Falconer [1]. A first estimate on the Hausdorff measure of $F(u)$ is the following:

Theorem 1.46. *Let* u *be the solution of* (3), (4). *Then*

$$H^{N-1}(F(u) \cap D) < +\infty \tag{29}$$

for each D *with* $\overline{D} \subset \Omega$.

The main idea in the proof of the above result, due to Phillips[ ], is to show first that if $0 \in F(u)$ and $B_{4r}(0) \subset \Omega$, then

$$H^{N-1}(F(u) \cap B_r(0)) \leqslant C \varliminf_{\varepsilon\to 0} \frac{|\{0 < u \leqslant \overline{M}\varepsilon^{2/(1+q)}\} \cap B_{2r}(0)|}{\varepsilon} \tag{30}$$

for some positive constants $\overline{M}$ and C. Finally

$$\frac{|\{0 < u \leqslant \overline{M}\,\varepsilon^{2/(1-q)}\} \cap B_{2r}(0)|}{\varepsilon} = O(1) \quad \text{as} \quad \varepsilon \to 0 \tag{31}$$

and thus the limit infimum is finite, which leads to the conclusion because, in the statement, D can be assumed to be a ball. The proof of this deep result is lengthy and will not be given here. Some direct consequences of Theorem 1.46 are the following:

i) $|F(u)| = 0$ and $|\tilde{S}(u)| = |\text{interior of } \tilde{S}(u)|$

ii) $\tilde{S}(u)$ has locally finite perimeter in Ω .

We recall that if $E \subset \Omega$ is a Borel set, then E is said to be <u>of finite perimeter</u> in Ω if the indicatrix function $\mathbb{1}_E$ has a bounded variation; in this case its perimeter P(E) is defined by $P(E) = V_\Omega(\mathbb{1}_E)$, see Federer[1] or Giusti [1]. We also recall that if E is a compact manifold with smooth boundary ∂E, then P(E) is the surface area of ∂E in the classical sense.

Much additional information on the free boundary $F(u)$ can be obtained through adequate Hausdorff measure estimates. This is the case of the estimate

$$\mathcal{H}^{N-1}(\mathcal{F}(u) \cap B_R(x_0)) \leqslant \mathcal{H}^{N-1}[\{u > 0\} \cap \partial B_R(x_0)] \tag{32}$$

which will allow us to prove that $\mathcal{F}(u)$ has a nonnegative mean curvature (and even that $\mathcal{F}(u)$ is convex if $N = 2$), when u is the solution of (1), (2) and $\partial\Omega$ is assumed to have a nonnegative mean curvature. In order to explain the utility of estimate (32) we shall state it more precisely

Theorem 1.47. *Let* u *be the solution of* (3),(4), *corresponding to* $h \equiv 1$. *Assume that*

$$\partial\Omega \text{ has a nonnegative mean curvature .} \tag{33}$$

Then, for any subdomain D *of* Ω *with piecewise smooth boundary* $\partial D \subset \Omega$ *and with* $\mathcal{H}^{N-1}(\mathcal{F}(u) \cap \partial D) = 0$ *we have*

$$\int_{D \cap \mathcal{F}(u)} d\mathcal{H}^{N-1} \leqslant \int_{\partial D \cap \{u>0\}} d\mathcal{H}^{N-1} . \tag{34}$$

Before giving the proof of the above theorem, we shall give a first application.

Theorem 1.48. *Assume* (33) , *and let* u *be a solution of* (3), (4) *corresponding to* $h \equiv 1$. *Then every* C^2*-portion of* $\mathcal{F}(u)$ *has a nonnegative mean curvature.*

Proof. Suppose that a smooth portion of $\mathcal{F}(u)$ is given by $x_N = \varphi(x')$, where $x' = (x_1, \ldots, x_{N-1})$ varies in a ball $\overline{B}_R$, and $u > 0$ if $x_N < \varphi(x')$. Take $\partial D \cap \{x_N < \varphi(x')\}$ where $x_N = \varphi(x') - \varepsilon\zeta(x')$, $\zeta' \geqslant 0$, $\zeta \in C_0^2(\overline{B}_R)$, $\varepsilon > 0$. Then (34) yields

$$\int_{\overline{B}_R} [1 + |\nabla\varphi|^2]^{1/2} \leqslant \int_{\overline{B}_R} [1 + |\nabla\varphi - \delta\nabla\zeta|^2]^{1/2}.$$

From this, we deduce that $\nabla \cdot \nabla\varphi/[1 + |\nabla\varphi|^2]^{1/2} \geqslant 0$, and the assertion follows (see, e.g., Gilbarg-Trudinger [1] p. 356). □

Now we return to the proof of estimate (34). This will carried out by using, in a fundamental way, the auxiliary function

$$\psi(u) = \int_0^u \frac{ds}{(2F(s))^{1/2}} \quad , \tag{35}$$

where $F(t) = \int_0^t f(s)ds = [1/(q+1)]u^{q+1}$. (Note that $\psi = \psi_1$, where ψ_1 is given in (33) of the Section 1.1). If u is any solution of (1), after some easy computations we obtain that in the set $\{u > 0\}$ we have

$$\Delta\psi(u) = \phi(u) \frac{1 - |\nabla\psi(u)|^2}{\psi(u)} , \tag{36}$$

where

$$\phi(u) = \frac{f(u)\ \psi(u)}{(2F(u))^{1/2}} \tag{37}$$

where $f(u) = u^q$, $0 < q < 1$. Note that $\phi(t)$ is a positive C^1 function of t away from $t = 0$, whereas near $t = 0$, $\phi(t) \sim$ constant > 0. Now, we recall that by Theorem 1.24.

$$|\nabla u|^2 \leqslant 2F(u) \quad \text{in} \quad \Omega .$$

Then

$$|\nabla\psi(u)| \leqslant 1 \tag{38}$$

and, in consequence,

$$\Delta\psi(u) \geqslant 0 \quad \text{in} \quad \{u > 0\} . \tag{39}$$

That (38) and (39) lead to estimate (32) can be proved, formally, in the following way. Assume $\mathcal{F}(u)$ smooth and let $x_0 \in \mathcal{F}(u)$. By the Green's formula

$$\int_{\{u>0\} \cap B_R(x_0)} \Delta\psi(u) = \int_{\mathcal{F}(u) \cap B_R(x_0)} \nabla\psi(u) \cdot \nu \, d\mathcal{H}^{N-1} +$$

$$+ \int_{\{u>0\} \cap \partial B_R(x_0)} \nabla\psi(u) \cdot \nu \, d\mathcal{H}^{N-1} .$$

Using (38),(39) and assuming that $|\nabla\psi(u)| \to 1$ if $\operatorname{dis}(x,\mathcal{F}(u)) \to 0$, we obtain (32).□

Now we shall make the above observation rigorous . We shall need the following estimate on the Hausdorff measure, due to Alt-Phillips [1] .

Theorem 1.49. *Let* u *be the solution of* (3),(4) *with* $h \equiv 1$. *Then*

$$\Delta\psi(u) = d\lambda + \mathbb{1}_{\{u>0\}}\phi(u)\,\frac{1-|\Delta\psi(u)|^2}{\psi(u)}, \tag{40}$$

where $d\lambda$ *is absolutely continuous with respect to* $dH^{N-1} \llcorner F(u)$ (L means "restriction to"). *More precisely,*

$$d\lambda = dH^{N-1} \llcorner F(u)_{red} + \theta(x)dH^{N-1}\llcorner F(u)_{sin}\ ,\ 0 \leqslant \theta(x) \leqslant C, \tag{41}$$

($F(u)_{red}$ *and* $F(u)_{sin}$ *defined as in Federer* [1]. *Chap.4*), *and*

$$\mathbb{1}_{\{u>0\}}\phi(u)\,\frac{1-|\Delta\psi(u)|^2}{\psi(u)} \in L^1(\Omega). \ \square \tag{42}$$

Now we shall integrate by parts correctly:

Proposition 1.50. *Let* D *be as in Theorem* 1.47. *Then*

$$\int_D \Delta\psi(u) = \int_{\partial D\cap\{u>0\}} \Delta\psi(u)\cdot \nu\, dH^{N-1} \tag{43}$$

Proof. The equality (43) is just the Green's formula for a function $\psi(u(x))$ whose Laplacian is a measure. We shall establish it by approximation. Let $\psi(u)_\varepsilon$ be a mollification of $\psi(u)$. Then, since $\Delta\psi$ is a measure

$$\Delta(\psi(u)_\varepsilon) \to \Delta\psi(u) \quad \text{as measures.} \tag{44}$$

Moreover , $\Delta\psi(u)(\partial D) = 0$ due to (40),(41) and the assumption $H^{N-1}(F(u)\cap\partial D) = 0$. Then, from (44),

$$\int_D \Delta(\psi(u)_\varepsilon) \to \int_D \Delta\psi(u) \tag{45}$$

By the Green's formula for smooth functions we have

$$\int_D \Delta(\psi(u)_\varepsilon) = \int_{\partial D} \nabla(\psi(u)_\varepsilon)\cdot\nu\, dH^{N-1}. \tag{46}$$

For any small $\delta > 0$, let V be any open neighborhood of $F(u)\cap\partial D$ with $\int_{V\cap\partial D} dH^{N-1} < \delta$. On $(\partial D - V)\cap\{u>0\}$ we have $\nabla(\psi(u)_\varepsilon) \to \nabla\psi(u)$

uniformly as $\varepsilon \to 0$. On the other hand,

$(\partial D - V) \cap \{u = 0\}$ is compactly contained in $\operatorname{int}\{u = 0\}$,

hence, if ε is small enough,

$$(\nabla \psi(u)_\varepsilon) = 0 \quad \text{on} \quad (\partial D - V) \cap \{u = 0\} .$$

From (38) we deduce that $|\nabla\psi(u)_\varepsilon| \leq 1$, and, therefore, if ε is small enough

$$\left| \int_{V \cap \partial D} \nabla(\psi(u)_\varepsilon) \cdot \nu \, dH^{N-1} \right| < \delta .$$

We now break the integral on the right-hand side of (46) into $\partial D \cap V$ and $\partial D - V$ and we obtain, letting $\varepsilon \to 0$ and then $\delta \to 0$,

$$\int_{\partial D} \nabla(\psi(u)_\varepsilon) \cdot \nu \, d\, H^{N-1} \to \int_{\partial D \cap \{u>0\}} \nabla\psi(u) \cdot d\, H^{N-1} \tag{47}$$

Taking $\varepsilon \to 0$ in (46) and using (45) and (47), the assertion (43) is proved. □

Finally the proof of Theorem 1.47 is obtained from (38),(39) and the following combination of Theorem 1.49 and Proposition 1.50. □

<u>Corollary 1.51.</u> *If* D *is as in Theorem* 1.47 *then*

$$\int_{D \cap F(u)_{red}} d\, H^{N-1} + \int_{\partial D \cap F(u)_{sing}} \theta \, d\, H^{N-1} = \tag{48}$$

$$= - \int_{D \cap \{u>0\}} \Delta\psi(u) dx + \int_{\partial D \cap \{u>0\}} \nabla \psi(u) \cdot \nu \, d\, H^{N-1} . \ \square$$

We shall end this subsection by recalling an improvement of Theorem 1.47 (both of them due to Friedman -Phillips [1]). Let Π be an open halfspace with $H^{N-1}(\partial\Pi \cap F(u)) = 0$ and denote by $\vec{\nu}$ the outward normal along $\partial\Pi$. Let D be a convex domain in Π with piecewise smooth boundary, such that $\partial D \cap \partial\Pi \neq \emptyset$ in an open neighborhood of $\partial D \cap \Pi$; then, recalling that $F(u)_{red}$ is the set of points of $F(u)$ where the exterior normal to $\tilde{S}(u)$ exists, we can state:

<u>Theorem 1.52</u> *Under the foregoing assumptions*

$$\int_{\{u>0\} \cap \partial G \cap \partial\Pi} d\, H^{N-1} \leq \int_{F(u)_{red} \cap D} d\, H^{N-1} , \tag{49}$$

with strict inequality if $|\{u > 0\} \cap D| > 0$. □

1.4c. Regularity of the free boundary and dependence with respect to a parameter in the equation.

The first regularity result for the free boundary $F(u)$ of the solution of (2),(3) is due to Alt-Phillips[1] and says that if u satisfies some "flatness condition" at the origin in the direction e_N, then $F(u)$ is, locally, a $C^{1,\alpha}$ function.

Theorem 1.53. (a) *Suppose* u *is a solution of* (3) *in* $B_1(0)$ *and* $0 \in F(u)$. *There exist positive constants,* α, $\beta \geq 1$, σ_0, τ_0 *and* C *depending on* N *and* q *such that* $\{x : x_N > \sigma\rho\} \cap B_\rho(0) \subset N(u)$ *with* $\sigma \leq \sigma_0$ *and* $\rho \leq \tau_0\, \sigma^\beta$ *implies that* $B_{\rho/4}(0) \cap F(u)$ *is a graph of a* $C^{1,\alpha}$ *function* φ *in the direction* e_N ; *moreover, if* $x' = (x_1, \ldots, x_{N-1})$,

$$|\nabla\varphi(x')| \leq C\,\sigma \quad \text{for} \quad |x'| \leq \frac{\rho}{8}, \tag{50}$$

$$|\nabla\varphi(x_1') - \nabla\varphi(x_2')| \leq C\,\sigma \left|\frac{x_1' - x_2'}{\rho}\right|^\alpha \quad \text{for} \quad |x_1'|,|x_2'| \leq \frac{\rho}{8}. \tag{51}$$

(b) *If* N = 2 *and if*

$$\limsup_{r\to 0} \frac{|N(u) \cap B_r(0)|}{|B_r(0)|} > 0 \tag{52}$$

then there exists a $\rho > 0$ *such that* $B_\rho(0) \cap F(u)$ *is a* $C^{1,\alpha}$ *graph.* □

Some more information on the graph φ giving the free boundary $F(u)$ can be obtained when the solution u is constant on $\partial\Omega$. Before giving such a result we shall first obtain some estimates on the location of $F(u)$ when there is a parameter λ in the equation. We shall use the notation

$$\Omega_\delta = \{x \in \Omega : d(x,\partial\Omega) > \delta\} \qquad , \delta > 0$$

Theorem 1.54. *Let* Ω *be a regular bounded open set of* $\mathbb{R}^N$, $\lambda > 0$. *Let* f *be a continuous nondecreasing function such that* $f(0) = 0$ *and*

$$\int_{0^+} \frac{ds}{F(s)^{1/2}} < +\infty$$

with F *primitive of* f, $F(0) = 0$. *Then, if* u *is the solution of*

$$- \Delta u + \lambda f(u) = 0 \quad \textit{in} \quad \Omega \tag{53}$$

$$u = 1 \qquad\qquad \textit{on} \quad \partial\Omega , \tag{54}$$

the null set N(u) *must satisfy the estimate*

$$\Omega_{\frac{\gamma}{\sqrt{\lambda}} + \frac{c_1}{\lambda}} \subset N(u) \subset \Omega_{\frac{\gamma}{\sqrt{\lambda}} - \frac{c_2}{\lambda}} \tag{55}$$

for some positive constants γ, c_1 *and* c_2 *independent of* λ *and for* λ *large enough.*

<u>Proof</u>. First we establish the crude estimate

$$\Omega_{\frac{K_2}{\sqrt{\lambda}}} \subset N(u) \subset \Omega_{\frac{K_1}{\sqrt{\lambda}}} , \tag{56}$$

where K_1, K_2 are positive constants independent of λ . Indeed, to prove the second inclusion it suffices to show that

$$\| \nabla u \|_\infty \leqslant K_1 \sqrt{\lambda} \tag{57}$$

for some constant K_1. Since the function $w(x) = u(x/\sqrt{\lambda})$ satisfies

$$- \Delta w + f(w) = 0 \quad \text{in} \quad \Omega^\lambda = \{x/\sqrt{\lambda} : x \in \Omega\}$$

and by the standard regularity $f(w) \in L^\infty(\Omega^\lambda)$, we have that $\| \nabla w \|_\infty \leqslant K_1$ (see, e.g., Chapter 4), for some constant K_1, and (57) holds. We also recall that the first inclusion in (56) was obtained in Theorem 1.9 (note that $\psi_{\lambda/N}(M) = K_2/\sqrt{\lambda}$ for some suitable constant K_2).

Now we shall prove the first inclusion in (55). Let $y \in \partial\Omega$ and let $B_R(x_0)$ be a ball in Ω with $y \in \partial B_R(x_0)$. Let U be the solution of (53),(54) on B_R instead of on Ω . By the comparison results, $u \leqslant U$ a.e. and by the uniqueness U is a nondecreasing radially symmetric function $U = U(r)$, $r = |x - x_0|$. The function U satisfies

$$- U''(r) - \frac{N-1}{r} U'(r) + \lambda f(U(r)) = 0 ,$$

and so the function

$$Z(s) = U(R - \frac{\gamma_0}{\sqrt{\lambda}} + \frac{s}{\sqrt{\lambda}}) \quad (\ \gamma_0 \text{ to be determined})$$

verifies

$$- Z'' - \frac{(N-1)}{\rho\sqrt{\lambda} + s} Z' + f(Z) = 0 \quad , \rho = R - \gamma_0/\sqrt{\lambda} \ . \tag{58}$$

Since $U'(r) \geqslant 0$, the support of $Z(0)$ consists of one interval, namely $0 \leqslant S \leqslant \gamma_0$. From (56) applied to U in $B_R(x_0)$, we have $\gamma_0 \leqslant K_2$, K_2 independent of λ. Multiplying both sides of (58) by $Z'(s)$, we get

$$\frac{1}{2} [(Z'(s))^2]' + \frac{(N-1)}{\rho\sqrt{\lambda} + s} (Z'(s))^2 = [F(Z(s))]'.$$

Hence

$$[(Z'(s))^2]' + \frac{C}{\sqrt{\lambda}} (Z'(s))^2 \geqslant \frac{1}{2} [F(Z(s))]' ,$$

where $C > 0$ is independent of λ. From this we obtain

$$[(Z')^2 e^{\frac{Cs}{\sqrt{\lambda}}}]' \geqslant \frac{1}{2} e^{\frac{Cs}{\sqrt{\lambda}}} [F(Z)]' .$$

Integrating and using the relations $Z'(0) = 0$ and $F(Z(0)) = 0$, we get

$$(Z'(s))^2 \geqslant \frac{1}{2} e^{-\frac{Cs}{\sqrt{\lambda}}} \int_0^s e^{\frac{Cs}{\sqrt{\lambda}}} [F(Z(t))]' dt =$$

$$= \frac{1}{2} F(Z(s)) - \frac{C}{\sqrt{\lambda}} \int_0^s e^{-\frac{C}{\sqrt{\lambda}}(s-t)} F(Z(t)) dt.$$

Recalling that $Z'(t) \geqslant 0$ we get

$$Z'(s) \geqslant \left(1 - \frac{C}{\sqrt{\lambda}}\right)^{1/2} \left[\frac{1}{2} F(Z(s))\right]^{1/2} .$$

Now we consider $\xi(s)$ solution of the Cauchy Problem

$$\xi'(s) = \left(1 - \frac{C}{\sqrt{\lambda}}\right)^{1/2} \left[\frac{1}{2} F(Z(s))\right]^{1/2}$$
$$\xi(0) = 0$$

Such a function ξ can be built in the following way. Consider the problem

$$\eta'(t) = \left[\frac{1}{2} F(\eta(t))\right]^{1/2} \qquad \text{for } t < 0$$
$$\eta(0) = 1.$$

There exists a unique solution of this system as long as $\eta(t) > 0$ and it determines a unique positive number γ such that $\eta(-\gamma) = 0$. Letting $\tilde{\xi}(r) = \eta(\gamma+r)$ we have

$$\tilde{\xi}'(r) = [\tfrac{1}{2} F(\tilde{\xi})]' \text{ for } 0 < r < \gamma, \quad \tilde{\xi}(0) = 0, \quad \tilde{\xi}(r) > 0 \text{ for } 0 < r < \gamma$$

and $\tilde{\xi}(\gamma) = 1$. Finally we take $\xi(s) = \tilde{\xi}(s(1 - C/\sqrt{\lambda})^{1/2})$. Since $U(R) = 1$ means $Z(\gamma_0) = 1$, we conclude that

$$\gamma_0 (1 - \frac{C}{\sqrt{\lambda}})^{1/2} \leqslant \gamma .$$

Recalling that $u \leqslant U$, we deduce that

$$N(u) \supset B_{R-\gamma_0/\sqrt{\lambda}}(x_0) \supset B_{R-\gamma/\sqrt{\lambda} - C/\lambda}(x_0) .$$

Thus, the first part of (55) follows. To prove the second part, we introduce the following shell $\tilde{\Omega}$: the inner boundary is a sphere $S_{\tilde{R}}$ in $\mathbb{R}^N - \Omega$ which contains a point $y \in \partial\Omega$, and the outer boundary is a sphere S_R containing $\overline{\Omega}$. Let V be the solution of (53),(54) for the shell $\tilde{\Omega}$. Again, by the comparison and uniqueness results, $V = V(r)$, $r = |x - x_0|$ for some $x_0 \in \mathbb{R}^N$, $V'(r) \leqslant 0$ and $V \leqslant u$ a.e. on Ω. The function

$$\tilde{Z}(s) = V(R + \frac{\tilde{\gamma}}{\sqrt{\lambda}} - \frac{s}{\sqrt{\lambda}})$$

can be analyzed similarly to $Z(s)$. Thus we find that

$$\tilde{Z}(s) \leqslant \tilde{\xi}(1 + C/\sqrt{\lambda})^{1/2}) , \quad \tilde{\gamma}(1 + C/\sqrt{\lambda})^{1/2} \geqslant \gamma ,$$

and this yields the second part of (55) if we let y vary over $\partial\Omega$ in the above construction. □

We end this subsection with a result due to Friedman -Phillips [1] improving Theorems 1.53 and 1.54, when $f(s) = s^q$. We shall assume now that $\partial\Omega$ have a $C^{1,\alpha}$ local parametrization in neighborhood of a point x^0 given by $x = \Phi(x')$ $(x' = (x_1,\ldots,x_{N-1})$, $x^0 = \Phi(0))$. We also denote by $\nu(x')$ the inner normal to $\partial\Omega$ at $\Phi(x')$.

Theorem 1.55. *There exist positive constants* σ_1, σ_2 *such that for any* λ *sufficiently large, if* u *satisfies* (53) *with* $f(s) = s^q$, $0 < q < 1$, *and* (54), *then* $F(u)$ *is a* C^{1,σ_1} *surface; furthermore, in terms of local coordinates* $x = \Phi(x')$, $F(u) \cap B_R(\Phi(0))$ *can be represented (for small enough* R *) in the form* $x = \Phi(x') + k(x',\lambda)\,\nu(x')$ *with* $k(x',\lambda)$ *satisfying, as a function of* x',

$$|\nabla_x k| \leqslant C/\lambda^{\sigma_1} \qquad (x' \in \overline{B}_R(x))$$

and

$$|k|_{C^{1,\sigma_2}(\overline{B}_R(0))} \leqslant C. \ \square$$

Remark 1.26. The above result improves Theorem 1.54 in the sense that now we know that the free boundary $F(u)$ is a smooth surface parallel to $\partial\Omega$ at a distance $C/\sqrt{\lambda} + O(1/\lambda)$, for some constant C independent of λ. $\square$

More information on the dependence of $F(u)$ with respect to λ is given in the next section.

1.4d. Geometrical properties of the free boundary.

From Theorems 1.48 and 1.55 it seems natural to expect some stronger geometrical properties such as, for instance, the convexity of the free boundary $F(u)$, u solution of (1) (2), assuming Ω to be a smooth convex domain. We shall prove that property, but only in the special case of two-dimensional domains Ω.

We first study the components of the null set N(u). Here by a component of N(u) we mean a maximal connected subset of N(u). Note that any component is necessarily a closed set.

Theorem 1.56. *Let* Ω *be a two-dimensional convex domain,* u *solution of* (3), (4) *with* $h \equiv 1$ *and let* T *be a component of* N(u) *with nonempty interior. Then* T *is a closed convex domain with* $C^{1,\alpha}$ *boundary and*

$$\text{dist}\,(T, N(u) - T) > 0. \tag{59}$$

Using now Theorem 1.55 we obtain:

Corollary 1.57. *Let* Ω *be a two-dimensional convex domain and* u *solution of* (53) *with* $f(s) = s^q$, $0 < q < 1$ *and* (54). *Then if* λ *is sufficiently large* , N(u) *is a convex domain with* $C^{1,\alpha}$ *boundary.*

Proof of Theorem 1.56. Let Q be an interior point of T and let l_1, l_2 be rays initiating at Q and forming an angle less than π . These rays intersect ∂T for the first time, say , at P_1 and P_2. We form the triangle $G = \{P_1, P_2, Q\}$ and claim that

$$\{u > 0\} \cap G = \emptyset \tag{60}$$

Indeed, otherwise Theorem 1.52 gives (assuming first that

$$\mathcal{H}^{N-1}(\overline{P_1P_2} \cap \mathcal{F}(u)) = 0)$$

$$\int_{\overline{P_1P_2} \cap \{u>0\}} d\,\mathcal{H}^1 < \int_{\mathcal{F}(u)_{red} \cap G} d\,\mathcal{H}^1 \quad . \tag{61}$$

On the other hand, from Corollary 1.51 and (39), we obtain a contradiction with (61).

If $\mathcal{H}^{N-1}(\overline{P_1P_2} \cap \mathcal{F}(u)) > 0$, since $\mathcal{H}^{N-1}_{loc}(\mathcal{F}(u)) < \infty$, then we can find $\tilde{P}_1$, $\tilde{P}_2$ with $\tilde{P}_i \in \overline{Q\,P_i}$ and $|\tilde{P}_i - P_i|$ arbitrarily small so that $\mathcal{H}^{N-1}(\overline{\tilde{P}_1\tilde{P}_2} \cap \mathcal{F}(u)) = 0$ and $\{\tilde{P}_1, \tilde{P}_2, Q\}$ still violates (59) (since $\{u > 0\}$ is open). The previous argument can then be applied to $\{\tilde{P}_1 . \tilde{P}_2, Q\}$ in order to derive a contradiction.

Having proved (60), we now denote by T the union of segments $\overline{Q\,P_1}$ when l_1 varies over all possible directions. From (60) it follows that T_0 is convex; in particular, ∂T_0 is Lipschitz continuous. Since $N = 2$, we can apply Theorem 1.53 (b). It then follows that ∂T_0 is in $C^{1,\alpha}$ and $u > 0$ in some $(\Omega - T_0)$-neighborhood of ∂T_0. Hence $T = T_0$ and (59) holds. □

The study of the free boundary for the equation (53) depending on a parameter λ is completed with the following result (like the above one, due also to Friedman-Phillips [1]).

Theorem 1.57. *Let* Ω *be a two-dimensional convex domain and* u *satisfy* (53) , *with* $f(s) = s^q$, $0 < q < 1$, *and* (54). *Then the null set* N(u) *is either a closed convex domain with* $C^{1,\alpha}$ *boundary , or a single point, or empty.* □

The convexity of the free boundary $\mathcal{F}(u)$ for the exterior problem has been shown in Caffarelli-Spruck [1](see also Kawohl [5]) with no restriction on the dimension of the space. More precisely, let G and Ω be convex sets in $\mathbb{R}^N$, $N \geq 1$, with G strictly contained in Ω, and let u be the unique solution of

$$- \Delta u + f(u) = 0 \quad \text{in} \quad \Omega - G \tag{62}$$

$$u = 1 \quad \text{on} \quad \partial G \text{ , } u = 0 \quad \text{on} \quad \partial\Omega \quad , \tag{63}$$

where f is a continuous nondecreasing function such that $f(0) = 0$. Then in the above references it is proved that the level surfaces $(\partial\{x : u(x)>t\},$ $t \in [0,1))$ are convex. We remark that the same conclusion is true for the solution u of

$$- \Delta u + f(u) = 0 \quad \text{in} \quad \mathbb{R}^N - G \tag{64}$$

$$u = 1 \quad \text{on} \quad \partial G \text{ ,} \tag{65}$$

assumed that f satisfies the hypothesis

$$\int_{0^+} \frac{ds}{F(s)^{1/2}} < + \infty \ .$$

Indeed, by Theorem 1.9 (or Theorem 1.18) we know that the support of u is bounded. So, there exists $R > 0$ such that $G \subset S(u) \subset B_R(0)$ and then, by uniqueness, u coincides with the solution of (62),(63) corresponding to $\Omega = B_R(0)$. It is also interesting to note that the proofs of Theorem 1.56 and of the above property are completely different. On this occasion, the convexity of the level sets $\{x : u(x) > t\}$ is obtained by showing that $u(z) \geq \min\{u(x), u(y)\}$ for every $x, y \in \Omega$ and $z = \lambda x + (1 - \lambda)y$, $\lambda \in (0,1)$.(See details in the cited reference, already compiled in the book by Friedman [3]). Finally we remark that problem (3),(4) with $h = 1$ cannot be reduced to an exterior problem like (62), (63). Indeed, if u satisfies (3) on a convex set Ω and with $u = 1$ on $\partial\Omega$ then the tentative change of variable $U = 1 - u$ reduces the problem to

$$\Delta U = - f(1 - U) \quad \text{in} \quad \Omega - G$$

$$U = 1 \quad \text{on} \quad \partial G \quad \text{and} \quad U = 0 \quad \text{on} \quad \partial\Omega \quad ,$$

where G is, for instance, a ball $G = B_R(x_0)$ contained in the set $N(u)$ (note that $U = 1$ on $N(u)$). The impossibility of applying the result of Caffarelli-Spruck [1] is due to the fact that now $\Delta U \leqslant 0$ in contrast with the case of equation (62).

A weaker property that the convexity of $F(u)$ is the starshapedness of this free boundary with respect to a point $x^0 \in \mathbb{R}^N$, assumed the domain to be starshaped with respect to that point. We shall prove this for nonnegative solutions of the x-dependent equation

$$- \Delta u + \theta(x)\, u^q = 0 \quad \text{in} \quad \mathbb{R}^N. \tag{66}$$

The above equation has been considered in Remark 1.8. We recall that if, for instance, $0 < q < 1$ and

$$\theta(0) < 0 \quad \text{and} \quad \liminf_{|x|\to\infty} \theta(x) > 0 \tag{67}$$

then by Proposition 1.11, any nonnegative solution of (66) has compact support.

Theorem 1.58. *Let* $\theta(x)$ *be a locally Lipschitz function satisfying* (67) *as well as*

$$x \cdot \nabla\theta(x) \geqslant 0 \tag{68}$$

$$\partial\{\theta = 0\} \text{ is a } C^{1,\alpha} \text{ surface.} \tag{69}$$

Then the support $S(u)$ *of any nonnegative solution of* (66) *consists of a single starlike component about the origin, and* $F(u)$ *is Lipschitz.*

We first need a preliminary result.

Lemma 1.59. *Under the hypotheses of the above theorem the function* $u^{1-q}(x)/|x|^2$ *is decreasing along rays from the origin and tends to zero at* $F(u)$.

Proof. Set $\varphi(x) = u^{1-q}(x)$ and $w = r\,\varphi_r - 2\varphi$, $r = |x|$. Then assuming $u \in C^3$ we have

$$\Delta w = r(\Delta\varphi)_r =$$

$$= (1-q)r\,\theta_r - \frac{q}{1-q}\Big[- \frac{|\nabla\varphi|^2}{\varphi^2} r\,\varphi_r + \sum_{j,k=1}^{N} 2\,\frac{\varphi_k}{\varphi}\, x_j\,\varphi_{jk}\Big]. \tag{70}$$

But

$$\sum_{j,k=1}^{N} x_j \varphi_{jk} = \sum_{k=1}^{N} (w_k + \varphi_k)$$

so that

$$\sum_{j,k=1}^{N} \frac{2\varphi_k}{\varphi} x_j \varphi_{jk} = \sum_{k=1}^{N} \frac{2\varphi_k w_k}{\varphi} + 2 \frac{|\nabla\varphi|^2}{\varphi} .$$

Moreover

$$\frac{|\nabla\varphi|^2}{\varphi^2} r \varphi_r = \frac{|\nabla\varphi|^2 w}{\varphi^2} + 2 \frac{|\nabla\varphi|^2}{\varphi} .$$

Inserting this equality into (70) we have

$$\Delta w = (1-q) r \theta_r - \frac{2q}{(1-q)} \sum_{k=1}^{N} \frac{\varphi_k w_k}{\varphi} + \frac{q}{(1-q)} \frac{|\nabla\varphi|^2 w}{\varphi^2} .$$

But by (68) $r \theta_r \geqslant 0$, and then

$$\Delta w + \frac{2q}{(1-q)} \sum_{k=1}^{N} \frac{\varphi_k w_k}{\varphi} - \frac{q}{(1-q)} \frac{|\nabla\varphi|^2}{\varphi} w \geqslant 0.$$

On the other hand, w tends to zero on $\mathcal{F}(u) \subset \partial S(u)$ due to the following inequality, proved in Spruck [1] by using the assumption (69)

$$|\nabla\varphi| = (1-q) \frac{|\nabla u|}{u^q} \leqslant C u^{\frac{1-q}{2}}$$

(note that such an inequality was proved in Theorem 1.24 for the Dirichlet problem). Then , by the maximum principle, we obtain that $w < 0$ on $S(u)$ and that gives the conclusion, since

$$\frac{d}{dr} \left(\frac{\varphi}{r^2}\right) = r^3 w < 0 \quad \text{in} \quad S(u).$$

Finally the restriction $u \in C^3$ can be easily removed to $u \in C^{2,\delta}$ by standard approximation arguments. □

Proof of Theorem 1.58. Lemma 1.59 already shows that $S(u)$ is a single starlike component about 0. To show that $\mathcal{F}(u)$ is Lipschitz it suffices to show that if we choose a new origin x_0 sufficiently close to 0, then

ϕ/r^2, (now $r = |x-x_0|$) is decreasing in a neighborhood of $\partial S(u)$. Let $\delta > 0$ and $\lambda > 0$ such that $\{x : 0 < u < \delta\} \subset \{x : \theta > \lambda\}$. Then by the proof of Lemma 1.59, $w < 0$ on $\partial\{x : 0 < u < \delta\}$. Let $|x_0|$ be so small that

$$\bar{w} = (x-x_0)\cdot \nabla\phi - 2\phi = w = |x_0|\ |\nabla\phi| \quad \text{on } \{u = \delta\}$$

By assumption, $x\cdot\nabla\theta > 0$ on $\{\theta \geqq 0\}$; hence, we may also assume $(x-x_0)\cdot \nabla\theta > 0$ for $|x_0|$ small. Repeating the argument of Lemma 1.59 we obtain that $\bar{w} < 0$ in $\{x : 0 < u < \delta\}$, which says that ϕ/r^2 is decreasing near $\partial S(u)$ as required. □

Remark 1.27. The starshapedness of the free boundary $\mathcal{F}(u)$ for solutions of the exterior problem (62), (63) has been obtained in Kawohl [4] by using a rearrangement method. (See also Kawohl [1] for another proof using the starshapedness function $S(\lambda,x) = u(\lambda x) - u(x)$, $\lambda \in (0,1)$ $x \in \Omega - \bar{G}$). □

1.5. BIBLIOGRAPHICAL NOTES

The study of the free boundary $\mathcal{F}(u)$ for solutions of nonlinear elliptic problems seems to have its origins in certain results on the compactness of the support for some special variational inequalities (Berkovitz-Pollard [1], Auchmuty-Beals [1], Brezis-Stampacchia [2]). After this, a systematic study for general second order variational inequalities on *unbounded* domains was made in Brezis [7], by means of the comparison principle.

Concerning semilinear elliptic equations, it seems that some particular solutions with compact support were first known in the study of stationary solutions of the porous media equation with absorption (Martinson-Pavlov [2]). Independently, and at the same time, a systematic study of solutions with compact support of semilinear equations in $\mathbb{R}^N$ was made by Benilan-Brezis-Crandall [1].

The treatment of quasilinear equations was started, from a physical point of view, by Martinson-Pavlov [1] and, mathematically, in a more general formulation, by Diaz-Herrero [1], [2].

The free boundary $\mathcal{F}(u)$ for *bounded* domains was first studied in the context of some particular problems in chemical engineering, (Aris [1]). Mathematically, the free boundary $\mathcal{F}(u)$ in bounded domains was first studied in the context of variational inequalities (Bensoussan-Brezis-Friedman [1]) and then extended to general (single or multivalued) second order nonlinear equations by Diaz [4]. We also mention in this context the early works of Bandle-Sperb-Stakgold [1], Brauner-Nicolaenko [1], Diaz-Hernandez [1] and Stakgold [2].

Section 1.1. The results of this section develop some ideas of Diaz [4]; in particular, that of deriving the existence and location estimates on $\mathcal{F}(u)$ from the existence of nontrivial solutions of the associated homogeneous Cauchy problem was already proposed. The study of the autonomous Cauchy problem is well known (the proof of the necessity in Theorem 1.4 follows Bandle-Sperb-Stakgold [1]). For the nonautonomous problem, it seems that only the case of homogeneous nonlinearities had been considered before (Diaz-Hernandez [1]). We point out that local super-solutions for general second order semilinear equations were already exhibited by Evans-Knerr [1] in the study of some parabolic problems. Nevertheless, such functions are not radially symmetric (see Lemma 2.39).

The ordinary differential equation (31), with other initial (or boundary) conditions, has been largely considered in other contexts such as, for instance, in the study of particular solutions of the porous media and other nonlinear parabolic equations (see the references to the works by Aronson, Atkinson, Gilding, Peletier and many others compiled in Diaz [7]); the study of removable singularities of nonlinear elliptic equations (see Veron [3] and the references therein); etc. Another nonautonomous ODE, amply treated in the literature, is the equation $u'' = a(t)u^q$ (see, for instance, the works of Fowler, Kiguradze, Taliaferro, Chanturia and others in the survey by Wong [1] and the recent treatment by Peletier-Tesei [1], [2]). We also mention here the results concerning nonoscillatory solutions of higher order ODEs of Kiguradze [1] and Svec [1], [2].

Returning to the problem in PDEs, we point out that Theorem 1.9 can also be applied to the minimal surfaces equation (see (20) of Section 4.1 and Remark 1.2). It also holds for functions g merely in $L^1_{loc}(\Omega)$ or $M(\Omega)$ (bounded measures on Ω). Thus sharp estimates on the location of $\mathcal{F}(u)$ when $g \in L^1$ were used by Gallouet-Morel [3] as a first step to derive an existence theorem for data $g \in L^1_{loc}(\mathbb{R}^N)$. Remark 1.7 follows Vazquez [1]. More recent references on the applications mentioned in Remark 1.8 are Aronson [1], Pozio-Tesei [1], [2], Peletier-Tesei [1], [2] and Badii-Diaz-Tesei [1]. Theorems 1.13 and 1.14 seem to be new (a complete proof of Theorem 1.13 is given in

Diaz [6]).

The boundary estimates given in Subsection 1.1c and, in particular, the property of nondiffusion of the support (Theorem 1.16), may be understood as an elliptic version of the waiting time property, well known for the porous media and other nonlinear parabolic equations (see, e.g., Knerr [1]). The results of this subsection seem to be new in the literature.

Theorem 1.18 is taken from Diaz-Herrero [2], where the previous results of Benilan-Brezis-Crandall [1] were generalized (in fact in Diaz-Herrero [2] the pseudo-Laplacian operator considered is the one given by (6) of Section 2.4). We also mention the approach of Veron [2] and Barbu [1] who considered semilinear equations as a second order evolution equation. Estimates on the support of the solution of the semilinear equations with $g = \delta$ (the Dirac delta) are due to Morel (personal communication). The compactness of the support may also be proved for other nonlinear equations not satisfying the assumptions of Theorem 1.18 but of a rather particular formulation. This is the case of the Thomas-Fermi model (see Lieb-Simon [1] and the results of Benilan and Brezis presented in Brezis [9]), as well as that of the vortex rings equation (Fraenkel-Berger [1]). A very complete survey of these and other variational problems with potentials can be found in Friedman [3]. Finally, we remark that Theorem 1.19 comes from an idea of Moet [1] for variational inequalities.

Section 1.2. The strong maximum principle given in Theorem 1.20 is due to Vazquez [5] and can be understood as a nonunique continuation property. A partial result, in this direction, may be derived asymptotically from the results of Bertsch-Kersner-Peletier [1] for the porous media equation with absorption.

The positivity of bounded solutions given in Theorem 1.23 is taken from Bandle-Sperb-Stakgold [1]. As indicated, their proof uses some gradient estimates called "the best maximum principle" by Payne. The proof is taken from Mossino [1]. Another positivity result can be found by adapting a result of Bandle [2] for parabolic semilinear equations.

Section 1.3. In Definition 1.1 we follow the usual notion of distribution function; decreasing and symmetric rearrangement of a function u by using $|u|$ instead of u. Nevertheless, for some purposes (see Theorem 2.22), sometimes it is interesting to work with the signed rearrangement of u (see details in Mossino [2]). For other different rearrangements, refer to the books by Bandle [1] and Kawohl [7]. Theorem 1.25 contains several basic properties of the symmetric rearrangement. Property (i) is easy to prove by using the Lebesgue-Stieltjes integral with respect to $d\mu$ (see, e.g., Hilden [1]). Properties (ii) and (iii) are due to Hardy and Littlewood and to Riesz, respectively (see a proof in Hardy-Littlewood-Polya [2]). Inequality (iv) was proved in Crandall-Tartar [1]. Finally, we comment on the important property (v). The observation that the Dirichlet integral $\int_\Omega |\nabla u|^2 dx$ diminishes under symmetrization, assumed $u = 0$ on $\partial\Omega$, was one of the starting points in the study of isoperimetric inequalities in mathematical physics (see Polya-Szego [1]). There are several proofs of property (v) according to the regularity assumed on u. Here we only mention the proof of Talenti [1] (by using the Fleming-Rishel formula) as well as those of Lieb [1] (for $p = 2$) and Berestycki-Lieb [1] (derived from some arguments which are more elementary). Other references relating to a more general inequality can be found in Kawohl [7].

It seems that the first application of rearrangement techniques to obtain *a priori* estimates of solutions of PDEs was given by Weinberger [1] for linear equations. A sharper result, containing the comparison $u^* \leqq v$ of Theorem 1.26, was proved by Talenti [1], [2] for linear equations and, later, by Talenti [3] for nonlinear equations (see also an alternative proof in Lions [1]). The general comparison $f(u^*) \underset{\sim}{<} f(v)$ of Theorem 1.26 was first shown by Chiti [1] and Lions [1] for the linear case and later by Maderna [1], Vazquez [3] and Mossino [2] for nonlinear equations. The proof given here is inspired by earlier ones and makes precise some delicate points as, for instance the proof of (53)).

An earlier and different proof of Theorem 1.28 for the semilinear equation and f Hölder continuous was given by Bandle-Sperb-Stakgold [1]. We remark that for the exterior problem such a result is not true, in general. In this case, the inequality similar to (14) involves some capacity terms (see Diaz [6]). Lemma 1.33 is due to Hardy-Littlewood-Polya [1] but the proof given here follows Bandle-Stakgold [1].

The results of Subsection 1.3a may be easily generalized to other equations, in the light of the works by Alvino-Trombetti [1], [2], [3], Talenti [4] and Trombetti -Vazquez [1]. We also mention the work by Alvino-Lions-Trombetti [1] in which it is proved, essentially, that if $u^* = v$ then necessarily u is radially symmetric (this is used in the proof of Theorem 1.36). A work related to the proof of Theorem 1.36, for quasilinear equations and in any dimension, is that by Mossino [3]. Theorem 1.38 seems to be new. We point out that assumption (71) holds if u is regular; this can be checked by applying the maximum principle to u_r.

Section 1.4. The careful study of solutions of free boundary problems near the free boundary was started by Caffarelli [1], [3] for the obstacle problem (taking, for instance, $q = 0$ in equation (3)) and, later, systematized by Alt-Caffarelli [1] for a problem in jet flow theory (for $q = -1$ in equation (3)). (See Friedman [3] for a very complete treatment of such a problem.) The results for the semilinear equation (1) are due to Phillips [2], Alt-Phillips [1] and Friedman-Phillips [1]. We refer the reader to those works for some more general formulations, remarks and bibliographical references.

Concerning Subsection 1.4d we remark that the convexity of N(u) for two-dimensional domains (Theorem 1.56, taken from Friedman-Phillips [1]) has recently been generalized by Caffarelli-Friedman [1] (strict convexity of a function of u) and by Korevaar-Lewis [1] and Kawohl [6] (for domains of arbitrary dimension). Many other geometrical properties, including star-shapedness, the exterior problem, quasilinear equations, among other topics, as well as an abundant literature, can be found in the recent monograph by Kawohl [7].

To end this subsection we mention that, to the author's knowledge, problems that seem to be unexplored included the following: the numerical approach to the free boundary $\mathcal{F}(u)$; some general stability results on $\mathcal{F}(u)$ with respect to f,g,h and Ω; and the study, near the free boundary, of degenerate quasilinear equations.

2 The free boundary in other second order nonlinear equations

The existence and properties of the free boundary $F(u)$ are examined for some second order nonlinear problems under formulations or assumptions not contemplated in the first chapter. Again, the comparison principle will be suitably applied.

The chapter starts with the consideration of several semilinear equations with nonmonotone perturbation, in Section 2.1. The extension of the results of Chapter 1 to general multivalued equations is made in Section 2.2, where the perturbation term in the equation is now assumed to be a general maximal monotone graph of $\mathbb{R}^2$. This allows the consideration of the obstacle problem, of the zero order reactions and of a general stationary porous media equation with absorption, among other examples.

In Section 2.3 the perturbation term has a singularity at the origin and is a decreasing function. A new phenomenology appears as a result of the lack of uniqueness.

Equations non-invariant by symmetries (or nonisotropic) are considered in Section 2.4, in which we examine the existence of the free boundary when originated by a suitable diffusion-convection balance. Also, some previous results are extended to fully nonlinear equations. The free boundary associated with solutions of the Hamilton-Jacobi-Bellman equations is also considered.

Finally, other boundary conditions are treated in Section 2.5 as well as the problem of thin obstacles, also called the Signorini problem.

2.1. EQUATIONS WITH A NONMONOTONE PERTURBATION TERM.

The main goal of this section is to point out how the free boundary $\mathcal{F}(u)$ studied in the last chapter also appears when the perturbation function f is not monotone or more precisely, when f may be decreasing for some real values. The main difficulty in the study of this problem comes from the fact that the comparison principle cannot be applied and so many different solutions may exist. In any case we shall restrict ourselves to the consideration of nonnegative solutions $u \geq 0$.

A first nonmonotone equation which can be easily studied corresponds to the case in which f is bounded from below by a nondecreasing function f_0. The main tool to be used is the following comparison result.

Theorem 2.1. *Let* $p > 1$, $g \in W^{-1,p'}(\Omega)$ *and* $h \in W^{1,p}(\Omega)$, $g \geq 0$, $h \geq 0$. *Let* f *be a real continuous function satisfying*

$$\left.\begin{array}{l}\exists f_0 \textit{ nondecreasing and continuous with } f_0(0) = 0 \textit{ and such that}\\ 0 \leq f_0(r) \leq f(r) \textit{ for every } r \geq 0.\end{array}\right\} \quad (1)$$

Let now $u_f \in W^{1,p}(\Omega)$ *be any nonnegative solution of the problem*

$$-\Delta_p u + f(u) = g \quad \textit{in} \quad \Omega \qquad (2)$$

$$u = h \quad \textit{on} \quad \partial\Omega \qquad (3)$$

Then $0 \leq u_f \leq u_{f_0}$ *in* $\overline{\Omega}$. □

With respect to the free boundary $\mathcal{F}(u)$ associated to any solution of (2),(3) we can now state the following result as a consequence of the above theorem (proved in Chapter 4) and Theorem 1.9.

Corollary 2.2. *Under the hypotheses of Theorem 2.1, if in addition*

$$\int_{0^+} \frac{ds}{F_0(s)^{1/p}} < \infty \quad , \qquad (4)$$

where $F_0(t) = \int_0^t f_0(s)ds$, *and if* u *is any nonnegative solution of* (2),(3), *then*

$$N(u) \supset \{x \in N(g) \cup N(h|_{\partial\Omega}) : d(x, S(g) \cup S(h|_{\partial\Omega})) \geq \varepsilon + L(\varepsilon)\} \qquad (5)$$

where $L(\varepsilon)$ *is given by* $L = \Psi_{1/N}(M)(\varepsilon))$, *with* $\Psi_{1/N}$ *defined from* f_0 *by* (33) *of Section* 1.1 *and* $M(\varepsilon)$ *is any bound of the supremum of* u_{f_0} *on the set* D_ε

of $N(g) \cup N(h|_{\partial\Omega})$ *defined as in Theorem* 1.9. □

In the rest of this section we shall study some more concrete nonmonotone problems for which assumptions (1) and (4) are verified, as well as some others in which the inequality (1) only holds in a neighbourhood of the origin. In these cases much more precise information will be obtained. In Subsection 2.1a the relations among the null set of two different solutions are considered. A nonlinear system in which (1) holds locally is studied in Subsection 2.1b. Finally, some equations in which f is even negative for some real values is analyzed in Subsection 2.1c (unidimensional case) and Subsection 2.1d (radially symmetric solutions in $\mathbb{R}^N$).

<u>2.1a. A nonmonotone semilinear equation in exothermic chemical reactions.</u> As we have already pointed out in the Introduction, the study of a single, irreversible, steady-state reaction taking place in a smooth bounded domain Ω of $\mathbb{R}^N$, leads to the semilinear problem

$$-\Delta u + \lambda f(u) = 0 \quad \text{in} \quad \Omega \tag{6}$$

$$u = 1 \quad \text{on} \quad \partial\Omega. \tag{7}$$

If the reaction is exothermic then f is not monotone and, then, some natural assumptions are the following (see Aris [1]).

$$f(t) = t^q f_1(t) \quad \text{is} \quad 0 \leqslant t < \infty \text{ , } 0 < q < 1. \tag{8}$$

$$m \leqslant f_1(t) \leqslant M \quad \text{for some} \quad 0 < m \leqslant M < \infty \tag{9}$$

$$|f_1''(t)| \leqslant K \quad \text{for some} \quad K. \tag{10}$$

The existence of classical solutions $u \in C^{2,\delta}(\Omega)$ of (6),(7) can be obtained by different methods. For instance, a solution can be found by minimizing the functional

$$J(v) \doteq \int_\Omega \left(\frac{1}{2}|\nabla v|^2 + F(v)\right)dx \quad , \quad F(t) = \int_0^t f(s)ds \tag{11}$$

on the convex set $K = \{v \in H^1(\Omega) : v = 1 \text{ on } \partial\Omega\}$. In the following we shall call the solutions of (6),(7) obtained in this way, <u>minimizer solutions</u>. They are also classical solutions, due to the regularity results (see Chapter 4). We also recall that the uniqueness of solutions is not true in general and that a sufficient condition in this sense is

$$f'(t) + \frac{f(t)}{1-t} > 0 \quad , \quad 0 < t \leqslant 1 \tag{12}$$

(see Chapter 4).

Althought the uniqueness of solutions is not assured, we can obtain some information about the difference between the solutions:

Theorem 2.3. a) *If* w_1, w_2 *are two solutions of* (6),(7) *corresponding to* λ_1 *and* λ_2 *respectively, and if* $\lambda_1 \leqslant \lambda_2$, $w_2 \leqslant w_1$ *then either* $w_2 \equiv w_1$ *and* $\lambda_1 = \lambda_2$ *or* $w_2 < w_1$ *on* $\{w_2 > 0\}$.

b) *If* u,v *are two solutions of* (6),(7) *with* $u \leqslant v, u \not\equiv v$, *then* $N(v) \subset \text{int } N(u)$, i.e.

$$d(N(v) , \mathbb{R}^N - N(u)) > 0 . \tag{13}$$

Proof. a) Suppose there exists a point $x_0 \in \Omega$ such that $0 < w_1(x_0) = w_2(x_0)$. If $\lambda_1 < \lambda_2$, then

$$\Delta(w_2 - w_1)(x_0) = (\lambda_2 - \lambda_1) f(w_2(x_0)) > 0$$

which contradicts the fact that $w_2 - w_1$ attains its maximum at x_0. If $\lambda_1 = \lambda_2$ then

$$\Delta(w_2 - w_1) = \lambda_2 c(x)(w_2 - w_1)$$

where

$$c(x) = \begin{cases} \dfrac{f(w_2) - f(w_1)}{w_2 - w_1} & \text{if } w_2 - w_1 \neq 0 \\ 0 & \text{if } w_2 - w_1 = 0. \end{cases}$$

By the strong maximum principle for linear equations (see e.g. Gilbarg-Trudinger [1]) $w_2 - w_1 \equiv 0$ in a neighbourhood of x_0. Hence the set $\{x : w_2 = w_1\}$ is open and, consequently , $w_2 \equiv w_1$.

b) Set $G_t = \{u < t\}$ for some small $t > 0$. By part a)

$$v \geqslant u + \delta \quad \text{on} \quad \partial G_t \text{ , for some } \delta > 0.$$

Suppose (13) is not true. Then, for any $\varepsilon > 0$ there is a unit vector e such that the function

$$v_\varepsilon(x) = v(x + \varepsilon e)$$

satisfies

$$v_\varepsilon(x_\varepsilon) = 0 < u(x_\varepsilon) \quad \text{at some point} \quad x_\varepsilon \in \partial G_t \tag{14}$$

Now, if ε is small enough (depending on δ) then

$$v_\varepsilon \geqslant u + \frac{\delta}{2} > u \quad \text{on} \quad \partial G_t.$$

On the other hand

$$\Delta u = f(u) \quad , \quad \Delta v_\varepsilon = f(v_\varepsilon) \quad \text{in} \quad G_t$$

and, in fact, by the assumptions (8),(9),(10), $f(s)$ is monotone increasing in s in the range of $u(x)$ and $v_\varepsilon(x)$, $x \in G_t$, provided that t is sufficiently small. Then, by comparison, we conclude that $v_\varepsilon \geqslant u$ in G_t, which contradicts (14). □

It is clear that our function f in consideration satisfies

$$f(t) \geqslant m\, t^q \quad , \quad 0 < q < 1$$

and so we are able to apply the Theorem 1.1 and Corollary 1.2, assuring in this way the existence of a null set $N(u)$, if λ is assumed large enough. The following theorem gives a sharp comparison result with respect to other nonlinear terms in the equation.

Theorem 2.4. *Suppose* f *and* $\tilde{f}$ *satisfy assumptions* (8),(9) *and* (10), *and let* $u,\tilde{u}$ *be minimizers of the problem* (6),(7) *for* f *and* $\tilde{f}$ *respectively. If*

$$\tilde{f}(t) > f(t) \quad \textit{for all} \quad 0 < t < 1 \tag{15}$$

then $\tilde{u} < u$ *in* $\{x \in \Omega : \tilde{u}(x) > 0\}$ *and* $d(N(u) : \mathbb{R}^N - N(\tilde{u})) > 0$.

Proof. Let $J(v)$ given in (11) , $\tilde{F}(t) = \int_0^t \tilde{f}(s)ds$ and

$$\tilde{J}(v) = \int_\Omega \left(\frac{1}{2} |\nabla v|^2 + \tilde{F}(v)\right)dx$$

We have

$$F(t) - F(s) < \tilde{F}(t) - F(s) \qquad \text{if} \qquad 0 \leqslant s < t \leqslant 1.$$

Consequently

$$\tilde{F}(\min(u,\tilde{u})) + F(\max(u,\tilde{u})) < \tilde{F}(\tilde{u}) + F(u) \tag{16}$$

at each point x for which $\tilde{u}(x) > u(x)$. If $\tilde{u}(x) \leqslant u(x)$ then equality holds in (16). Thus, provided that the set $\{\tilde{u} > u\}$ is non empty, we have

$$\int_\Omega \tilde{F}(\min(u,\tilde{u})) + \int_\Omega F(\max(u,\tilde{u})) < \int_\Omega \tilde{F}(\tilde{u}) + \int_\Omega F(u)$$

and hence

$$\tilde{J}(\min(u,\tilde{u})) + J(\max(u,\tilde{u})) < \tilde{J}(\tilde{u}) + J(u). \tag{17}$$

However, since $\max(u,\tilde{u})$ and $\min(u,\tilde{u})$ belong to K, we must have

$$\tilde{J}(\min(u,\tilde{u})) \geqslant \tilde{J}(\tilde{u}) \quad , \quad J(\max(u,\tilde{u})) \geqslant J(u)$$

contradicting (17). We have thus proved $u \geqslant \tilde{u}$. (Note that, in contrast with Theorem 2.1, f and $\tilde{f}$ are not assumed to be non-decreasing). We can now proceed as in Theorem 2.3 in order to deduce that $u > \tilde{u}$ in $\{x \in \Omega: \tilde{u} > 0\}$ and that $N(u) \subset \operatorname{int} N(\tilde{u})$. □

Finally, it is important to point out that the regularity result and dependence of the null set $N(u)$ with respect to the λ of Subsection 1.4c were established by Friedman-Phillips [1] for f satisfying the general conditions (8), (9) and (10). In addition, the convexity of the null set for two-dimensional convex domains (Theorem 1.56) as well as the classification of those null sets (Theorem 1.57) were also given without any monotonicity assumption on f.

2.1b. A nonlinear system.

Nonlinear systems involving monotone terms were considered in Remark 1.8 as applications of the results for an x-dependent scalar equation. Here we shall study a different nonlinear system of a different formulation involving nonmonotone perturbations, in the following way:

$$- \Delta u_i(x) + f_i(u(x)) = 0 \quad , \; i = 1,\dots n, \quad \text{in} \quad \mathbb{R}^N, \tag{18}$$

where $u = (u_1,\dots,u_n)$, $N \geq 2$ and the n functions $f_i : \mathbb{R}^n \to \mathbb{R}$ are the gradients of some function $F \in C^1(\mathbb{R}^n - \{0\})$, namely

$$f_i(u) = \frac{\partial F(u)}{\partial u_i} \quad \text{if} \quad u \neq 0 \quad \text{and} \quad f_i(u) = 0 \quad \text{if} \quad u = 0 \tag{19}$$

Nonlinear systems like (18) arise in several branches of mathematical physics (see,e.g, references in Brezis-Lieb[1]).

The existence of a nontrivial solution u of (18) was proved in Brezis-Lieb [1] under very general assumptions of F such as, for instance,

$$\lim_{|u|\to\infty} |u|^{-p}|F(u)| = 0 \quad , \quad p = 2^* = 2N/(N-2). \tag{20}$$

The solutions u are found in the class of functions

$$L = \{u \mid u \in L^1_{loc}(\mathbb{R}^N) \;,\; \nabla u \in L^2(\mathbb{R}^N) \;,\; F(u) \in L^1(\mathbb{R}^N) \quad \text{and}$$

$$\text{meas}\{|u| > t\} < \infty \quad \text{for all} \quad t > 0\} \; .$$

These solutions verify (18) in the sense of distributions ($\mathcal{D}'$) and , moreover, minimize the action

$$J(v) = \frac{1}{2} \sum_{i=1}^{n} \int_{\mathbb{R}^N} |\nabla v_i(x)|^2 dx + \int_{\mathbb{R}^N} F(v)dx$$

in the sense that $0 < J(u) \leq J(v)$ for all $v \in L \cap L^\infty_{loc}$, $v \not\equiv 0$ and v satisfying (18) in the sense of distributions.

If, in addition to (20), it is assumed that

$$f(u)\cdot u \geq C\, |u|^{q+1} \qquad \text{for all} \quad u \in \mathbb{R}^n \quad , |u| < \delta,\; q > 0 \quad , \tag{21}$$

then the above solutions u of (18) are more regular, as the mentioned authors have shown, because they must satisfy that $u \in [W^{2,s}_{loc}(\mathbb{R}^N)]^n$ for all $s < \infty$ (and so $u \in [C^{1,\alpha}_{loc}]^n, \forall\alpha < 1$) as well as

$$u \in [L^\infty(\mathbb{R}^N)]^n \quad \text{and} \quad \lim_{|x|\to\infty} u(x) = 0. \tag{22}$$

We remark that assumption (21) is a sort of vectorial version of hypothesis (1). Nevertheless, (21) is only assumed for small vectors u, in contrast with (1). The following result shows how the information given in (22) may be used, jointly with (21), to assure the existence of a free boundary.

Theorem 2.5. *Assume that* (21) *holds with* $0 < q < 1$ *and let* $u \in L$ *with* $f(u) \in L^1_{loc}$ *satisfying* (18) *in the sense of distributions. Then* $u(x)$ *has compact support in* $\mathbb{R}^N$.

Proof. First of all we note that

$$-\Delta|u|^2 = -2u \cdot \Delta u - 2|\nabla u|^2 \leqslant 2u \cdot (-f(u))$$

Then by (22) we have that

$$-\Delta|u|^2 + 2C|u|^{q+1} \leqslant 0 \quad \text{in} \quad \{|x| > R\}$$

for some large enough $R > 0$. Then, the function $w(x) = |u(x)|^2$ is a bounded nonnegative subsolution of the scalar equation

$$-\Delta w + 2C\, w^{\frac{q+1}{2}} = 0 \quad \text{in} \quad \{|x| > R\}. \tag{23}$$

Finally, the conclusion holds by using the comparison principle for equation (23) as well as Theorem 1.9 or 1.18. □

2.1c. Equilibrium solutions of a degenerate parabolic equation in biological population models.

In the above examples of this section the monotone perturbation f satisfied a "positivity" assumption

$$f(u)u \geqslant 0\ . \tag{24}$$

Now we shall consider a special, but very illustrative, example of a non-monotone problem in which (24) fails. More precisely, given $L > 0$ we shall study the one-dimensional semilinear problem

$$- u_{xx} + f(u) = 0 \qquad \text{in} \qquad \Omega = (-L,L) \tag{25}$$

$$u(\pm L) = 0 \tag{26}$$

where

$$f(u) = f_1(u^{1/m}) \ , \ f_1(r) = r(1-r)(a-r) \ , \ 0 < a < (m+1)/(m+3), \ m > 1. \tag{27}$$

Problem (25),(26) appears in the study of the equilibrium points of the degenerate parabolic equation

$$v_t - (v^m)_{xx} + f_1(v) = 0 \tag{27}$$

arising in some biological population models (see references in the work Aronson-Crandall-Peletier [1] from which are taken the results of this subsection).

First of all, we remark that, as in 2.1b, the majoration of f (hypotheses (1) and (4)) only holds for small values of u. Nevertheless, we shall see that for adequate values of L the free boundary $\mathcal{F}(u)$ may appear, modifying in a substantial way the structure of the set of equilibrium points of (27) in contrast with the nondegenerate case $(m = 1)$ for which $\mathcal{F}(u)$ cannot exist (see e.g. Smoller-Wasserman [1]).

In the analysis of (25)(26) we shall exploit the one-dimensionality of the problem by reducing it to a Cauchy problem. So, if $u(x) > 0$ in $(-L,L)$, then there exists $x_0 \in (-L,L)$ such that $0 < u(x) < u(x_0)$ for $x \in (-L,L)$ and clearly $u'(x_0) = 0$ (recall that any solution u of (25) is $u \in C^{2,\delta}$ and satisfies it in a classical sense). Conversely, let us seek conditions on $x_0 \in (-L,L)$ and $\mu \in \mathbb{R}^+$ which guarantee that the solution of the initial value problem

$$- u_{xx} + f(u) = 0 \tag{28}$$

$$u(x_0) = \mu^m \ , \ u'(x_0) = 0 \tag{29}$$

is also a positive solution of (25),(26). If $\mu = 1$ then $u \equiv 1$ is the unique solution of (28),(29) (recall that $f(1) = 0$). If $\mu > 1$, then $f(s)>0$ for $s > 1$ implies that any solution of (28),(29) is convex and hence cannot satisfy $u(\pm L) = 0$. Thus (28),(29) has no solutions satisfying the boundary conditions unless $\mu<1$. Consequently, we consider only $\mu \in (0,1)$.

To solve (28),(29) we integrate the equation in the usual way. For simplicity, we shall first recall that if $v = u^{1/m}$ then (28) leads to

$$-(v^m)_{xx} + f_1(v) = 0 \tag{30}$$

$$v(x_0) = \mu \quad , \; v'(x_0) = 0 \tag{31}$$

Now, we multiply (30) by $(v^m)'$ and integrate the result using (31). Then

$$\frac{1}{2}(v^m)_x^2 + m\,F_1(v) = m\,F_1(\mu) \tag{32}$$

where

$$F_1(r) = \int_0^r s^{m-1} f_1(s)ds = -\frac{1}{m+3}\, r^{(m+1)}[\, r^2-(1+a)\frac{m+3}{m+2}\, r + a\,\frac{m+3}{m+1}\,]. \tag{33}$$

Since $f_1 < 0$ on $(a,1)$, F_1 is strictly increasing on $(a,1)$. Thus, if $\mu > a$, we can integrate (32) to obtain

$$\left(\frac{m}{2}\right)^{1/2} \int_v^{\mu} \frac{\eta^{m-1}}{(F_1(\mu)-F_1(\eta))^{1/2}}\, d\eta = |x_0 - x| \; . \tag{34}$$

The integrand in (34) has a singular point at $\eta = \mu$, but $F_1(\mu) - F_1(\eta) \geqslant \delta(\mu - \eta)$ for some $\delta > 0$ and η near μ, so the singularity is integrable. Equation (34) defines v implicitly as a function of $|x_0 - x|$ as long as $v \leqslant \mu$. If $F_1(\mu) < 0$, then there exists a unique $\nu \in (0,a)$ such that $F_1(\nu) = F_1(\mu)$ and $F_1(\eta) < F_1(\mu)$ for $\eta \in (\nu,\mu)$. In this case (34) represents a periodic solution of (30) whose values lie in $[\nu,\mu]$. Thus, in order that (34) represents a positive solution of (25),(26) it is necessary that $F_1(\mu) \geqslant 0$.

The sign of F_1 is determined by the sign of

$$H(r) = r^2 - (1+a)\,\frac{m+3}{m+2}\; r + a\,\frac{m+3}{m+1}\; . \tag{35}$$

It can be easily checked that H has a unique root $\alpha \in (a,1)$ if and only if $H(1) < 0$ or $0 < a < \frac{m+1}{m+3}$. In this case,

$$F_1 < 0 \quad \text{on} \quad (0,\alpha) \quad \text{and} \quad F_1 > 0 \quad \text{on} \quad (\alpha,1),$$

hence we may restrict our attention to the range $\alpha \leqslant \mu < 1$.

For $\mu \in [\alpha,1)$ we have $F_1(\eta) < F_1(\mu)$ for all $\eta \in (0,\mu)$. Thus we can extend the integration in (34) down to $v = 0$. Define

$$\lambda(\mu) = \left(\frac{m}{2}\right)^{1/2} \int_0^{\mu} \frac{\eta^{m-1}}{(F_1(\mu)-F_1(\eta))^{1/2}}\, d\eta \quad , \quad \alpha \leqslant \mu < 1. \tag{36}$$

If $\mu = \alpha$ the integrand in (36) may have a second singularity at $\eta = 0$. However, $-F_1(\eta) \geqslant \delta^2 \eta^{m+1}$ for some $\delta > 0$ and $\eta > 0$ near 0, so $\eta^{m-1}(-F_1(\eta))^{-1/2} \leqslant \delta^{-1}\, \eta^{(m-3)/2}$ near $\eta = 0$. Since $m > 1$ this singularity is integrable and λ is well defined on $[\alpha,1)$.

For a positive solution u of (25),(26), $u = 0$ only at $\pm L$. Therefore

$$\lambda(\mu) = |x_0 - L| = |x_0 + L|$$

from which we conclude that $x_0 = 0$. To summarize , we have proven the following result:

<u>Proposition 2.6.</u> *Suppose* $0 < a < (m+1)/(m+3)$. *Then* u *is a positive solution of* (25),(26) *if and only if*

$$\left(\frac{m}{2}\right)^{1/2} \int_{u^{1/m}(x)}^{\mu} \frac{\eta^{m-1}}{(F_1(\mu)-F_1(\eta))^{1/2}}\, d\eta = |x|, \textit{for } |x| \leqslant L,$$

where $\mu \in [\alpha,1)$ *and* $L \in \mathbb{R}^+$ *are related by the equation*

$$\lambda(\mu) = L \tag{37}$$

and α *is the unique root of* H *in* $(a,1)$.□

In view of (37), there is a positive solution of (25)(26) for a given interval $(-L,L)$ if and only if L is in the range of λ ,i.e. $L \in \lambda([\alpha,1))$. When $L = \lambda(\mu)$ we write $u(x,\lambda)$ for the corresponding positive solution. The multiplicity of these positive solutions is the same as the multiplicity of the roots of $\lambda(\mu) = L$, which is determined by the shape of the graph of λ . The next result shows that the graph of $\lambda(\mu)$ always has the general features indicated in Figure 7.

<u>Proposition 2.7.</u> 1) $\lambda \in C[\alpha,1) \cap C^1(\alpha,1)$; 2) $\lambda(\mu) \to \infty$ *and* $\lambda'(\mu) \to \infty$ *as* $\mu \uparrow 1$; 3) $\lambda'(\mu) \to -\infty$ *as* $\mu \downarrow \alpha$ *and* 4) $\lambda'(\mu)$ *has a unique root* $\mu_0 \in (\alpha,1)$. □

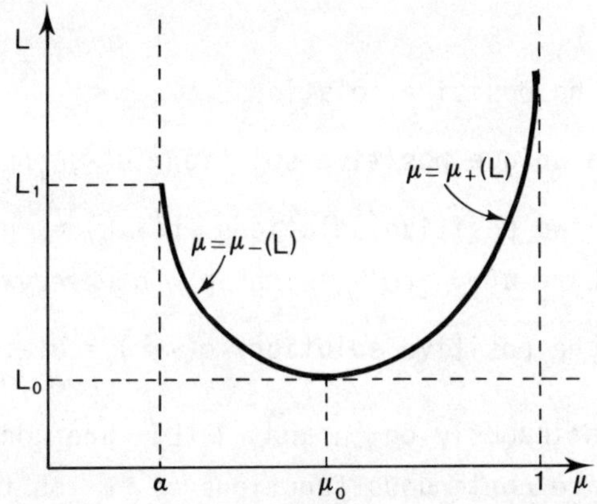

Figure 7.

The proof of Proposition 2.7 is rather technical, and we omit it (we refer to Aronson-Crandall-Peletier [1] for the proof). Some remarks about the interpretation of this result are the following. Define $L_0 = \lambda(\mu_0)$ and $L_1 = \lambda(\alpha)$. Clearly $L_0 > 0$ and by 2) $\lambda([\alpha,1)) = [L_0,+\infty)$. Moreover,

$$\lambda(\mu) = L \text{ has } \begin{cases} \text{no solutions for } 0 < L < L_0 \\ \text{one solution for } L = L_0 \text{ and for } L > L_1 \\ \text{two solutions for } L_0 < L \leq L_1. \end{cases}$$

It is interesting to note the dependence of L_1 on m. Let us write $F_1 = F_1(\eta,m)$, $\alpha = \alpha(m)$ and $L_1 = L_1(m)$. Then

$$L_1(m) = \left(\frac{m}{2}\right)^{1/2} \int_0^{\alpha(m)} \frac{\eta^{m-1}}{(-F_1(\eta:m))^{1/2}}\, d\eta$$

If $a \in (0,1/2)$, then $\alpha(m)$ is defined and continuous for $m \geq 1$. Moreover, as $m \to 1$, $F_1(\eta:m) \to F_1(\eta,1)$. Since $-F_1(\eta;1) = a\eta^2/2 + O(\eta^3)$ for sufficiently small $\eta > 0$, $(-F_1(\eta:1))^{-1/2}$ is not integrable at $\eta = 0$. It follows from Fatou's lemma that

$$\lim_{m \downarrow 1} L_1(m) = \infty .$$

Propositions 2.6 and 2.7 provide a complete characterization of the set of positive solutions of (25),(26). For $L_0 \leqslant L$ let $\mu_+(L)$ denote the largest solution of $L = \lambda(m)$ and for $L_0 \leqslant L \leqslant L_1$ let $\mu_-(L)$ be the smallest solution (so $\mu_+(L_0) = \mu_-(L_0)$). We distinguish the following cases

$0 < L < L_0$. There are no positive solutions.

$L = L_0$ There is a unique positive solution $u(\cdot,\mu_+(L_0))$.

$L_0 < L \leqslant L_1$ There are two positive solutions $p(\cdot,L) = u(\cdot,\mu_-(L))$ and $q(\cdot,L) = u(\cdot,\mu_+(L))$ with $p < q$ everywhere on $(-L,L)$.

$L > L_1$ There is one positive solution $q(\cdot,L) = u(\cdot,\mu_+(L))$.

Since $u(\cdot,\mu)$ depends continuously on μ and $\mu_\pm(L)$ are continuous on their domains, p and q are continuous functions of L on their domains. Note also that the nonexistence of the small positive solution p for $L > L_1$ is due to the nonlinearity of the diffusion ($m > 1$). (See Smoller-Wasserman [1] for the study of the linear case).

Another effect of the nonlinearity is that $u(\cdot,\alpha) = u(\cdot,\mu_-(L_1))$ generates families of nonnegative solutions of (25),(26) on intervals $(-L,L)$ with $L > L_1$. To show this, note that for $\mu \in (\alpha,1]$ we have $F_1(\mu) > 0$ so that, according to (32), $(v^m)'(\pm\lambda(\mu),\mu) \neq 0$. However $F_1(\alpha) = 0$, so $(v^m)'(\pm\lambda(\alpha),\alpha) = (v^m)'(\pm L_1,\alpha) = 0$. It follows that $u(x,\alpha)$ extended to 0 for $L \geqslant |x| \geqslant L_1$ is a solution of (25),(26) for $L > L_1$ and so is

$$U(x;h) = \begin{cases} u(x-h;\alpha) & \text{for } |x-h| \leqslant L_1 \\ 0 & \text{for } |x-h| > L_1 \end{cases}$$

provided $|h| \leqslant L - L_1$. More generally, we may piece several such solutions together if their support are disjoint. Let n be a positive integer and $L \geqslant nL_1$. For each n-vector $\zeta = (\xi_1,\dots \xi_n)$ which satisfies

$$-L \leqslant \xi_1 - L_1,\ \xi_i + L_1 \leqslant \xi_{i+1} - L_1, i = 1,\dots,n-1 \text{ and } \zeta_n + L_1 \leqslant L \tag{36}$$

the function

$$U(x;\xi) = \begin{cases} u(x-\xi_i,\alpha) & \text{for } |x-\xi_i| \leq L_1 \\ 0 & \text{for } |x-\xi_i| > L_1 \text{ for } i = 1...n \end{cases}$$

is a nonnegative solution of (25),(26) . We shall use $P_n(L)$ to denote the collection of functions $u(\cdot,\xi)$ where $\xi \in \mathbb{R}^n$ satisfies (36).

Clearly, a nonnegative solution of (25),(26) is either positive or belongs to some $P_n(L)$. We thus have:

Proposition 2.8. *For* $L > L_1$, *let* n *be the integral part* $[L/L_1]$ L/L_1. *Then, with* $P(L) = \bigcup_{j=1}^{m} P_j(n)$ *we have that the set* $E^*(L)$ *of all the nonnegative and non-trivial solutions of* (25),(26) *is given by* $E^*(L) = \{q(\cdot,L)\} \cup P(L)$.

Remark 2.1. If $L/L_1 > [L/L_1] = n$ then P_n is a true n-parameter family while if $n = L/L_1$, P_n contains only $U(x,\zeta)$, $\xi_i = (i-1)L_1 + L_{1/2}$. □

Combining Propositions 2.6, 2.7 and 2.8 we obtain the complete description of the set $E(L)$ of nonnegative equilibrium solutions of the parabolic equation (27) with homogeneous Dirichlet conditions.

Theorem 2.9. $E(L) = \{0\}$ *if* $0 < L < L_0$, $E(L) = \{0,p(\cdot,L), q(\cdot;L)\}$ *if* $L_0 \leq L \leq L_1$, *and finally* $E(L) = \{0,q(\cdot,L)\} \cup P(L)$ *if* $L_1 < L$. □

2.1d. Non-negative radial solutions of a nonmonotone semilinear equation in $\mathbb{R}^N$.

An interesting result about the compactness of the support of the radial solutions of the non-monotone equation

$$\begin{cases} -\Delta u + f(u) = 0 \quad \text{in} \quad \mathbb{R}^N & (37) \\ u(x) \to 0 \quad \text{as} \quad |x| \to \infty & (38) \end{cases}$$

is given in Peletier-Serrin [1] as a byproduct of the discussion of the uniqueness of radial solutions of (37),(38) . The basic assumptions made on the function f are the following:

f is locally Lipschitz continuous in $(0,+\infty)$, (39)

$\int_0^\delta f(s)ds < 0$ for some $\delta > 0$, (40)

$\lim_{s\to 0} f(s) = 0$ and $f(0) = 0$. (41)

In order to show the compactness of the support of the radial solution, they make some additional assumptions on the primitive F of f, $F(t) = \int_0^t f(s)ds$. Let

$\alpha = \inf \{u > 0 : f(u) < 0\}$

$\beta = \inf \{u > 0 : F(u) < 0\}$.

In the following, we shall always assume

$$\beta > 0 \tag{42}$$

Note that $\alpha > 0$ implies $\beta > 0$. Moreover the condition $\beta > 0$ can be satisfied even when f is not everywhere non-positive near $u = 0$,($\alpha = 0$). Of course, for such a possibility f must be rather special near $u = 0$, the behaviour of

$$f(u) = \begin{cases} -u^2\sin\frac{1}{u} & \text{when} \quad \sin\frac{1}{u} > 0 \\ -u\sin\frac{1}{u} & \text{when} \quad \sin\frac{1}{u} \leqslant 0 \end{cases}$$

is an example of this situation. We also remark that the function f given in the study of (25),(26) satisfied $\alpha > 0$. Since the solutions can have compact support, the uniqueness of radial solutions of (37),(38) is studied in the mentioned work in the class of non-negative and non-trivial functions. As in the above situations, here we shall restrict ourselves to the question of the study of the support of solutions. Among the radial solutions , (37),(38) can be formulated as

$$-u'' - \frac{(N-1)}{r}u' + f(u) = 0 \qquad r > 0 \tag{43}$$

$$u'(0) = 0 \ , \quad \lim_{r\to\infty} u(r) = 0 \quad , \ u \geqslant 0 \quad \text{on} \quad r \geqslant 0. \tag{44}$$

The main result about the compactenss of the support is the following :

Theorem 2.10. *Assume the hypothesis (42) and let* u *be a solution of* (43),(44). *Then, the condition*

$$\int_{0^+} \frac{ds}{F(s)^{1/2}} < \infty \tag{45}$$

is sufficient for u *to have compact support. Moreover, if* $\alpha > 0$, *condition* (45) *is also necessary.*

Remark 2.2. Condition (45) has been already used in Section 1.2a. to find sufficient conditions for the existence of the free boundary when f is assumed non-decreasing and $f(0) = 0$. □

We need some preparatory Lemmas.

Lemma 2.11. *Let* u *be a solution of equation* (43) *on a finite interval* $[r_0,r_1] \subset [0,+\infty)$ *and let* $u > 0$ *on* (r_0,r_1). *Then*

$$[\frac{1}{2}u'(r_1)^2 - F(u_1)]-[\frac{1}{2}u'(r_0)^2 - F(u_0)] = -(N-1)\int_{r_0}^{r_1} \frac{u'(r)^2}{r}\,dr, \tag{46}$$

where $u_i = u(r_i)$, $i = 0,1$.

Proof. Multiply equation (45) by u', and integrate over (r_0,r_1).(Note that in the one-dimensional case (e.g. equation (25)) the right hand side of (46) is identically zero). □

Lemma 2.12. *Let* u *be a nonnegative solution of* (43),(44). *Then* $u' \leq 0$ *on* $[0,\infty)$. *Equality holds at a point* $\bar{r}$ *if and only if* $u(\bar{r}) = 0$. *In consequence, either* $u > 0$ *on* $[0,\infty)$ *or* $u > 0$ *on* $[0,a)$, *and* $u = 0$ *on* $[a,+\infty)$ *for some* $a > 0$.

Proof. Let ξ be a critical point of $u(r)$, with $\xi \in (0,\infty)$. Then there are three cases to consider : i) $f(u(\xi)) < 0$, ii) $f(u(\xi)) > 0$ and iii) $f(u(\xi)) = 0$. In case (i) the equation shows that $u''(\xi) < 0$, so ξ is a strict maximum, in case (ii) we get $u''(\xi) > 0$ hence ξ is a strict minimum. Finally, if $f(u(\xi)) = 0$, and if $u(\xi) > 0$, by the uniqueness of the initial value problem for ordinary differential equations (applied at the initial point $r = \xi$) we see that

$$u(r) = u(\xi) = \text{constant}, \quad r \in [0,\infty).$$

This violates the condition $u \to 0$ as $r \to \infty$ and so (iii) cannot occur if $u(\xi) > 0$. Thus the isolated critical points in $(0,\infty)$ are either strict local maxima or strict local minima. We now treat two further cases. First, suppose that there are no maximum points in $(0,\infty)$. Then, obviously (since $u(0) \geqslant 0$ and $u \to 0$ as $r \to \infty$), we have u everywhere nonincreasing, indeed with $u'(r) \leqslant 0$. Second, suppose (for contradiction) that there is <u>at least</u> one maximum point $\bar{\xi}$ in $(0,\infty)$. Hence, to the left of $\bar{\xi}$ we have $u'(r) > 0$. Since $u'(0) = 0$, there is a first point r_0 to the left of $\bar{\xi}$ where $u'(r_0) = 0$. Clearly $u(r_0) < u(\bar{\xi})$. To the right of $\bar{\xi}$ it is evident that there is a last point r_1 where $u(r_1) = u(r_0)$; see Figure 8

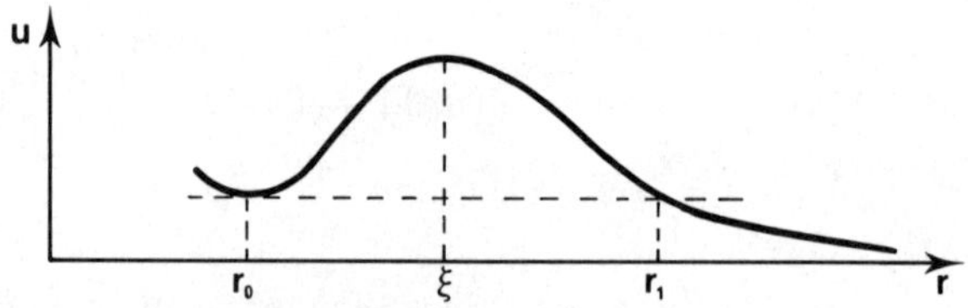

<u>Figure 8.</u>

Then applying Lemma 2.11 on (r_0,r_1) we get

$$\frac{1}{2}\, u'(r_1)^2 = -(N-1) \int_{r_0}^{r_1} \frac{u'(r)^2}{r}\, dr < 0$$

This is impossible, and so, as $u \geqslant 0$ and $\lim_{r\to a} u(r) = 0$ then $u' \leqslant 0$ for all $r > 0$. □

<u>Remark 2.3.</u> Note the important difference between the behaviour of solutions of equation (37) for $N = 1$ and $N > 1$. Indeed, Lemma 2.13 is not true if $N = 1$ (see Proposition 2.8). □

<u>Lemma 2.13.</u> *Let* u *be a nontrivial nonnegative solution of* (43),(44). *Then* $\lim_{r\to\infty} u'(t) = 0$ *and*

$$\frac{1}{2}\, u'(r)^2 - F(u(r)) = (N-1) \int_r^{\infty} \frac{u'(s)^2}{s}\, ds \quad , \; r \geqslant 0 \tag{47}$$

Proof. By Lemma 2.12 it suffices to assume $u > 0$ on $[0,a)$ for some maximal $a > 0$ which may be finite or infinite. If $a < +\infty$ then (47) follows by setting $r_1 = a$ in (46). Next suppose $a = \infty$. We let $r_1 \to \infty$ in (46). The right-hand side converges to some negative limit, or to $-\infty$. This implies that $\frac{1}{2} u'(r_1)^2 + F(u(r_1))$ converges, through possibly to $-\infty$. However, as $r_1 \to \infty$, $u(r_1) \to 0$ and, hence, $F(u(r_1)) \to 0$. Thus $\frac{1}{2} u'(r_1)^2$ converges to some nonnegative limit V^2. Again by (44), we have $V = 0$ and then

$$\lim_{r \to \infty} \{ \frac{1}{2} u'(r)^2 - F(u(r)) \} = 0$$

which proves (47) by using (46). □

Remark 2.3. Note that by (47)

$$- F(u(0)) = (N-1) \int_0^\infty \frac{u'(r)^2 dr}{r} > 0.$$

Since $F \geqslant 0$ on $(0,\beta)$, this implies that $u(0) > \beta$. □

Lemma 2.14. *Assume* $\beta > 0$. *Let* u *be a nontrivial nonnegative solution of* (43), (44) *and let* $R > 0$ *such that* $u(R) = U \in (0,\beta)$. *Then, if* $r(u)$ *is the inverse of* u *on* $(0,U]$, *we have*

$$r(u) \leqslant r(U) + 2^{-1/2} \int_u^U (F(s))^{-1/2} ds \qquad 0 < u \leqslant U. \tag{48}$$

Proof. By Lemma 2.13

$$\frac{1}{2} u'(r)^2 - F(u(r)) \geqslant 0 \quad \text{for every } r \geqslant 0.$$

But if $r > R$ then $u(r) \in [0,\beta)$ and hence $F(u(r)) \geqslant 0$ and

$$\frac{1}{2} u'(r)^2 \geqslant F(u(r)) \qquad \text{for} \quad r \geqslant R.$$

Since $u' \leqslant 0$ then

$$u'(r) \leqslant -2^{1/2} (F(u(r)))^{1/2} \tag{49}$$

By Lemma 2.12 there exists a number $a \leqslant +\infty$ such that $u(r) > 0$

when $r < a$ and $u(r) = 0$ when $r \geq a$. Thus, as $R < a$, we obtain from (49)

$$(F(u(r)))^{-1/2}\, u'(r) \leq -\, 2^{1/2} \quad \text{for } R \leq r < a.$$

Integrating this inequality from R to $r \in (R,a)$ we obtain the conclusion. □

Proof of Theorem 2.10. That the condition (45) is sufficient for u to have compact support is an immediate consequence of (48). To prove that (45) is necessary we assume that u has compact support $[0,a]$ with $a<\infty$. By Lemma 2.13 we have

$$u'(r)^2 \leq 2|F(u(r))| + C \int_r^a u'(s)^2 ds \qquad a - \delta < r < a$$

where $C \equiv 2(N-1)/(a-\delta)$ and $\delta \in (0,a)$ will be chosen later. Applying Gronwall's Lemma we deduce that

$$u'(r)^2 \leq 2|F(u(r))| + 2C \int_r^a |F(u(s))| e^{C(s-r)} ds\ . \tag{50}$$

Now we choose δ so small that $u(r) \in (0,\alpha)$ for $a - \delta < r < a$. Then

$$\frac{d}{dr} F(u(r)) = f(u(r))u'(r) < 0$$

and hence $F(u(s)) \leq F(u(r))$ if $s \in (r,a)$. Using this in (50) we obtain

$$u'(r)^2 \leq 2F(u(r))\, [1 + C \int_r^a e^{C(s-r)} ds\} \leq 2F(u(r))e^{C\delta}. \tag{51}$$

Dividing by F in (51), taking the square root and integrating over $(a-\delta,a-\varepsilon)$, $0 < \varepsilon < \delta$ we have

$$\int_{u(a-\varepsilon)}^{u(a-\delta)} (F(u))^{-1/2} du \leq 2^{1/2}\ \delta e^{\delta(N-1)/(a-\delta)}.$$

Finally, as $u(a-\varepsilon) \to 0$ as $\varepsilon \to 0$ we conclude that (45) holds. □

Remark 2.4. The uniqueness of nonnegative radial solutions is proved in Peletier-Serrin [1]under a hypothesis (slightly) stronger than (42). □

2.2. VARIATIONAL INEQUALITIES AND MULTIVALUED EQUATIONS.

The main goal of this section is to extend the results of Chapter 1 to some multivalued equations which can be understood as limit equations of some quasilinear equations. A possible motivation for the study of such a type of equations may come from the three following examples:

i) The obstacle problem. Given a function $\psi \in H^1(\Omega)$ (called "obstacle"), an already classical problem appearing in many different contexts is to find w minimizing the functional

$$J(v) = \frac{1}{2}\int_\Omega |\nabla v|^2 dx - \int_\Omega g\ v\ dx \tag{52}$$

on the convex set $K = \{v - h \in H_0^1(\Omega)\ ,\ v \geq \psi \text{ on } \Omega\}$, where g and h are data of the problem, for instance, $g \in L^2(\Omega)$, $h \in H^1(\Omega)$, $h \geq \psi$. It is easy to see that any solution u of this problem is characterized by the inequality

$$\int_\Omega \nabla w \cdot \nabla(v - w)dx \geq \int_\Omega g(v - w)dx \qquad \forall v \in K\ ,\ w \in K. \tag{53}$$

Finally, under additional assumptions on ψ, $\partial\Omega$ and h, we have that $u \in H^2(\Omega)$ and, then, u satisfies the "complementary formulation"

$$-\Delta w \geq g\ ,\quad w \geq \psi \quad \text{and} \quad (-\Delta w - g)(\psi - w) = 0 \text{ a.e. on } \Omega \tag{54}$$

$$w = h \quad \text{on} \quad \partial\Omega. \tag{55}$$

We also note that if $\psi \in H^2(\Omega)$ the above problem reduces to the case of zero obstacle. Indeed, the function $u = w - \psi$ satisfies

$$-\Delta u \geq g\ ,\quad u \geq 0\ ,\quad (-\Delta u - g)u = 0 \quad \text{a.e. on} \quad \Omega \tag{56}$$

$$u = h \quad \text{on} \quad \partial\Omega \tag{57}$$

where g and h are now given by $g + \Delta\psi$, $h + \psi$, from the functions g and h of the original formulation. (For an exhaustive treatment of the above variational inequalities, as well as of its variants, see the books by Duvaut-Lions [1], Kinderlehrer -Stampacchia [2] and Friedman [3]). We finally remark that situations in which the operator Δ is replaced by another different (linear or nonlinear) differential operator arise frequently in the applications (see the references mentioned).

ii) Zero order reactions and equations with discontinuous perturbations. Semilinear equations

$$- \Delta u + f(u) = 0 \quad \text{in} \quad \Omega \tag{58}$$

involving discontinuous functions f appear, for instance, in the study of a single, irreversible steady state reaction of zero order (see the Introduction). In this special case f is discontinuous at $r = 0$, and it is given by

$$f(r) = 0 \quad \text{if} \quad r \leqslant 0 \quad \text{and} \quad f(r) = \lambda \quad \text{if} \quad r > 0, \lambda > 0. \tag{59}$$

Other equations with discontinuous terms are discussed, for instance, in Stuart [1]Chang[1] and Frank-Wendt [1].

iii) Stationary states in porous media with absorption. The study of stationary solutions of the porous media equation with an absorption term leads to the problem

$$- \Delta\phi(v) + f(v) = g \quad \text{in} \quad \Omega \tag{60}$$

$$v = 0 \quad \text{on} \quad \partial\Omega \quad . \tag{61}$$

Here the functions ϕ and f are assumed to be continuous, non-decreasing and $f(0) = \phi(0) = 0$. When ϕ is strictly increasing the function $u = \phi(v)$ satisfies

$$- \Delta u + \tilde{f}(u) = g \quad \text{in} \quad \Omega \tag{62}$$

$$u = 0 \quad \text{on} \quad \partial\Omega \tag{63}$$

where $\tilde{f} = f \circ \phi^{-1}$, but if ϕ is not strictly increasing $\tilde{f}$ is a multi-valued function.□

Note that in examples (ii) and (i) the free boundary $\mathcal{F}(u)$ makes sense, and that for the obstacle problem the free boundary $\mathcal{F}_\psi(u)$ is defined as the common boundary of the sets $S_\psi(w) = \{x \in \bar{\Omega}: w(x) > \psi(x)\}$ (the continuation set) and $N_\psi(w) = \{x \in \bar{\Omega}: w(x) = \psi(x)\}$ (the coincidence set). If $\psi = 0$ then $\mathcal{F}_0(u)$ coincides with the usual free boundary.

The three above problems can be reformulated in an unified way by using the notion of <u>maximal monotone graphs</u> of $\mathbb{R}^2$. This class of graphs β of $\mathbb{R}^2$ can be characterized through a nondecreasing real function b to which we add some vertical segments at every jump of b, i.e.

$$\beta(r) = [b(r-), b(r+)] \quad \text{if} \quad -\infty < b(r-) \leqslant b(r+) < -\infty$$

$$\beta(r) = (-\infty, b(r-)] \quad \text{if} \quad -\infty = b(r-) < b(r+) < +\infty$$

$$\beta(r) = [b(r-), +\infty) \quad \text{if} \quad -\infty < b(r-) < b(r+) = +\infty .$$

Of course, the graph of any continuous non-decreasing real function defined on the whole $\mathbb{R}$ is a maximal monotone graph, which in fact is identified with the function itself. Using this terminology it is easy to check that the above examples correspond to special cases of the multivalued equation

$$-\Delta_p u + \beta(u) \ni g \quad \text{in} \quad \Omega \tag{64}$$

$$u = h \quad \text{on} \quad \partial\Omega , \tag{65}$$

where β are the maximal monotone graph of $\mathbb{R}^2$ given, respectively by (see Figure 9)

$$\beta(r) = \emptyset \text{ (the empty set)}, \quad \beta(0) = (-\infty, 0] \text{ and } \beta(r) = \{0\} \text{ if } r > 0 \tag{66}$$

$$\beta(r) = \{0\} \text{ if } r < 0 , \quad \beta(0) = [0,\lambda] , \beta(r) = \{\lambda\} \text{ if } r > 0 , \lambda > 0 \tag{67}$$

and

$$\beta(r) = f \circ \phi^{-1}(r). \tag{68}$$

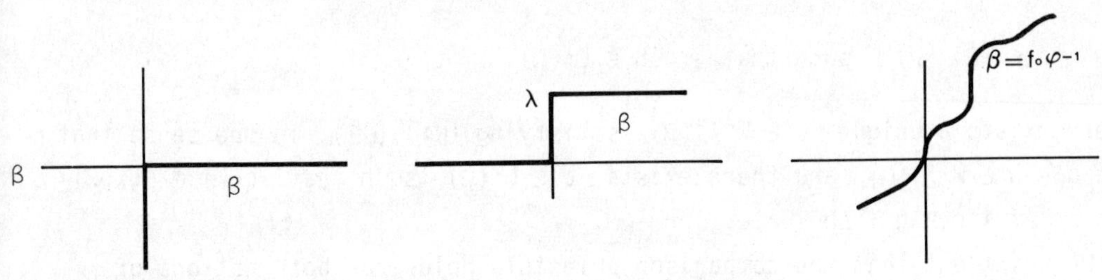

<u>Figure 9.</u>

Prior to the study of the free boundary we recall that, as in the case of single-valued equations, the solutions of equation (64) may be defined in several senses according to the regularity of the data g and h as well as to the boundedness or unboundedness of the domain Ω. A discussion of this topic is given in Chapter 4 but in any case we recall here that if, for instance, Ω is bounded and we assume

$$g \in W^{-1,p'}(\Omega) \quad , \text{ i.e, } \quad g = g_0 + \sum_{i=1}^{N} \frac{\partial g_i}{\partial x_i} \quad , \quad g_0, g_i \in L^{p'}(\Omega) \tag{69}$$

and

$$h \in W^{1,p}(\Omega) \quad \text{with} \quad j(h) \in L^1(\Omega) \; , \tag{70}$$

then there exists a unique $u \in W^{1,p}(\Omega)$ verifying (64),(65) in the sense that u minimizes the functional

$$J(u) = \int_\Omega [|\nabla u|^p + j(u) - g_0 u + \sum_{i=1}^{N} g_i \frac{\partial u}{\partial x_i}]dx$$

on the set $K = \{v : v - h \in W_0^{1,p}(\Omega)$ such that $j(v) \in L^1(\Omega)\}$. Here j is a "primitive" of the maximal monotone graph β in the sense that $j : \mathbb{R} \to (-\infty, +\infty]$ is a convex, l.s.c. function with $j \not\equiv +\infty$ and such that β is the subdifferential of j, $\beta = \partial j$ (see definition in Chapter 4). Under additional assumptions ($g_i \equiv 0$ and $D(\beta) = \mathbb{R}$) this solution, in fact, satisfies the equation (64) in a.e. $x \in \Omega$.

Another solution of the problem (64),(65) may be carried out in the $L^1(\Omega)$ space. So if, for instance, $p = 2$ and Ω is bounded, assuming

$$g \in L^1(\Omega)$$

and

$$h \in W^{1,1}(\Omega) \quad \text{such that} \quad \Delta h \in L^1(\Omega)$$

there exists a unique $u \in W^{1,1}(\Omega)$ satisfying (64),(65) in the sense that $z = u - h \in W_0^{1,1}(\Omega)$ and there exists $c \in L^1(\Omega)$ such that $c(x) \in \beta(z(x))$ and $-\Delta z + c = g + \Delta h$ a.e. on Ω.

We point out that the comparison principle holds for both notions of solutions. With respect to the boundedness of solutions of (64),(65) we send the reader to the exposition and references in Chapter 4.

After these preliminaries we mention that results on the existence and location of the free boundary are given in subsection 2.2a, where solutions with compact support are also considered. In subsection 2.2b we use symmetric rearrangement to derive another sufficient condition for the existence of $\mathcal{F}(u)$. Finally, in subsection 2.2c some qualitative properties of the free boundary are given.

2.2a. Existence and location of the free boundary. Solutions with compact support.

Let β be a maximal monotone graph of $\mathbb{R}^2$ such that $0 \in \beta(0)$, and let u be a function satisfying the equation (69) in a.e. $x \in \Omega$. It is clear that on the null set $N(u)$ we must have

$$\beta(0) \ni g(x) \tag{71}$$

and in consequence (71) must be satisfied on a subset of Ω positively measured. Due to the assumption $0 \in \beta(0)$, (71) holds if, for instance, the null set $N(g)$ is positively measured. As we shall show, it turns out that in this case, a sufficient condition for the existence of $\mathcal{F}(u)$ is that β satisfies an integral condition similar to the one given in Section 1.1. Nevertheless, (71) shows that if β is multivalued at the origin then it is possible to have solutions with a non-empty null set $N(u)$, corresponding to equations in which $g \neq 0$ on $N(u)$, and so, new results with respect to the single valued case are possible. We shall start by considering the first of the above situations.

If the null set $N(g)$ is not empty, using the same arguments as in Theorem 1.9, the existence of the null set $N(u)$ for solutions u of (69),(70) is related to the existence of nontrivial solutions of the homogeneous Cauchy problem

$$-\frac{1}{r^{N-1}}\frac{d}{dr}\left(r^{N-1}\left|\frac{du}{dr}\right|^{p-1}\frac{du}{dr}\right) + \beta(u) \ni 0 \tag{72}$$

$$u(0) = u'(0) = 0. \tag{73}$$

The analysis made in Section 1.1a for continuous non-decreasing functions f, instead of β, remains true after some minor changes in the notation. Indeed, first of all we note that the auxiliary function $\psi_\mu(\tau)$ defined in (33) of Section 1.1a may be understood in the following sense:

$$\psi_\mu(\tau) = \left(\frac{p-1}{p\mu}\right)^{1/p} \int_0^\tau \frac{ds}{j(s)^{1/p}} \tag{74}$$

where now j is a "primitive' of the maximal monotone graph β, $\beta = \partial j$. Introducing the sections of β given by the following real non-decreasing functions

$$\beta^0(r) = \{s_0 \in \beta(r) : |s_0| \leq |s| \quad \forall s \in \beta(r)\}$$

$$\beta^+(r) \text{ (resp. } \beta^-(r)) = \{s_+ \in \beta(r) \text{ (resp. } s_- \in \beta(r)): s_+ \geq s \text{ (resp. } s_- \leq s)$$
$$\forall s \in \beta(r)\}$$

then it turns out that $\beta^0(r) = \beta^+(r) = \beta^-(r)$ a.e. $r \in \mathbb{R}$ (recall that by a well-known result of the measure theory the set of $r \in \mathbb{R}$ in which β is multivalued must be necessarily numerable) and so, a primitive j of β is given by

$$j(r) = \int_0^r \beta^0(s)ds \quad \text{if} \quad r \in D(\beta) \quad \text{and} \quad j(r) = +\infty \text{ if } \quad r \notin D(\beta)$$

(here $D(\beta) = \{r \in \mathbb{R} : \beta(r) \neq \emptyset\}$).

Since the function ψ_μ can be well-defined, the statement and proof of Theorem 1.5 can be translated word for word to the Cauchy problem (23) (24) . We only remark that the function η defined by $\eta(\tau,\mu) = \psi_\mu^{-1}(\tau)$ is not, in general, of class C^2, but it is $\eta \in C^1([0,\psi_\mu(+\infty))) \cap W^{2,1}((0,\psi_\mu(+\infty))$ and this suffices to apply the arguments of the proof of Theorem 1.5. Analogously the function

$$\phi(r) = \left(\frac{p\mu}{p-1}\right)^{p/(p-1)} j(\eta(r))^{(p-1)/p}$$

is still an l.s.c convex function and its subdifferential $\partial\phi$ satisfies $\partial\phi(r) = \mu\beta^0(r)$ for a.e. $r \in D(\beta)$.

Using the same arguments as in Chapter 1 we have

Theorem 2.15. *Let* $p > 1$ *and assume*

$$\max \left\{ \int_{0^-} \frac{ds}{j(s)^{1/p}} , \int_{0^+} \frac{ds}{j(s)^{1/p}} \right\} < +\infty \tag{75}$$

Then if u *is the solution of (69),(70) and* $|u| \leq M$ *on* $N(g)$ *(for instance,* $u \in L^\infty(\Omega)$ *and* $M = \|u\|_\infty$*), then we have the following estimate for the null set* $N(u)$:

$$N(u) \supset \{x \in N(g) \cup N(h|_{\partial\Omega}) : d(x, S(g) \cup S(h|_{\partial\Omega})) \geq L\}$$

where

$$L = \max\{-\psi_{1/N}(-M) , \psi_{1/N}(M)\} . \square \tag{76}$$

Remark 2.5. In the above statement the behaviour of β on $(-\infty,0)$ is taken into account because now it is not natural to assume that β is odd. In any case, if the data g and h are of prescribed sign as occurs, for instance, in the example of zero order chemical reactions, where $g \geq 0$, $h \geq 0$, the same holds for the solution u (in that case, $u \geq 0$). Then, the behaviour of β on $(-\infty,0)$ has no interest and (75) may be weakened to the assumption

$$\int_{0^+} \frac{ds}{j(s)^{1/p}} < +\infty . \square \tag{77}$$

Remark 2.6. Assumption (75) depends only on the behaviour of β near the origin. So, if β is single-valued except in some points r_i, $r_i \neq 0$ where it is multivalued, the existence of the free boundary depends on the values of β near 0 and not on those near r_i (Figure 9). Nevertheless, if β is multivalued at $r = 0$ the situation is different. Indeed, if $[0,\beta^+(0)] \subset \beta(0)$ then $j(r) \geq \beta^+(0)r$ for $r > 0$ small enough and then

$$\int_0^\varepsilon \frac{ds}{j(s)^{1/p}} \leq \frac{1}{\beta^+(0)^{1/p}} \int_0^\varepsilon \frac{ds}{s^{1/p}} = \frac{p}{(p-1)\beta^+(0)^{1/p}} \varepsilon^{(p-1)/p} .$$

In consequence, if β is multivalued at 0 and $0 < \beta^+(0) < +\infty$, the hypothesis (77) holds _for every $p > 1$_ (the same for (75), assumed $-\infty < \beta^-(0) < 0 < \beta^+(0) < +\infty$). Note that if β is the graph (66) associated to the obstacle problem then $\beta = \partial j$ with $j(r) = +\infty$ if $r < 0$ and $j(r) = 0$ if $r > 0$. In consequence (77) does not hold although β is multivalued at the origin, $\beta(0) = (-\infty, 0\,]$. □

Remark 2.7. The special example of zero order reactions (with β given by (67)) can be treated directly. In fact Lemma 1.5 remains true for the limit case $q = 0$ and so the function

$$u(r) = C\, r^{\frac{p}{p-1}}$$

satisfies that

$$\mathbb{L}u = -\Delta_p u + \beta(u) = [\lambda - C^{(p-1)} \frac{p^{(p-1)} N}{(p-1)^{(p-1)}}]$$

In particular, if

$$C = K_{N,\lambda} = \frac{(p-1)}{p} [\frac{\lambda}{N}]^{1/(p-1)}$$

$\mathbb{L}u = 0$ and, in consequence, the estimate given in Theorem 2.15, can be improved by taking

$$L = (M/K_{N,\lambda})^{(p-1)/p} .$$ □

We point out that the extensions of Theorem 1.9 given in the subsection 1.1b still remain true for the multivalued equation. It is the same with the results of subsection 1.1c relative to the non-diffusion of the support. The following is a statement similar to Theorem 1.16, for the particular case of β as in the zero order reaction equation.

Theorem 2.16. *Let* $p > 1$ *and* β *given by (67). Let* $h \equiv 0$ *and* $g \in L^\infty(\Omega)$ *such that* $0 \leq g(x) < \lambda - \varepsilon$ a.e. $x \in \Omega$ *for some* $\varepsilon > 0$. *Then the solution* u *of (69),(70) satisfies that* $N(u) = N(g)$.

Proof. From the assumption made, there exists $C_0 \in (0, K_{N,\lambda})$ such that

$$0 \leq g(x) \leq \lambda - C_0^{(p-1)} \frac{p^{(p-1)} N}{(p-1)^{(p-1)}} .$$

Then, it suffices to take the function $C_0|x - x_0|^{p/(p-1)}$ as a local supersolution, as in Theorems 1.15 and 1.16.□

Remark 2.8. The condition (75) is also a necessary condition for the existence of a nonempty null set $N(u)$. This may be obtained from a strong maximum principle similar to that in Theorem 1.20 and this holds for multivalued equations like (69) (see Vazquez [5]).□

If β is multivalued at the origin the null set $N(u)$ of solutions of (69)(70) may exist although $g(x)$ does not vanish on Ω. The following result proves that the necessary condition given in (71) is "almost sufficient".

Theorem 2.17. *Let* $p > 1$ *and* β *be a maximal monotone graph such that* $\beta(0) = [\beta^-(0), \beta^+(0)]$, $-\infty \leq \beta^-(0) < 0 < \beta^+(0) \leq +\infty$. *Let* $u \in L^\infty(\Omega)$ *with* $\|u\|_\infty \leq M$ *satisfying the equation (69) in* a.e. $x \in \Omega$ *as well as the Dirichlet condition (70). Then if we define*

$$N_\varepsilon(g) = \{x \in \Omega : \beta^-(0) + \varepsilon \leq g(x) \leq \beta^+(0) - \varepsilon\}$$

we have the following estimate for the null set $N(u)$:

$$N(u) \supset \{x \in N_\varepsilon(g) \cup N(h|_{\partial\Omega}) : d(x, (\mathbb{R}^N - N_\varepsilon(g)) \cup S(h|_{\partial\Omega})) \geq L\}$$

where

$$L = [\frac{Mp}{p-1}]^{\frac{(p-1)}{p}} (\frac{N}{\varepsilon})^{\frac{1}{p}} .$$

Proof. Without loss of generality we can assume that $D(\beta) - \{0\} \neq \emptyset$. We take, for instance, $\beta^+(0) < +\infty$. Let $x_0 \in N_\varepsilon(g) \cup N(h|_{\partial\Omega})$ and let $R = d(x_0, (\mathbb{R}^N - N_\varepsilon(g)) \cup S(h|_{\partial\Omega}))$. Let $\tilde{\Omega} = \Omega \cap B_R(x_0)$ and define

$$\overline{u}(x) = \overline{u}(x : x_0) = C|x - x_0|^{\frac{p}{p-1}} .$$

Then on $\tilde{\Omega}$ we have

$$- \Delta_p \bar{u} + \beta^-(\bar{u}) \geq - C^{(p-1)} \frac{p^{(p-1)}N}{(p-1)^{(p-1)}} + \beta^+(0) \geq - \varepsilon + \beta^+(0) \geq g(x)$$

if $C \leq C_{\varepsilon,N} \equiv \frac{(p-1)}{p} [\frac{\varepsilon}{N}]^{1/(p-1)}$. On the other hand, as in Theorem 1.9, we have $\bar{u} \geq u$ on $\partial\tilde{\Omega}$ if , for instance ,

$$C R^{\frac{p}{p-1}} \geq M .$$

Taking $C = C_{\varepsilon,N}$ and applying the comparison principle on $\tilde{\Omega}$ we deduce that $u \leq \bar{u}$ on $\tilde{\Omega}$. Now if $\beta^-(0) = -\infty$, $D(\beta) \subset [0,\infty)$ and so , $0 \leq u(x)$ a.e. $x \in \Omega$. In particular , $0 \leq u(x_0) \leq \bar{u}(x : x_0) = 0$. If $-\infty < \beta^-(0)$ then, by the same argument, we also prove that $-\bar{u}(x) \leq u(x)$ and so $x_0 \in N(u)$. □

Remark 2.9. The above result applies to the obstacle problem, in which case β is given by (66). If, for instance, we consider the case of $p = 2$ and an obstacle $\psi \in H^2(\Omega)$ then the coincidence set $\{u = \psi\}$ is not empty if the set where $g(x) + \Delta\psi(x) \leq -\varepsilon$ is sufficiently large for some $\varepsilon > 0$. We point out that the local supersolution in the proof of Theorem 2.17 was chosen as $u = u_0(r)$, $r = |x - x_0|$, with u_0 satisfying the homogeneous Cauchy problem

$$\frac{1}{r^{N-1}} \frac{d}{dr} (r^{N-1} |\frac{du}{dr}|^{p-1} \frac{du}{dr}) = \varepsilon \quad , \quad u'(0) = 0 \quad , \quad u(0) = 0.$$

This argument can be applied to other differential operators. For instance, for the equation

$$- \Delta u + \alpha u + \beta(u) \ni g \quad , \quad \alpha > 0$$

the local supersolution is given as the solution of the associated homogeneous Cauchy problem and it can be explicitly defined in terms of modified Bessel functions (see Nagai [1]and Yamada [1], [2]). □

Theorem 2.17 allows us to obtain , as an application, some estimates on the location of the level sets $\{x \in \Omega : u(x) \geq t\}$ (and analogously $\{x \in \Omega : u(x) \leq t\}$ for solutions of quasilinear equations):

Theorem 2.18. *Let* $u \in L^\infty(\Omega)$, *with* $0 \leq u \leq M$,*be the solution of the problem*

$$- \Delta_p u + f(u) = g \quad \textit{in} \quad \Omega \tag{78}$$

$$u = h \quad \textit{on} \quad \partial\Omega , \tag{79}$$

where f *is a nondecreasing real function and* g *and* h *are nonnegative functions. Then, given* $0 < t < M$, *we have the estimate*

$$\{x \in \Omega : u(x) > t\} \supset \{x \in N_{\lambda,\varepsilon}(g) \cup N(h-\lambda|_{\partial\Omega}) : d(x, (\mathbb{R}^N - N_{\lambda,\varepsilon}(g)) \cup S(h-\lambda|_{\partial}) \geq L\}$$

where

$$N_{\lambda,\varepsilon}(g) = \{x \in \Omega : g(x) \geq \varepsilon + f(\lambda)\} \quad , \quad \varepsilon > 0$$

and

$$L = [\frac{\lambda p}{p-1}]^{\frac{(p-1)}{p}} (\frac{N}{\varepsilon})^{\frac{1}{p}}$$

Proof. Consider the obstacle problem

$$v \leq M , \quad - \Delta_p v + f(v) \leq g \quad , (-\Delta_p v + f(v) - g)(v - M) = 0 \quad \text{in} \quad \Omega \tag{80}$$

$$v = h \quad \text{on} \quad \partial\Omega . \tag{81}$$

By the results of Chapter 4 such a problem admits a unique solution v which must coincide with u because u already satisfies (80),(81). For $\lambda \in (0,M)$ let now u_λ be the solution of the new obstacle problem

$$v \leq \lambda , \quad - \Delta_p v + f(v) \leq g \quad , (- \Delta_p v + f(v) - g)(v - \lambda) = 0 \quad \text{in} \quad \Omega \tag{82}$$

$$v = h \quad \text{on} \quad \partial\Omega . \tag{83}$$

Again, by the comparison results with respect to the obstacle, we have that $u_\lambda \leq u$ a.e. in Ω. On the other hand, u_λ satisfies

$$- \Delta_p u_\lambda + f(u_\lambda) + \beta(u_\lambda - \lambda) \ni g$$

where β is the maximal monotone graph given by

$$\beta(r) = \{0\} \quad \text{if} \quad r \leq 0 \quad , \quad \beta(0) = [0,+\infty) \; , \quad \beta(r) = \emptyset \quad \text{if} \quad r > 0.$$

Then $w = u_\lambda - \lambda$ satisfies

$$- \Delta_p w + \beta(w) \ni \tilde{g} \quad \text{in} \quad \Omega$$

$$w = h - \lambda \qquad \text{on} \quad \partial\Omega ,$$

where $\tilde{g}(x) = g(x) - f(u_\lambda(x))$. The conclusion follows from the Theorem 2.17 by noting that $\beta^-(0) = 0$, $\beta^+(0) = +\infty$ and that

$$N_\varepsilon(\tilde{g}) = \{x \in \Omega : \ \varepsilon \leqslant g(x)\} \supset N_{\lambda,\varepsilon}(g). \square$$

Now we shall consider the case of <u>Ω unbounded</u> . It is clear that Theorem 2.17 remains true and so the existence of the null set is assured if the set $N_\varepsilon(g) \cup N(h|_{\partial\Omega})$ is also unbounded. For instance, if the support of $h|_{\partial\Omega}$ is compact and

$$\beta^-(0) + \varepsilon \leqslant g(x) \leqslant \beta^+(0) - \varepsilon \qquad \text{for} \qquad |x| \geqslant R \tag{84}$$

then we conclude that the solution u of (69),(70) has compact support if R is large enough. As in subsection 1.1d, it is also possible to construct global supersolutions having compact support. Indeed, it is enough to take $\bar{u} = G(|x-x_0|)$ with G defined by (80) of Section 1.1 and changing it on $R_0 \leqslant r \leqslant R_1$ by $G(r) = K_3 r^{p/p-1}$, for a suitable K_3 . (See details in Brezis [7]). An analogous version of Theorem 1.19 is also possible (Moet [1]).

It is not difficult to see that the necessary condition $g \in \beta(0)$ is not sufficient to conclude the existence of the free boundary $\mathcal{F}(u)$ or, for instance, the compactness of the support $S(u)$. Even the condition $\beta^-(0) < g(x) < \beta^+(0)$ for $|x| \geqslant R$ is not enough . To see this it suffices to note that $u(x)\, e^{-\rho x}$ $(\rho > 1)$ verifies on $\Omega = (0,+\infty)$ the equation

$$- u'' + \beta(u) \ni g$$

with $\beta(r) = \emptyset$ if $r < 0$, $\beta(0) = [0,+\infty)$, $\beta(r) = r$ if $r > 0$ and where $g = (1 - \rho^2)e^{-\rho x}$. Nevertheless, another sufficient condition , weaker than (84) , may be given for $\beta^-(0) < g(x) < \beta^+(0)$ for $|x| \geqslant R$ assumed that g does not converge too fast to $\beta^-(0)$ or $\beta^+(0)$ when $|x| \to +\infty$.

This was proved in Benilan-Brezis-Crandall [1] when in (64) $p = 2$ and $\Omega = \mathbb{R}^N$ giving at the same time an existence result (see also Brezis [7] for a similar, but weaker, result).

Theorem 2.19. *Let* $\beta(0) = [\beta^-(0), \beta^+(0)]$, $-\infty < \beta^-(0) < 0 < \beta^+(0) < +\infty$ *and let* $g \in L^1_{loc}(\mathbb{R}^N)$. *Suppose* $R > 0$ *and that there are nonnegative functions* $\varepsilon_\pm \in L^1_{loc}([0,\infty))$ *such that*

$$\beta^-(0) + \varepsilon_-(|x|) \leqslant g(x) \leqslant \beta^+(0) - \varepsilon_+(|x|) \text{ a.e. for } |x| \geqslant R \quad (85)$$

and satisfying

$$\int_0^\infty r^{N-1} \varepsilon_\pm(r)dr = +\infty . \quad (86)$$

Then there exists a solution $u \in L^1(\mathbb{R}^N)$, *with compact support, of the equation*

$$-\Delta u + \beta(u) \ni g \quad \text{in } \mathbb{R}^N . \quad (87)$$

If $N = 1$ *or* $N = 2$ *and* $\beta^\pm(0) \in \text{int}\,\beta(\mathbb{R})$, *then*

$$\int_0^\infty r\, \varepsilon_\pm(r)dr = \infty \text{ if } N = 2, \text{or } \int_0^\infty r \log(1+r)\varepsilon_\pm(r)dr = \infty \text{ if } N = 1 \quad (88)$$

are sufficient to imply that (87) *has a solution* $u \in L^1(\mathbb{R}^N)$ *with compact support.*

Proof. We shall only reproduce here the easier case $N \geqslant 3$. (For the complete proof see the mentioned reference). From (85) we have $\beta^+(0) - \beta^-(0) \geqslant \varepsilon_+ + \varepsilon_-$, so ε_+ and ε_- are bounded. Since ε_+ is bounded, $\min(\varepsilon_+(x), \beta^+(0))$ satisfies the same integral condition (86) and $g^+(x) \leqslant \beta^+(0) - \min(\varepsilon_+(x), \beta^+(0))$. Dealing similarly with the minus case and using the comparison principle (see Chapter 4), we can suppose: $g \geqslant 0$ and there is an $R > 0$ such that $g > \beta^+(0)$ on $|x| < R$ while $g(x) = \beta^+(0) - \varepsilon_+(|x|)$ on $|x| > R$. Let $g_n = g$ on $|x| \leqslant n$ and $g_n = 0$ on $|x| > n$. By Theorem 2, the solution $u_n \in L^1(\mathbb{R}^N)$ of $-\Delta u + \beta(u) \ni g_n$ in $\mathbb{R}^N$ has compact support (see also Remark 1.12). Moreover, u_n and $w_n = g + \Delta u_n$ are nondecreasing in n since g_n is nondecreasing in n. Since

$$- \Delta u_n + (w_n - \beta^+(0)) = g_n - \beta^+(0) \leqslant (g_n - \beta^+(0))^+ ,$$

then, by comparison, we have that $u_n \leqslant \overline{u}$, $w_n \leqslant \overline{w}$, where $\overline{u}$ is the solution of $- \Delta u + \tilde{\beta}(u) = (g - \beta^+(0))^+$, $\tilde{\beta}(u) = \beta(u) - \beta^+(0)$ and $\overline{w} = (g - \beta^+(0))^+ + \Delta\overline{u}$. (The existence of such functions $\overline{u}, \overline{w} \in L^1_{loc}(\mathbb{R}^N)$ can be proved as the limit of some coercive semilinear equations in $\mathbb{R}^N$ since $(g - \beta^+(0))^+ \in L^1(\mathbb{R}^N)$; see the mentioned reference). Thus, $u_n \uparrow u \leqslant \overline{u}$ and $w_n \uparrow w \leqslant \overline{w}$ for some functions $u, w \in L^1_{loc}(\mathbb{R}^N)$. By the maximal monotonicity of β, $w \in \beta(u)$ a.e. and $g + \Delta u = w$ in $\mathcal{D}'(\mathbb{R}^N)$, so u is a solution of (87). Thus, it is enough to bound the supports of the u_n uniformly in n.

We make one further reduction. By the comparison principle we may assume that $\beta(\mathbb{R}) \subset [-A,A]$ for some $A > 0$ (argument as in Theorem 2.1). On $|x| > R$, $(g - \beta^+(0))^+ = (- \varepsilon_+(|x|))^+ = 0$, so $(g - \beta^+(0))^+ + \Delta\overline{u} \in \beta(\overline{u})$ implies $\Delta\overline{u} \in L^\infty(\{x : |x| > R\})$ and $\overline{u} \in W^{1,1}_{loc}(\{x : |x| > R\})$. By standard arguments we conclude that $\overline{u} \in W^{2,s}_{loc}(\{x : |x| > R\})$ for $1 \leqslant s < \infty$. Choosing $p > N$, the Sobolev embedding theorem implies $\overline{u} \in C^1(\{x : |x| > R\})$. Now, choose $R_0 > R$ and set $M = \sup_{|x|=R_0} \{|\overline{u}(x)|\}$. We define the auxiliary function $\eta(r;r_0)$ for $r > r_0$ by

$$\eta(r;r_0) = \int_{r_0}^{r} \left(\int_{t}^{r} \left(\frac{s}{t}\right)^{N-1} \varepsilon_+(s)ds\right)dt.$$

Then, by (86), $\lim_{r\to\infty} \eta(r:r_0) = +\infty$, since

$$\eta(r:r_0) = \frac{1}{N-2} \int_{r_0}^{r} \left(\frac{s^{N-1}}{r^{N-2}} - s\right) \varepsilon_+(s)ds .$$

Finally, choose $\overline{R} > R_0$ so that $M = \eta(\overline{R};R_0)$ and let

$$v(x) = \begin{cases} \eta(\overline{R},|x|) & \text{if} \quad R_0 \leqslant |x| \leqslant \overline{R} \\ 0 & \text{of} \quad |x| > \overline{R}. \end{cases}$$

We have $v \in C^1(\{x : |x| > R\})$, $v = M \geqslant \overline{u} \geqslant u_n$ if $|x| = R$ and $- \Delta u + \beta^+(0) = \beta^+(0) - \varepsilon_+$ on $R_0 < |x| < \overline{R}$, and $- \Delta v = 0$ on $|x| > \overline{R}$. Then if $z = \beta^+(0)$ in $R_0 < |x| < R_1$ and $z = \beta^+(0) - \varepsilon_+$ on $|x| > \overline{R}$, then $z \in \beta(v)$ and $- \Delta v - z = \beta^+(0) - \varepsilon_+ \geqslant g_n$. Then, by the comparison principle, $v \geqslant u_n$ on $|x| \geqslant R_0$ and so $\operatorname{supp} u_n \subset \{x : |x| \leqslant \overline{R}\}$. □

Remark 2.10. The hypotheses of Theorem 2.19 cannot be weakened (see Benilan-Brezis-Crandall [1]). □

2.2b. Rearrangement and multivalued equations.

The symmetric rearrangement of a function can be used in order to study different properties of solutions of multivalued equations. A first result in this direction corresponds to an analogous version of Theorem 1.26 of subsection 1.3a relative to single valued equations. For simplicity we shall restrict ourselves to the consideration of the semilinear multivalued equation

$$- Lu + \beta(u) \ni g_1 \quad \text{in} \quad \Omega \tag{89}$$

$$u = 0 \quad \text{on} \quad \partial\Omega \tag{90}$$

where L is the second order elliptic operator given by

$$Lu = \sum_{i,j=1}^{N} D_j(a_{ij}(x)D_i u) - a(x)u \quad ,$$

with $a_{ij} \in C^1(\overline{\Omega})$, $a \in L^\infty(\Omega)$, $a \geqslant 0$, and

$$\sum_{i,j} a_{ij}(x)\, \xi_i \xi_j \geqslant |\xi|^2 \quad \text{for every} \quad \xi \in \mathbb{R}^N \ , \ \xi \neq 0. \tag{91}$$

Here β represents a maximal monotone graph of $\mathbb{R}^2$ with $0 \in \beta(0)$ and the main goal is to "compare" the solution u of (89), (90) with the solution v of the simplified problem

$$- \Delta v + \beta(v) \ni g_2 \quad \text{in} \quad \Omega^* \tag{92}$$

$$v = 0 \quad \text{on} \quad \partial\Omega^* \ , \tag{93}$$

where Ω^* is a ball of measure meas Ω . We shall take advantage of this opportunity to work with L^1-solutions in contrast with subsection 1.3a where we considered variational solutions. We recall that the existence of the (unique) function $u \in W_0^{1,1}(\Omega)$ (resp. $v \in W_0^{1,1}(\Omega^*)$) satisfying the existence of $c_1 \in L^1(\Omega)$ (resp $c_2 \in L^1(\Omega^*)$) with $c_1(x) \in \beta(u(x))$ (resp. $c_2(x) \in \beta(v(x))$) for a.e. x and such that $-Lu + c_1 = g_1$ (resp. $-\Delta v + c_2 = g_2$) for a.e. x, is a consequence of the main result of Brezis-Strauss [1](see also Chapter 4), when it is assumed $g_1 \in L^1(\Omega)$ (resp $g_2 \in L^1(\Omega^*)$).

Theorem 2.20. *Let u and v be nonnegative L^1-solutions of (89)(90) and (92)(93) respectively. Assume that g_2 is a nonincreasing radial function and that $g_1^* \lesssim g_2$ (in the sense of Definition 1.2). Then $c_1^* \lesssim c_2$. Moreover, for any convex nondecreasing real function Φ we have that*

$$\int_{\Omega} \Phi(c_1(x))dx \leqslant \int_{\Omega_*} \Phi(c_2(x))dx . \tag{94}$$

Sketch of the proof. We first assume that g_1 and g_2 are bounded functions. Let $\varepsilon > 0$ fixed and $\mu > 0$. Consider u_μ^ε and v_μ^ε to be the solutions of the (single valued) problems (we drop for a while ε)

$$- Lu_\mu + \beta_\mu^\varepsilon(u_\mu) = g_1 \quad \text{in } \Omega$$

$$u_\mu = 0 \quad \text{on } \partial\Omega$$

and

$$- \Delta v_\mu + \beta_\mu^\varepsilon(v_\mu) = g_2 \quad \text{in } \Omega^*$$

$$v_\mu = 0 \quad \text{on } \partial\Omega^* ,$$

where β_μ^ε is the Lipschitz continuous nondecreasing function defined by $\beta_\mu^\varepsilon(r) = \varepsilon r + \beta_\mu(r)$, β_μ being the Yosida-approximation of β, ($\beta_\mu = (I - (I + \mu\beta)^{-1})(1/\mu), \mu > 0$). In this context Theorem 1.26 can be applied, and so $\beta_\mu^\varepsilon(u_\mu^*) \lesssim \beta_\mu^\varepsilon(v_\mu)$. Finally the proof ends by using the convergence result of Brezis-Strauss[1]. Indeed, there it is shown that $\beta_\mu^\varepsilon(u_\mu) \to \varepsilon u^\varepsilon + c_1^\varepsilon$ in $L^1(\Omega)$ (resp. $\beta_\mu^\varepsilon(v_\mu) \to \varepsilon v^\varepsilon + c_2$ in $L^1(\Omega^*)$) when $\mu \to 0$ for some adequate functions u^ε, v^ε, c_1^ε and c_2^ε with $c_1^\varepsilon \in \beta(u^\varepsilon)$, $c_2^\varepsilon \in \beta(v^\varepsilon)$ a.e. Finally, if g_1 and g_2 are in L^1 we take g_1^ε and g_2^ε bounded functions satisfying $(g_1^\varepsilon)^* \lesssim g_2^\varepsilon$, g_2^ε being a nonincreasing radial function, and again, by the mentioned work, we know that $u^\varepsilon \to u$, $v^\varepsilon \to v$ in L^1 and that $c_1^\varepsilon \to c_1$ and $c_2^\varepsilon \to c_2$ in L^1, when $\varepsilon \downarrow 0$. □

Remark 2.11. It is also possible to state Theorem 2.20 for quasilinear multivalued equations like (69). Indeed, it suffices to apply the convergence result due to Benilan [2]and the accretiveness of the operator Δ_p in L^1 (see also Chapter 4). In any case, the L^1-setting requires some further assumptions in the nonlinearities than the variational one. For instance, for general diffusion equations like (3) of Section 1.3, we need the monotone dependence of $A(x,u,\xi)$ with respect to ξ. □

As in subsection 1.3a, from the above result we can derive an isoperimetric inequality for the null set of solutions of multivalued equations. This is of interest in the special case of zero order reactions.

Corollary 2.21. *Let* L *be as above and let* β *a maximal monotone graph of* $\mathbb{R}^2$ *with* $0 \in \beta(0)$. *For* $k > 0$, *let* u *be the solution of*

$$- Lu + \beta(u) \ni 0 \quad \text{in} \quad \Omega$$

$$u = k \quad \text{on} \quad \partial\Omega$$

Finally, let Ω^* *be a ball of the same measure as* Ω , *and let* v *be the solution of*

$$- \Delta v + \beta(v) \ni 0 \quad \text{in} \quad \Omega^*$$

$$v = k \quad \text{on} \quad \partial\Omega^*.$$

Then, if $v > 0$ *on* Ω^**, the null set* N(u) *is empty. If, moreover,* $\beta^{-1}(0) = 0$, *then,* meas N(u) $\leqslant$ meas N(v). □

Now we return to the context of Theorem 2.20. As was already pointed out in Remark 1.16, the explicit inequality $u^* \leqslant v$ is not true,in general Nevertheless we shall prove another explicit inequality , very useful in order to have a new sufficient condition for the ocurrence of the free boundary in multivalued equations. The idea is to compare u^* (u the nonnegative solution of (80,(90)) with the radially symmetric function z, solution of the problem

$$- \Delta z + \beta^{\#}(z) \ni g_2 \quad \text{in} \quad \Omega^* \tag{95}$$

$$z = 0 \quad \text{on} \quad \partial\Omega^* \tag{96}$$

where $\beta^{\#}$ is the maximal monotone graph of $\mathbb{R}^2$ defined from β by

$$\beta^{\#}(r) = \{\beta^{+}(0)\} \text{ if } r > 0, \quad \beta^{\#}(0) = (-\infty, \beta^{+}(0)], \quad \beta^{\#}(r) = \emptyset \text{ if } r < 0 \qquad (97)$$

(note that now $D(\beta^{\#}) = D(\beta) \cap [0,\infty)$).

Theorem 2.22. *Let* u *and* z *be nonnegative* L^1*-solutions of* (89),(90) *and* (95)(96) *respectively. Assume that* g_2 *is a nonincreasing radial function and that* $g_1^{*} \lesssim g_2$ *. Then* $u^{*} \leqslant z$ *a.e. in* Ω^{*}.

Remark 2.12. If β is in fact a nondecreasing continuous function with $\beta(0) = 0$ and $g_1 \geqslant 0$ then Theorem 2.22 reduces to Remark 1.16 .However, if β is a maximal monotone graph, multivalued at the origin, then nonnegative solutions u of (89),(90) may exist even for g_1 changing in sign (this is the case if e.g., $D(\beta) \subseteq [0+\infty)$). In any case, z is the solution of the variational inequality

$$z \geqslant 0, \quad -\Delta z \geqslant g_2 - \beta^{+}(0) \quad \text{and} \quad (-\Delta z - g_2 + \beta^{+}(0))z = 0 \quad \text{in} \quad \Omega^{*} \qquad (98)$$

$$z = 0 \qquad \text{on} \qquad \partial\Omega^{*}. \qquad (99)$$

Again, note that $z \geqslant 0$ even if g_2 is not a positive function and that the coincidence set $\{z = 0\}$ may be non empty.□

As in subsection 1.3a, the proof of Theorem 2.22 is given in several steps relative to variational solutions.

Lemma 2.23. *Let* $g_1 \in L^2(\Omega)$ *and* $u \in H_0^1(\Omega)$ *be the (nonnegative) solution of* (89),(90). *Then the decreasing rearrangement* $\tilde{u}$ *of* u *satisfies*

$$-\frac{d\tilde{u}}{ds}(s) \leqslant \left(\frac{1}{N\omega_N^{1/N} s^{(N-1)/N}}\right)^2 \left[\int_0^s (\tilde{g}_1(\theta) - \beta^{+}(0))d\theta\right] \qquad (100)$$

a.e. $s \in (0,|\{u > 0\}|)$, *where* $\tilde{g}_1$ *is the signed decreasing rearrangement of* g_1.

Proof. From the assumptions, there exists $c_1 \in L^1(\Omega)$ (in fact in $L^2(\Omega)$) such that $c_1(x) \in \beta(u(x))$ a.e. $x \in \Omega$ and verifying $-Lu + c_1 = g_1$ in Ω . Then, for every $w \in H_0^1(\Omega) \cap L^{\infty}(\Omega)$ we have

$$\sum_{i,j=1}^{N} \int_{\Omega} a_{ij}(x) \frac{\partial u}{\partial x_i} \frac{\partial w}{\partial x_j} - \int_{\Omega} auw + \int_{\Omega} c_1 w = \int_{\Omega} g_1 w$$

Taking $w = T_{t,h}(u)$, with $T_{t,h}$ defined in the proof of Lemma 1.29, and remarking that $a(x) \geqslant 0$ and $c_1 T_{t,h}(u) \geqslant \beta^{+}(0)\, T_{t,h}(u)$ for a.e.

where s_0 *is the unique solution of* $\chi(s_0)=0$ *in* $]|\{g_2-\beta^+(0)>0\}|,|\Omega^*|[$ (χ given by (101)).

Proof. Inequality (103) follows from the comparison $u^* \leq z$. On the other hand, if $g_2 \leq \beta^+(0)$ then the unique solution of (95),(96) (i.e. of the variational inequality (98),(99)) is $z \equiv 0$. Now, the function χ is concave because $\frac{d\chi}{ds} = g_2 - \beta^+(0)$ is nonincreasing. Moreover, χ is strictly increasing in $(0,|\{g_2 - \beta^+(0) > 0\}|)$, constant in $(|\{g_2 - \beta^+(0) > 0\}|, |\{g_2 - \beta^+(0) \geq 0\}|)$, and strictly decreasing in $(|\{g_2-\beta^+(0) \geq 0\}|, |\Omega^*|)$; $\chi(0) = 0$ and $\chi(|\Omega^*|) = \int_\Omega (g_2-\beta^+(0))dx$. Then, if $\int_{\Omega^*}(g_2 - \beta^+(0))dx \geq 0$ (and $g_2 \neq \beta^+(0)$) we have $\chi > 0$ on $]0, |\Omega^*|[$ and then by (102) $\frac{d\tilde{z}}{ds}(s) < 0$ on $]0,|\Omega^*|[$ which implies that meas $N(z) = 0$. Finally, assume that $\int_{\Omega^*}(g_2 - \beta^+(0))dx < 0$ and $|\{g_2 - \beta^+(0) > 0\}| > 0$. Now it is clear that $|\{z > 0\}| > 0$ because, otherwise, $g_2 - \beta^+(0)$ must be non positive on Ω^*. By the above description of χ, there exists a unique $s_0 \in (|\{g_2 - \beta^+(0) \geq 0\}|, |\Omega^*|)$ such that $\chi(s) > 0$ if $0 < s < s_0$, $\chi(s_0) = 0$ and $\chi(s) < 0$ if $s_0 < s < |\Omega^*|$. But from (102) and the monotonicity of $\tilde{z}(s)$ we conclude that $s_0=|\{z>0\}|$ □

The above result is a slight generalization of the work Bandle-Mossino [1] concerning the obstacle problem (i.e. β given by (66)). Taking $g_2= g_1^*$ and using the equimeasurability of g_1^* and g_1, we find the following sufficient condition for the existence of the free boundary $F(u)$

Corollary 2.24. *Let* $u \geq 0$ *be the solution of (89),(90). Then if*

$$\int_\Omega (g_1(x) - \beta^+(0))dx < 0 \tag{104}$$

the null set $N(u)$ *has a positive measure. If in addition* $|\{g_1-\beta^+(0) > 0\}| > 0$ *then the existence of the free boundary* $F(u)$ *is assured.* □

Remark 2.13. It is interesting to compare Corollary 2.24 with Theorem 2.17 where a different criterion is given for the existence of the free boundary. Essentially, each result is of a different nature. Hypothesis (104) is a global assumption and, in contrast, Theorem 2.17 is a local result. In spite of this, there are some cases in which Theorem 2.17 cannot be applied; however, these are included in Corollary 2.24.

$x \in \Omega$, then we obtain the same conclusion as in Lemma 1.29, but now replacing the term $f(u)$ by the constant $\beta^+(0)$. Note that now $\tilde{g}_1(s) = \inf\{t \in \mathbb{R} : \text{meas}\{x : g_1(x) > t\} < s\}$. In particular, if we define the function

$$\chi(r) = \int_0^r (\tilde{g}_1(\theta) - \beta^+(0))d\theta \tag{101}$$

then the above result proves that $\chi(\mu(t)) \geq 0$ for a.e. $t \in (0, \text{ess.sup}\, u)$, where $\mu(t)$ is the distribution function of u. Arguing as in the proof of Lemma 1.31 we have that in fact $\chi \geq 0$ on $(0, |\{u > 0\}|]$. Finally, inequality (100) is proved using Lemma 1.30 and reproducing again the proof of Lemma 1.31.□

Proof of Theorem 2.22. We first prove it for $g_i \in L^2$ and so for u and z being variational solutions. It is easy to check that the (unique) solution z of (95),(96) is a radial symmetric function, with $z \in C^1(]0,|\Omega|])$ and such that

$$-\frac{d\tilde{z}}{ds}(s) = \left(\frac{1}{N\omega_N^{1/N} s^{(N-1)/N}}\right)^2 \left[\int_0^s (g_2(\theta) - \beta^+(0))d\theta\right] \tag{102}$$

for every $s \in]0, |\{z > 0\}|]$. (See Lemma 1.32). The conclusion now follows by using that $g_1^* \leq g_2$, $\tilde{u}(|\Omega|) = \tilde{z}(|\Omega|) = 0$ and integrating in (100) and (102). Finally, the conclusion for L^1-solutions is obtained by approximation arguments as in Theorem 2.20.□

An important consequence of the above theorem is the following isoperimetric inequality for the null set (or coincidence sets) of solutions of the mentioned problems.

Theorem 2.23. *Under the assumptions of Theorem 2.22 we have*

$$\text{meas}\, N(u) \geq \text{meas}\, N(z) \tag{103}$$

Moreover

$$\text{meas}\, N(z) = |\Omega^*| \quad \textit{if} \quad g_2 \leq \beta^+(0),$$

$$\text{meas}\, N(z) = 0 \quad \textit{if} \quad \int_\Omega (g_2 - \beta^+(0))dx \geq 0 \ \textit{and} \ g_2 \neq \beta^+(0),$$

and

$$\text{meas}\, N(z) = |\Omega^*| - s_0 > 0 \quad \text{otherwise},$$

They correspond, for instance, to the case when the function $g_1(x) - \beta^+(0)$ oscillates with a high frequency and small amplitude. Obviously, the corresponding result for nonpositive solutions u of (89)(90) is also true, but the case of u changing in sign seems to be delicate because the limitation of the techniques of the rearrangement. (Part v) of Theorem 1.25 is not true for the signed rearrangement when u changes in sign; see Mossino [2]. □

Remark 2.14. Theorems 2.22 and 2.23 are also true for quasilinear multivalued equations as in (69). On the other hand, the mentioned result also holds for the general obstacle problem (54), assuming $\psi \in H^2(\Omega) \cap H_0^1(\Omega)$; see Bandle-Mossino [1]. See also Maderna-Salsa [2] for the case of other non-zero obstacles ψ. □

2.2c. Further results.

Many of the results of Section 1.4 remain true for the obstacle problem as well as for zero order reactions. The main reason is that both problems can be formulated as a semilinear equation $-\Delta u + f(u) = 0$ with $f(r) = \lambda \mathbb{1}_{[r>0]}$ and so they correspond to the limiting case of $f(r) = \lambda r^q$ with $0 < q < 1$, $q \downarrow 0$. We note that the solution of this limit case satisfies

$$- \Delta U \geq - \lambda , \quad U \geq 0 , \quad U(- \Delta U + \lambda) = 0 \quad \text{a.e. in} \quad \Omega \tag{105}$$

$$U = 1 \qquad \text{on} \quad \partial\Omega \tag{106}$$

Note also that the non-homogeneous obstacle problem

$$- \Delta u \geq 0 \quad , \quad u \geq \psi \quad , \quad (u - \psi) \Delta u = 0 \quad \text{in} \quad \Omega \tag{107}$$

$$u - \psi = 1 \qquad \text{on} \quad \partial\Omega \tag{108}$$

can be reduced to the homogeneous one assuming that ψ satisfies

$$\Delta\psi = - \lambda \quad , \quad \lambda > 0 . \tag{109}$$

Indeed, it suffices to take $u = U + \psi$.

In this way, the behaviour of solutions of the zero order reactions or of the obstacle problem (105),(106) can be studied as in subsection 1.4a by making there $q = 0$. The same happens with the Lebesgue and Hausdorff measure results of the free boundary given in subsection 1.4a. Finally, the convexity of the coincidence set $\{u = \psi\}$, with u solution of (107) (108), remains also true (and with the same proof as in Subsection 1.4c) assuming that Ω is a convex set in $\mathbb{R}^2$ and ψ satisfies (109).

Remark 2.15. It is important to point out that a very exhaustive qualitative study of solutions of variational inequalities (even under formulations more general than (107),(108)) was made by L.A.Caffarelli in [1] and [3] (see also the recompilation made in the book Friedmann [3]) In fact these results were the starting point for the study of semilinear equations.□

Remark 2.16. The nature and geometrical properties of the coincidence set in variational inequalities have been considered in the literature by many different authors (see the panoramic exposition of Kinderlehrer [1]). Most of those results relate to domains Ω of $\mathbb{R}^2$ and give a very complete description of this set. For $N > 2$ the starshapedness of the coincidence sets (for convex domains Ω and concave obstacles ψ) was proved by Kawohl [1] and Sakaguchi [1]. The convexity of this set for the problem (107),(108) has been recently obtained in Kawohl [6] for $N \geq 2$.□

We shall end this section by studying the convergence of the free boundary $\mathcal{F}(u)$ for the obstacle problem. First of all we recall the notion of Hausdorff distance $d(A,B)$ between two sets A,B:

$$d(A,B) = \inf\{\varepsilon > 0 : A \subset B^{\varepsilon} \text{ and } B \subset A^{\varepsilon}\} \quad ,$$

where we have used the notation

$$D^{\varepsilon} = \{x \in \mathbb{R}^N : d(x,D) < \varepsilon\}$$

for any set $D \subset \mathbb{R}^N$ and $\varepsilon > 0$. (We also recall that the set $\mathcal{H}(K)$ of all compact subsets of the compact set K of $\mathbb{R}^N$ becomes a compact metric space with the Hausdorff distance: see , e.g. Dellacherie [1]).Now we are concerned with the solutions of the variational inequality

$$-\Delta u \geq g \quad , u \geq 0 \quad , (-\Delta u - g)u = 0 \quad \text{in} \quad \Omega \tag{110}$$

$$u = 0 \qquad \text{on} \quad \partial\Omega . \tag{111}$$

By the existence results (see Chapter 4) We know that if $g \in L^p(\Omega)$ the solution u is $H^1_0(\Omega) \cap W^{2,p}(\Omega)$.

Theorem 2.25. *Let* ω *be an open smooth subset,* $\omega \subset\subset \Omega$ *. Let* g_ν *and* g_0 *in* $C^{0,1}(\Omega)$ *satisfying*

$$g_\nu \to g_0 \quad in \quad L^p(\Omega) \text{ , for some } p > N \text{ , when } \nu \to 0 \tag{112}$$

$$g_\nu \leq -\mu < 0 \quad in \quad \omega \text{ , for every } \nu \geq 0 \text{ and some } \mu > 0. \tag{113}$$

Let u_ν *(resp.* u_0 *) be the solutions of (110), (111) corresponding to* $g = g_\nu$ *(resp.* $g = g_0$*) and assume that*

$$\overline{\text{int } (N(u_0) \cap \omega)} = N(u_0) \cap \bar{\omega}. \tag{114}$$

Then $F(u_\nu) \cap \bar{\omega} \to F(u_0) \cap \bar{\omega}$ *in Hausdorff distance, and if* $\mathbf{1}_\nu$ *(resp.* $\mathbf{1}_0$*) denotes the characteristic function of* $N(u_\nu)$ *(resp.* $N(u_0)$*) then* $\mathbf{1}_\nu \to \mathbf{1}_0$ *in* $L^s(\omega)$ *for any* $1 \leq s < \infty$.

Proof. Due to (112) and using the Sobolev imbedding it is easy to see that the solutions u_ν , u_0 are such that

$$u_\nu \to u_0 \text{ in } C^0(\bar{\omega}) \text{ and } -\Delta u_\nu - g_\nu \to -\Delta u_0 - g_0 \quad \text{in } \mathcal{D}'(\omega) \tag{115}$$

(see details in Rodrigues [1]). Now consider an arbitrary small open ball $B_\varepsilon \subset \omega$ such that $B_\varepsilon \cap F(u_\nu) = \emptyset$ for infinitely many $\nu > 0$. So we have $u_\nu = 0$ or $u_\nu > 0$ in B_ε . In the first case, it is clear that $u_0 = 0$ in B_ε . In the second one, u_ν verifies the equation

$$-\Delta u_\nu = g_\nu \quad \text{in } B_\varepsilon . \tag{116}$$

By (115) , the same equation holds for u_0 . But $g_0 \neq 0$ in ω and then $\text{int } N(u_0) \cap B_\varepsilon = \emptyset$ and, by (114), this implies that B_ε lies in the noncoincidence set $\{u_0 > 0\}$. In any case $B_\varepsilon \cap F(u_0) = \emptyset$ and, therefore, $\inf\{\varepsilon : (F(u_0) \cap \bar{\omega}) \subset (F(u_0) \cap \bar{\omega})^\varepsilon\}$ tends to zero as $\nu \downarrow 0$. On the other hand , let $\varepsilon > 0$ be the radius of any open ball $B_\varepsilon \subset \omega$ such that $B_\varepsilon \cap F(u_0) = \emptyset$. If $u_0 > 0$ in B_ε , by (115) we find that $u_\nu > 0$ in $B_{\varepsilon'}$ and $B_{\varepsilon'} \cap F(u_\nu) = \emptyset$ for any $0 < \varepsilon' < \varepsilon$ and for all ν small enough. If $u_0 = 0$ in B_ε and $x_\nu \in B_{\varepsilon'} \cap F(u_\nu)$ for infinitely many $\nu > 0$, by the "nondegeneracy property" (Theorem 1.44 of Chapter 1 with $q = 0$, or Friedman [3]p. 154) we know that there exists infinitely many $y_\nu \in \partial B_\varepsilon$ verifying

$$u_\nu(y_\nu) \geq C(\varepsilon - \varepsilon')^2 > 0$$

which contradicts (115). Therefore, $B_\varepsilon \cap F(u_\nu) = \emptyset$ for all ν small enough and it follows that $\inf\{\varepsilon : (F(u_\nu) \cap \bar{\omega}) \subset (F(u_0) \cap \bar{\omega})^\varepsilon\} \to 0$ as $\nu \downarrow 0$. Finally, to prove the second assertion we recall that by the analogous version of Corollary 1.45 of Chapter 1 for $q = 0$ (see also Friedmann [3] p. 156) $\text{meas}(F(u_0) \cap \bar{\omega}) = 0$. Therefore, $\text{meas}(F(u_0) \cap \bar{\omega})^\varepsilon \to 0$ as $\varepsilon \downarrow 0$. But by the Hausdorff convergence of $F(u_\nu) \cap \bar{\omega}$ to $F(u_0) \cap \bar{\omega}$ we have $(F(u_\nu) \cap \bar{\omega}) \subset (F(u_0) \cap \omega)$ for all ν small enough. In consequence, for any $1 \leq s < \infty$

$$\int_\omega |\mathbf{1}_\nu - \mathbf{1}_0|^s = \int_\omega |\mathbf{1}_\nu - \mathbf{1}_0| \leq \text{meas}(F(u_0) \cap \bar{\omega})^\varepsilon$$

and the theorem is proved.□

Remark 2.17. The above result is taken from Rodrigues [1]in which some quasilinear operators are also considered. The proof adapts an argument of Alt-Caffarelli [1]. Other stability results for the free boundary were given in Cafarelli [4](see also Rodrigues [1]).□

Remark 2.18. Theorem 2.25 remains true for nonhomogeneous boundary conditions such as(106).Note that the above proof has, essentialy, a local character. In particular, it can be applied to the variational inequality (105),(106) associated to zero order reactions. So, if $g_\nu = \lambda_\nu \to g_0 = \lambda_0$, $\lambda_0 > 0$ then we have the convergence of the respective free boundaries in the Hausdorff distance. Note that a stronger result was given in Theorem 1.54 (also valid for the limiting case $q = 0$). A careful study of the dependence of the null set $N(u)$ with respect to λ in zero order reactions and general nonhomogeneous boundary conditions was made in Frank-Wendt [1] . The asymptotic behaviour of the free boundary as $\lambda \to +\infty$ is also considered there.□

2.3. A SINGULAR EQUATION.

Another class of nonlinear equations giving rise to the free boundary $\mathcal{F}(u)$ may be illustrated by the following special problem

$$-\Delta u + \lambda u^{-k} = 0 \quad \text{in} \quad \Omega \tag{1}$$

$$u = 1 \quad \text{on} \quad \partial\Omega \; . \tag{2}$$

where $\lambda > 0$, $k \in (0,1)$ and Ω is assumed to be an open regular bounded set of $\mathbb{R}^N$. As we shall detail later many pathologies may occur (lack of comparison principle, nonuniqueness, low regularity etc), due to the presence in the equation of the nonlinear singular term u^{-k} (recall that the case $k=0$ corresponds, formally, to the obstacle problem). Nevertheless, the main reason to expect the existence of the free boundary is that such a nonlinearity satisfies the fundamental assumption for the existence of local super and subsolutions given in Section 1.1 (see Theorem 1.5).

The problem (1),(2) appears as the limiting case of some models in heterogeneous chemical catalyst kinetics (Langmuir-Hinshelwood model) where the equation is

$$-\Delta u + \lambda u^m \left(\frac{\varepsilon+1}{\varepsilon+u}\right)^{m+k} = 0 \quad \text{in} \quad \Omega \; , \tag{3}$$

with $k > 0$, $m \geqslant 1$, $\lambda > 0$ and ε small, $\varepsilon > 0$ (Aris [1]), as well as in models in enzyme kinetics

$$-\Delta u + \lambda \frac{u^m}{\varepsilon + u^{m+k}} = 0 \quad \text{in} \quad \Omega \; , \tag{4}$$

(Banks [1]). In subsection 2.3a we recall different results about the solutions of (1) and (4) of interest in the study of the free boundary, which is made in subsection 2.3b.

In the following, and for the sake of simplicity, we shall identify (nonnegative) solution of (1),(2) with functions satisfying the boundary condition (2) and the equation (1) in the sense that $\Delta u = \lambda u^{-k} \mathbb{1}_{\tilde{S}(u)}$ in $\mathcal{D}'(\Omega)$ where $\tilde{S}(u) = \{x \in \Omega : u(x) > 0\}$.

2.3a. On the variational and other limiting solutions.

Perhaps, the most natural way to attack problem (1),(2) is by the minimization of the functional

$$J_0(v) = \frac{1}{2}\int_\Omega |\nabla v|^2 dx + \frac{\lambda}{1-k}\int_\Omega v^{1-k}\, dx \tag{5}$$

in the convex set $K = \{v \in H^1(\Omega),\ v \geq 0$ a.e. on Ω, $v = 1$ on $\partial\Omega\}$. (note that here we are only interested in nonnegative solutions and that the boundary condition may be replaced, in general, by $u = h$ on $\partial\Omega$, $h \in H^1(\Omega)$). The existence of, at least, one $u \in K$ that minimizes J_0 on K is standard, even though J_0 is not a convex functional (see Chapter 4). A more delicate question is to prove that u satisfies the equation (1). As a first step, we shall see that at least u is subharmonic, that is,

$$\int_\Omega u\, \Delta\zeta \geq 0 \quad \text{for all} \quad \zeta \in C_0^\infty(\Omega),\ \zeta \geq 0 .$$

Since $u \in H^1(\Omega)$, by the comparison principle for the Laplace operator, it is sufficient to show that u is below any harmonic replacement.

Lemma 2.26. *Let $x_0 \in \Omega$ and $B_r(x_0) \subset \Omega$. Consider $H_r(x)$ defined by $H_r(x) = u(x)$ on $\Omega - B_r(x_0)$ and such that $\Delta H_r(x) = 0$ in $B_r(x_0)$. Then $u \leq H_r$ on $B_r(x_0)$.*

Proof. Let $w = \min(H_r, u)$. Then $w \in K$ and, if we denote by B_r to $B_r(x_0)$, we have

$$J_0(w) - J_0(u) = \frac{1}{2}\int_{B_r} (|\nabla w|^2 - |\nabla u|^2)dx + \frac{\lambda}{1-k}\int_{B_r} (w^{1-k} - u^{1-k})dx .$$

To estimate such a difference, we note that $w - u \in H^1(B_r)$ and, by a well-known result (see e.g. Kinderlehrer-Stampacchia [2]. Lemma A.4) we have that $\nabla(w-u) = 0$ a.e. on the set $\{x \in B_r : w - u = 0\}$. Thus,

$$\int_{B_r} \nabla w \cdot \nabla(w-u)dx = \int_{B_r} \nabla H_r \cdot \nabla(w-u)dx = 0$$

because H_r is harmonic and $w - u \in H_0^1(B_r)$. Then

$$\int_{B_r} (|\nabla w|^2 - |\nabla u|^2)dx = \int_{B_r} \nabla(w+u)\cdot\nabla(w-u)dx = -\int_{B_r} |\nabla(w-u)|^2 dx.$$

Since $w \leq u$ we deduce that $J_0(w) - J_0(u) \leq 0$ with equality if and only if $w = u$. Then, as u minimizes J_0 we have $w = u$ and the assertion follows. □

With the help of the above result we can already examine the one-dimensional case in order to show, in this special case, how the free boundary may exist if the "sizes" of the domain and of the solutions are in an adequate balance. Let u be a variational solution of

$$- u'' + \lambda u^{-k} = 0 \quad , \quad \text{in} \quad \Omega = (0,R) \tag{6}$$

$$u(0) = 0 \ , \ u(R) = h \quad , \quad h > 0 \ , \tag{7}$$

with $\lambda > 0$ and $0 < k < 1$. Since $u \in H^1(0,R)$, $u(x)$ is absolutely continuous on $[0,R]$. Furthermore, since u is subharmonic, u is convex. Then

$$u(x) = 0 \ \text{on} \ [0,s] \ \text{for some} \ s \, ; \ u(x) > 0 \ \text{and} \ -u'' + \lambda u^{-k} = 0 \ \text{in} \ (s,R).$$

Now we shall prove that if h is small (or λ, or R is large) the free boundary $F(u)$ exists i.e. $0 < s < R$. Indeed, if this is not the case, i.e. if $s = 0$, then since u is subharmonic, $u \leqslant h$ on $(0,R)$ and $u'' \geqslant \lambda h^{-k}$, which implies

$$u(x) \geqslant \frac{\lambda h^{-k}}{2} x^2$$

because $u(0) = 0$ and u is nondecreasing. Then

$$h = u(R) \geqslant \frac{\lambda h^{-k}}{2} R^2,$$

which is impossible if

$$h^{1+k} < \frac{\lambda R^2}{2} \, . \tag{8}$$

We remark that some particular solutions of (6),(7) may be constructed as the solutions of the homogeneous Cauchy problem given by (7) and $u(0) = u'(0) = 0$. Indeed, if $0 < k < 1$, the function $f(r) = \lambda r^{-k}$ satisfies the integral condition

$$\int_{0^+} \frac{ds}{[F(s)]^{1/2}} < + \infty \tag{9}$$

with $F(r) = [\lambda/(1-k)]r^{1-k}$ and then, as in Lemma 1.3, we conclude that for every $s > 0$ the function

$$u(x) = 0 \text{ if } 0 < x < s\ ,\ u(x) = \frac{2}{1+k}\left(\frac{2(1-k)}{\lambda}\right)^{1/2}(x-s)^{\frac{1+k}{2}} \text{ if } x > s \tag{10}$$

is such a solution. Note that condition (8) may be improved for this special solution.

Returning to the N-dimensional formulation (1),(2), we point out that the study of the regularity of solutions becomes a difficult question if we assume that the free boundary $\mathcal{F}(u)$ does exist. For instance, the example given in (10) shows that the expected optimal regularity is $u \in C^{1,(1-k)/(1+k)}(\Omega)$. Such a regularity was proved in Phillips [1] by exploiting the fact that the functional $J_0(\cdot)$ preserves minimizers under a certain scaling:

$$J_{0,r}(v) = s^{\left(N + \frac{4}{1+k} - 2\right)} J_{0,r/s}(v^s) \tag{11}$$

where, here $J_{0,r}$ denotes the functional (5) on the ball B_r and for $v \in H^1(B_r)$, v^s is given by $v^s(x) = v(sx)/s^{2/(1+k)}$. The study of the growth of solutions near the free boundary is also used there. By a similar program to the one mentioned in subsection 4.1, the following result is proved in Phillips [1] and [2](see also Giaquinta-Giusti [1]):

<u>Theorem 2.27.</u> *Let* $h \in H^1(\Omega)$, $h \geq 0$. *Then any variational solution of equation* (1) *satisfying* $u = h$ *at* $\partial\Omega$ *is such that* $u \in C^{1,(1-k)/(1+k)}(\Omega)$. *Moreover, if* $\tilde{S}(u) = \{x \in \Omega : u(x) > 0\}$, *then* $u \in C^\infty(\tilde{S}(u))$, $-\Delta u + \lambda u^{-k} = 0$ *in* $\tilde{S}(u)$ *and* $u = 0$, $\nabla u = 0$ *on* $\mathcal{F}(u)$. *Finally* $u \in W^{2,p}_{loc}(\Omega)$ *for* $p < (1+k)/2k$ *and* $\Delta u = \lambda u^{-k}\, 1\!\!1_{\tilde{S}(u)}$ *in* $\mathcal{D}'(\Omega)$. □

In order to give sufficient conditions for the existence of the free boundary $\mathcal{F}(u)$ we shall use the approximate equation (4), with $m = 1$. Equation (3) may also be taken for this purpose . First of all we recall an existence result which is proved by means of the classical method of super and subsolutions:

<u>Theorem 2.28.</u> *Let* $\varepsilon > 0$ *and* $k > 0$ *fixed. Consider the problem*

$$- \Delta u + \lambda \frac{u}{\varepsilon + u^{1+k}} = 0 \qquad \textit{in} \quad \Omega \tag{12}$$

$$u = 1 \qquad \textit{on} \quad \partial\Omega \tag{13}$$

Then : i) $\forall\lambda > 0$ there exists at least one classical solution u_ε of (12), (13) satisfying that $u_\varepsilon \in C^\infty(\bar{\Omega})$ and $0 < u \leqslant 1$; ii) $\forall\lambda > 0$, there exists a maximal solution $\bar{u}_\varepsilon$ and a minimal solution $\underline{u}_\varepsilon$. Moreover, for $\lambda > 0$ fixed $\bar{u}_\varepsilon$ and $\underline{u}_\varepsilon$ are monotonically decreasing as $\varepsilon > 0$. □

The proof of this theorem consists in defining the operator T by

$$\Delta(T\psi) = \lambda \frac{T\psi}{\varepsilon + \psi^{1+k}} \quad \text{in} \quad \Omega \ , \quad T\psi = 1 \quad \text{on} \quad \partial\Omega$$

and verifying that T is a monotone increasing mapping. Finally, we apply the method of super and subsolutions (see Section 4.1.) starting with the subsolution $\underline{u} \equiv 0$ and the supersolution $\bar{u} \equiv 1$.

Problem (12),(13) admits a variational formulation which also supplies solutions, a priori being not necessarily coincident with any of these exhibited in Theorem 2.28. To explain this, we introduce the functional

$$J_\varepsilon(v) = \frac{1}{2} \int_\Omega |\nabla v|^2 + \lambda \int_\Omega F_\varepsilon(v)dx \tag{14}$$

where

$$F_\varepsilon(t) = \int_0^t f_\varepsilon(s)ds \quad , \quad f_\varepsilon(s) = \frac{s}{\varepsilon + s^{1+k}} \quad , \ \varepsilon > 0 \tag{15}$$

We have:

<u>Theorem 2.29.</u> *Let $\varepsilon > 0, \lambda > 0$ and $k > 0$ fixed. Then there exists at least one function $u_\varepsilon^J \in H^1(\Omega)$ which minimizes the functional J_ε on the convex set $K = \{v \in H_0^1(\Omega) : 0 \leqslant v \leqslant 1$ a.e., $v = 1$ on $\partial\Omega\}$. Moreover, $u_\varepsilon^J > 0$ and $u_\varepsilon^J \in C^\infty(\Omega)$ and satisfies the equation (12).* □

The proof of the above theorem is quite standard and may be obtained by the arguments of Section 4.1. (Note that again $F_\varepsilon(t)$ is not convex but the application $v \to \int_\Omega F_\varepsilon(v)dx$ is continuous in $L^2(\Omega)$). We also remark that, from theorems 2.28 and 2.29, we deduce that $\underline{u}_\varepsilon \leq u^J_\varepsilon \leq \overline{u}$ but, in general, it is not known if any of the inequalities is, in fact, an identity, except if λ is near 0 or $+\infty$ for ε and k fixed. In any case, it is not difficult to show that $\underline{u}_\varepsilon$ and $\overline{u}^\varepsilon$ are relative minimum points of J_ε.

In order to show the convergence of solutions of (12) as $\varepsilon \downarrow 0$, we recall the Pohozaev identity, already used in Theorem 1.36.

Lemma 2.30. *Let Ω be a regular domain of $\mathbb{R}^N$ of outer normal unit vector $\vec{n}$ at the boundary $\partial\Omega$. Let $\vec{r}$ be the radius vector of a boundary point relative to some fixed point of Ω. Let $v \in H^2(\Omega) \cap C^0(\Omega)$ be a nontrivial solution of the problem*

$$- \Delta v + f(v) = 0 \quad \text{in} \quad \Omega \tag{16}$$

$$v = 0 \quad \text{on} \quad \partial\Omega \tag{17}$$

where f is a real continuous function. Then, the following identities hold:

$$\int_\Omega (\nabla v)^2 dx + \frac{1}{N-2} \int_{\partial\Omega} \vec{r}\cdot\vec{n} \left(\frac{\partial v}{\partial n}\right)^2 ds = - \frac{2N}{N-2} \int_\Omega F(v)dx \quad , \text{if } N \geq 3 \tag{18}$$

and

$$\int_{\partial\Omega} \vec{r}\cdot\vec{n} \left(\frac{\partial v}{\partial n}\right)^2 ds = -4 \int_\Omega F(v)dx \quad , \text{if } N = 2 \tag{19}$$

with $F(t) = \int_0^t f(s)ds.$ □

The proof is obtained by multiplying both sides of (16) by $\Sigma\, x_i \frac{\partial u}{\partial x_i}$ and integrating by parts (see Pohozaev [1]). As an application we have:

Theorem 2.31. *Assume Ω to be a regular bounded convex open set. Then, there exist positive constants C_1 and C_2 (independent of ε) such that, for any solution u_ε of (12), (13), we have $|\nabla u_\varepsilon|_{L^2} \leq C_1$ and $|\Delta u_\varepsilon|_{L^1} \leq C_2$*

Proof. We set $v_\varepsilon = 1 - u_\varepsilon$. Then v_ε satisfies (16), (17) with

$$f(s) = - \lambda \frac{(1 - s)}{\varepsilon + (1-s)^k}$$

and, therefore,

$$- \int_\Omega F(v_\varepsilon)dx = - \int_\Omega \left(\int_0^{v_\varepsilon(x)} f(t)dt\right) \leq \lambda \int_\Omega \left(\int_0^1 (1-t)^{-k} dt\right) = \lambda \frac{|\Omega|}{(1-k)}.$$

Now, let $N \geq 3$. Since Ω is starshaped we can assume $\vec{r} \cdot \vec{n} > 0$ a.e., and the boundedness of $|\nabla u_\varepsilon|_{L^2}$ follows from (18). When $N = 2$, using the fact that there exists $\theta > 0$ such that $\vec{r} \cdot \vec{n} > \theta$ a.e. on $\partial\Omega$, from (19) we deduce that $\frac{\partial v_\varepsilon}{\partial n}$ is bounded in $L^2(\partial\Omega)$. Multiplying equation (12) by u_ε and using Green's formula, we get

$$\int_\Omega (\nabla u_\varepsilon)\, dx = \int_{\partial\Omega} \frac{\partial u_\varepsilon}{\partial n} u_\varepsilon ds - \lambda \int_\Omega \frac{u_\varepsilon^2}{\varepsilon + u_\varepsilon^k}\, dx$$

and the right-hand side is bounded; hence $|\nabla u_\varepsilon|_{L^2}$ is bounded. Finally, from the equation (11) we deduce that $\nabla u_\varepsilon \geq 0$, and then it suffices to use Green's formula and remark that $\frac{\partial u_\varepsilon}{\partial n}$ is bounded in $L^1(\Omega)$. □

Letting $\varepsilon \downarrow 0$ we obtain:

<u>Theorem 2.32.</u> *Let $\bar{u}_\varepsilon$, $\underline{u}_\varepsilon$ and u_ε^J be the maximal, the minimal and any variational solution of (12), (13). Then there exist $\bar{u}_0$, $\underline{u}_0$ and u_0^J nonnegative functions in $H^1(\Omega)$ such that: $\bar{u}_\varepsilon \to \bar{u}_0$ and $\underline{u}_\varepsilon \to \underline{u}_0$ strongly in $L^p(\Omega)$ for every $1 \leq p \leq \infty$ and weakly in $H^1(\Omega)$, $\bar{u}_0^{-k} \in L^1(\tilde{S}(\bar{u}_0))$, $\underline{u}_0^{-k} \in L^1(\tilde{S}(\underline{u}_0))$ and $\bar{u}_0$, $\underline{u}_0$ satisfies the boundary condition (2) as well as the equation (1) in $L^1(\tilde{S}(\underline{u}_0))$ and $L^1(\tilde{S}(\bar{u}_0))$ respectively. Finally, $u_\varepsilon^J \to u_0^J$ strongly in $H^1(\Omega)$ and u_0^J minimizes the functional J_0 given in (5) on the convex set $K = \{v \in H^1(\Omega) : v \geq 0$ a.e. in Ω, $v = 1$ on $\partial\Omega\}$ In particular, $u_0^J \in C^{1,(1-k)/(1+k)}(\Omega) \cap W^{2,p}_{loc}(\Omega)$ for every $1 \leq p < \infty$.*

<u>Proof</u>. Since $\bar{u}_\varepsilon$ and $\underline{u}_\varepsilon$ are decreasing as $\varepsilon \to 0$, the convergence in $L^p(\Omega)$ to $\bar{u}_0$ and $\underline{u}_0$ respectively is a consequence of the Beppo-Levi Theorem on monotone convergence. Moreover, if u_ε is $\bar{u}^\varepsilon$ or $\underline{u}_\varepsilon$ and u_0 represents $\bar{u}^0$ or $\underline{u}_0$, then we have that

$$[u_\varepsilon/(\varepsilon + u_\varepsilon^k)] \to u_0^{-k} \quad \text{a.e. in} \quad \tilde{S}(u_0).$$

As Δu_ε is bounded in $L^1(\Omega)$, the same follows for $u_\varepsilon/(\varepsilon + u_\varepsilon^k)$ in $\tilde{S}(u_0)$ and, by Fatou's Lemma, $u_0^{-k} \in L^1(\tilde{S}(u_0))$. On the other hand, by Theorem 2.31 as $\varepsilon \to 0$, one can extract a subsequence, also noted by u_ε, such that u_ε converges to u_0 weakly in $H^1(\Omega)$, and so in fact this is the same for the whole sequence, and u_0 satisfies (12) and (13). Finally, let u_ε^J be the variational solution of (12), (13). We know that one can extract a subsequence, again noted by u_ε^J, such that $u_\varepsilon^J \to u_0^J$ weakly in $H^1(\Omega)$. Moreover, by the Lebesgue's theorem, $F_\varepsilon(u_\varepsilon^J) \to F_0(u_0^J)$ (F_ε given in (14))in $L^1(\Omega)$ and then, from $J_\varepsilon(u_\varepsilon^J) \leq J_\varepsilon(v)$ $\forall v \in K$, we deduce that $J_0(u_0^J) \leq J_0(v)$ $\forall v \in K$ and u_0^J realizes the minimum of J_0. The strong convergence in $H^1(\Omega)$ is then a consequence of the fact that $\inf J_\varepsilon \to \inf J_0$ when $\varepsilon \to 0$. □

The above convergence may be made more precise under some circumstances. For instance, by adapting an argument of Brauner-Eckhaus-Garbey-Van Harten [1], it is possible to identfy the limit solution $\underline{u}$, assumed adequate regularity on the eventual free boundary ($\underline{u}$). This is certainly verified in the following special case:

Proposition 2.33. *Let* $\Omega = B_R(x_0)$ *and let* $\underline{u}_0 = \lim \underline{u}_\varepsilon$ *,where* $\underline{u}_\varepsilon$ *is the minimal solution of (12),(13) on* $B_R(x_0)$. *Then* $\underline{u}_0$ *is the minimal solution of (1),(2) among the set of radially symmetric solutions.*

Sketch of the proof. First of all we note that the radial symmetry of $\underline{u}_0$ is deduced from this same property for each $\underline{u}_\varepsilon$. Now let $\tilde{u}_0$ be the minimal solution among the radially symmetric solutions of (1),(2). By using an asymptotic argument (see the above reference for a related result)for every $\varepsilon > 0$ it is possible to construct a function $\tilde{u}_\varepsilon$, solution of (12),(13) on $B_R(x_0)$ and such that $\tilde{u}_\varepsilon \to \tilde{u}_0$ in $C^1(B_R(x_0))$. Finally, as $\tilde{u}_\varepsilon \geq \underline{u}_\varepsilon$ we deduce that $\tilde{u}_0 \geq \underline{u}_0$ and then $\tilde{u}_0 = \underline{u}_0$. □

2.3b. On the existence of the free boundary.

A first result on the existence of the free boundary $\mathcal{F}(u)$ can be obtained via bifurcation theory. Indeed, the limit problem (1),(2) can be reformulated as

$$- \Delta w = \lambda \frac{1}{(1-w)^k} \quad \text{in} \quad \Omega \tag{20}$$

$$w = 0 \quad \text{on} \quad \partial\Omega \tag{21}$$

by making $w = 1-u$. Then, if we define the open set B by

$$B = \{v \in W^{2,s}(\Omega) : \exists \theta > 0 \ , \ v(x) < 1- \theta \} \ , \quad s > N/2,$$

it is possible to adapt an abstract result of Crandall-Rabinowitz ([1] Proposition 2.16 and 2.17), and we get

Lemma 2.34. *There exists a* $\lambda_* > 0$ ($\lambda_* \leq \mu_0$, μ_0 *smallest eigenvalue of* $-\Delta$) *such that, if* $\lambda > \lambda_*$ *there exists no solution of* (20),(21) *in the open set* B . *Moreover, if* $\lambda = \lambda_*$, *there is a unique solution in* B . □

By the above lemma, if $\lambda > \lambda^*$, any solution u of (1),(2) with $u \in W^{2,s}(\Omega)$, $s > N/2$, must be zero in some subset of Ω . We shall improve this result by studying first the case in which Ω is a ball. Again, the study of such a special case becomes much more complicated than in previous situations due to the nonuniqueness of solutions.

First of all we remark that it is not difficult to check that the computations of Lemma 1.5 still remains true for $q = -k \in (-1,0)$. So , the function

$$u(x) = C\,|x - x_0|^{\frac{2}{1+k}} \tag{22}$$

is such that u is a solution of the equation (1) on $B_R(x_0)$ if $C = K_{N,\lambda}$ and that u is a supersolution (resp. a subsolution) of (1), if $C < K_{N,\lambda}$ (resp $C > K_{N,\lambda}$). In particular, u is a supersolution of (1),(2) on $B_R(x_0)$ if

$$\frac{1}{R^{2(1+k)}} \leq C \leq K_{N,\lambda} \quad , \quad K_{N,\lambda} = \left[\frac{\lambda(1+k)^2}{2(N+k(N-2))} \right]^{\frac{1}{1+k}} \quad , \tag{23}$$

and in fact u is a solution of (1),(2) if $C = K_{N,\lambda} = R^{-2/(1+k)}$. However, this information is not satisfactory because of the nonuniqueness. A more carefully study was made in Brauner-Nicolaenko [1],[2] and Misiti-Guyot [1] whose main results are compiled in the following. First of all we restrict ourselves to the consideration of $\Omega = B_1(0)$ and then we ask for nonnegative functions $u = u(r)$ satisfying the following conditions that, for short, we shall call Problem P : there exists $r_0 \in (0,1)$ such that

$$u \equiv 0 \quad \text{on} \quad r < r_0 \tag{24}$$

$$-\Delta u + \frac{\lambda}{u^k} = 0 \quad \text{on} \quad r_0 < r < 1 \quad , \quad 0 < k < 1, \tag{25}$$

$$u(1) = 1 \,, \tag{26}$$

$$\lim_{r \to r_0^-} u(r) = \lim_{r \to r_0^+} u(r) = 0 \tag{27}$$

$$\lim_{r \to r_0^-} u'(r) = \lim_{r \to r_0^+} u'(r) = 0. \tag{28}$$

In order to present the main result for the above problem, we need to introduce some notation :

$$\beta = \frac{2}{1+k} \,, \quad \lambda_c = \beta(\beta + N - 2) \,,$$

$$A(k,N) = \frac{4+(N-2)(1+k)}{(4+2(N-2)(1+k))^{1/2}} = \frac{\beta}{\sqrt{\lambda_c}} + \frac{\sqrt{\lambda_c}}{\beta}$$

$$B(k,N) = A^2(k,N) - 4(1+k).$$

Finally, in the plane (k,N) we define the three following regions:

$$R_1 = \{(k,N) : B(k,N) < 0\} = \{(k,N) \mid N^-(k) < N < N^+(k)\}$$

$$R_2 = \{(k,N) : \frac{3k-1}{\sqrt{2(1-k)}} > \sqrt{B(k,N)}\} = \{(k,N) \mid k > 1/3 \text{ and } 1 < N \leq N^-(k) \text{ or }$$

$$N^+(k) \leqslant N < 2 + \frac{2}{1+k}\}$$

$$R_3 = \{(k,N) : \sqrt{B(k,N)} \geqslant \frac{3k-1}{2(1-k)}\} \,,$$

being

$$N^{\pm}(k) = \frac{6k+2}{k+1} \pm 4\left(\frac{k}{1+k}\right)^{1/2} .$$

An illustration of these regions is given in the Figure 10.

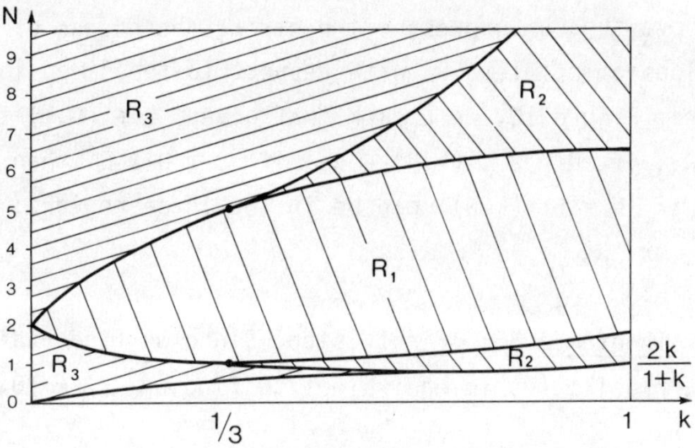

Figure 10

We have

Theorem 2.35. (i) *Let* $(k,N) \in R_1$. *Then there exist* $\lambda_1 < \lambda_c$ *and* $\lambda_2 > \lambda_c$ *such that* : $\forall \lambda < \lambda_1$ *problem* (P) *has no solution ; if* $\lambda \in [\lambda_1, \lambda_2] - \{\lambda_c\}$ *problem* (P) *has a finite number of solutions; if* $\lambda = \lambda_c$ *there is an infinitely countable number of solutions, and finally, if* $\lambda > \lambda_2$ *then there is an unique solution of* (P).

(ii) *Let* $(k,N) \in R_2$. *Then there exists* $\lambda_1 < \lambda_c$ *such that* : $\forall \lambda < \lambda_1$ *problem* (P) *has no solution; if* $\lambda = \lambda_1$ *there is a unique solution; if* $\lambda \in]\lambda_1, \lambda_c[$ (P) *has two solutions; and if* $\lambda \geq \lambda_c$ *then there is a unique solution of* (P).

(iii) *Let, finally,* $(k,N) \in R_3$. *Then, for every* $\lambda \leq \lambda_c$ (P) *has no solution and, for every* $\lambda > \lambda_c$, *there is a unique solution.* □

The following figure illustrates the dependence of r with respect to λ in the three above cases.

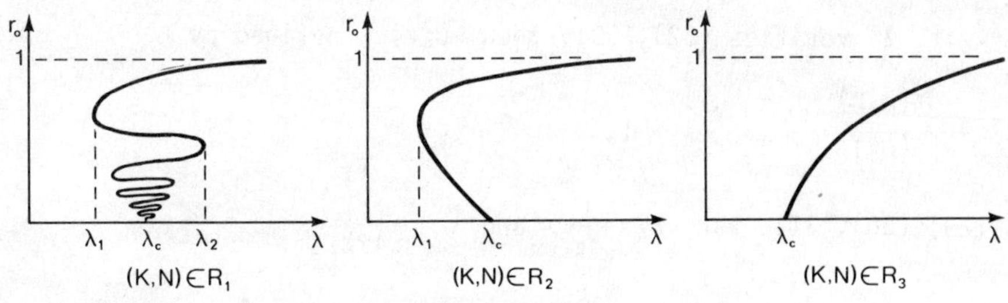

Figure 11.

Remark 2.19. The above theorem expresses the strong dependence of the null set $N(u)$ of solutions of problem P with respect to the dimension N of the space. So, from Figure 10, we remark that given $k \in (0,1)$ then, if $N = 1$, $(k,N) \in R_3$; if $N = 2$, $(k,N) \in R_2$; if $3 \leq N \leq 5$ then $(k,N) \in R_1$ or R_3; if $N = 6$ (k,N) can be in the three regions; and if $N \geq 7$ $(k,N) \in R_2$ or R_3. □

Idea of the proof of Theorem 2.35. First step. Since we find radially symmetric solutions, problem (P) is equivalent to finding $r_0 \in (0,1)$ and $u = u(r) \geq 0$ such that

$$- u''(r) - \frac{N-1}{r} u'(r) + \frac{\lambda}{u^k(r)} = 0 \quad \text{in} \quad (r_0,1) \tag{29}$$

$$u'(r_0) = u(r_0) = 0 \tag{30}$$

$$u(1) = 1. \tag{31}$$

But, if u satisfies (29),(30), (31) then the function z defined by

$$z(\rho) = \frac{1}{(\lambda\, r_0^2)^{1+k}} u(r_0\rho)$$

satisfies

$$z''(\rho) + \frac{N-1}{\rho} z'(\rho) - \frac{1}{z^k(\rho)} = 0 \quad \text{in} \quad (1,1/r_0) \tag{32}$$

$$z(1) = z'(1) = 0 \tag{33}$$

$$z(1/r_0) = \frac{1}{\lambda\, r_0^2} \quad . \tag{34}$$

Conversely, if z verifies (32),(33), then $u(r)$, defined by

$$u(r) = \frac{z(\rho_0 r)}{z(\rho_0)} \quad , \quad \rho_0 > 1,$$

satisfies (29),(30),(31) for $r_0 = \dfrac{1}{\rho_0}$ and $\lambda = \dfrac{\rho_0^2}{z^{1+k}(\rho_0)}$.

Second step. The Cauchy Problem given by equation (32) and initial conditions (33) has a unique solution $z \in C^{1,(1-k)/(1+k)}([1,\infty)) \cap C^{\infty}((1,\infty))$ as can be shown by considering the nonlinear integrodifferential equation

$$\frac{1}{2} z'(\rho)^2 + \int_1^{\rho} \frac{N-1}{\tau} z'(\tau)^2 d\tau = \frac{1}{1-k} z^{1-k}(\rho). \tag{35}$$

Then, by the change of unknown function, if suffices to study the values (and multiplicity) of the function $\rho \to \frac{\rho^2}{z^{1+k}(\rho)}$. In this way, for $r_0 = \frac{1}{\rho}$ fixed, we may know the number of solutions of (29),(30),(31).

Third step. Defining the function $v(\rho)$ by

$$z(\rho) = M \rho^{2/(1+k)} v(\rho) , \qquad M = \left(\frac{1}{\lambda_\tau}\right)^{1/(1+k)}$$

and making

$$\rho = e^{s/\sqrt{\lambda_c}} \qquad \text{and} \qquad w(s) = v(\rho) \quad , \tag{36}$$

it is easy to check that w(s) is the unique nonnegative solution of the new Cauchy Problem

$$w''(s) + A(k,N)\, w'(s) + w(s) - \frac{1}{w^k(s)} = 0 \quad \text{in} \quad (0,\infty) \tag{37}$$

$$w(0) = w'(0) = 0 . \tag{38}$$

On the other hand, we have

$$\frac{\rho^2}{z^{1+k}(\rho)} = \frac{\lambda_c}{w^{1+k}(s)}$$

and so, it suffices to study the values taken by the function w(s) with their multiplicity. It is shown that $\lim_{s\to\infty} w'(s) = 0$ and $\lim_{s\to-\infty} w(s) = 0$ and finally the conclusions of the theorem are obtained by discussing in the phase plane (w,w') or by reasoning directly on the equation (37) with results of the same kind than Sturm's theorems. (See details in the mentioned references).□

Remark 2.20. Note that for $\Omega = B_1(0)$ the only solution u of the form (22) corresponds to $C = 1$ and then $\lambda = \lambda_c$, i.e. $u = u_c(r) = r^{2/(1+k)}$. It is also important to remark that when λ is near λ_c problem (1),(2) may admit strictly positive solutions, as well as solutions with a non empty null set, according to the parameters $k \in (0,1)$ and N. In any case, it turns out that (λ_c, u_c) is an endpoint of the bifurcation diagram for positive solutions of (1),(2). For very precise bifurcation diagrams of positive (and nonnegative) solutions we refer the reader to Brauner-Nicolaenko [2]and Misiti-Guyot [1]. □

Now, we consider the case of general domains Ω. In contrast to Chapter 1, now we cannot apply the comparison principle directly and so other arguments are needed. First of all we recall an useful comparison-matching lemma due to Berestycki-Lions [1].(See also Il'in-Kalashnikov-Oleinik [1]). The result may be stated in a general setting which has many different applications (see the mentioned works):Let Ω be a regular domain in $\mathbb{R}^N$ and let Ω_1 be a subdomain of Ω such that $\partial\Omega_1$ is regular and $\overline{\Omega}_1 \subset \Omega$. We denote by $\Omega_2 = \Omega - \Omega_1$ and by ν the unit outward normal to Ω_1. Let L be the operator defined by

$$Lu = \sum_{i,j=1}^{N} D_j(a_{ij}(x)D_i u) + \sum_{j=1}^{N} b_j(x)\, D_j u + c(x)u$$

where $a_{ij} = a_{ji} \in L^\infty(\Omega)$, $b_i \in L^\infty(\Omega)$, $c \in L^\infty(\Omega)$. We denote by ν the conormal associated with L (i.e. $\nu^i = \sum_j a_{ij}(x)n_j$).

Lemma 2.36. *Let us assume that* f *is a measurable function on* $\Omega \times \mathbb{R} \times \mathbb{R}^N$ *such that if* $v \in H^1(\Omega)$, $f(x,v,\nabla v) \in L^1(\Omega)$. *Furthermore, we assume the existence of* $u_i \in H^2(\Omega_i)$ *satisfying*

$$-Lu_i \geq f_i \text{ a.e. in } \Omega_1, \qquad i = 1,2, \tag{39}$$

$$u_1 = u_2 \text{ on } \partial\Omega_1 \text{ and } \frac{\partial u_1}{\partial \nu} \geq \frac{\partial u_2}{\partial \nu} \text{ on } \partial\Omega_1, \tag{40}$$

with $f_i = f(x,u_i\nabla u_i)$. *Then, the function* u *defined by* $u = u_i$ *on* Ω_i *belongs to* $H^1(\Omega)$ *and satisfies*

$$- Lu \geq f(x,u,\nabla u) \text{ in } \quad D'(\Omega).$$

Proof. Let $\Phi \in D_+(\Omega)$. We have

$$< - Lu,\Phi > = \int_\Omega \{ \sum_{i,j} a_{ij} D_i u D_j \Phi - \sum_j b_j D_j u \Phi - c u \Phi \} dx =$$

$$= \int_{\Omega_1} (- Lu_1)\Phi dx + \int_{\Omega_2} (- Lu_2) \Phi dx + \int_{\partial\Omega_1} (\frac{\partial u_1}{\partial\nu} - \frac{\partial u_2}{\partial\nu}) \Phi ds$$

Then, using (39) ,(40) and the fact that $u \in H^1(\Omega)$ (see e.g. Gilbarg-Trudinger [1]), we conclude that

$$< - Lu,\Phi > \geq \int_\Omega f(x,u,\nabla u) \Phi dx. \square$$

As an elementary application of the above lemma we can easily construct supersolutions from solutions on subdomains.

Corollary 2.37. *Let* $B_R(x_0) \subset \Omega$ *and let* u_ε^R *be any solution of (12),(13) in* $B_R(x_0)$. *Then the function*

$$u_\varepsilon(x,R) = \begin{cases} u_\varepsilon^R(x) & x \in B_R(x_0) \\ 1 & x \in \Omega - B_R(x_0) \end{cases} \tag{41}$$

is a supersolution of the problem (12),(13) in Ω .

Proof. Take $L = \Delta$, $\Omega_1 = B_R(x_0)$, $u_1 = u_\varepsilon^R$ and $u_2 = 1$ and note that $\frac{\partial u_1}{\partial\nu} \geq 0$ because, by the maximum principle , $u_\varepsilon^R(x) < 1$ on $B_R(x_0)$. $\square$

Concerning the existence and location of the free boundary for solutions of (1),(2) in general domains we have

Theorem 2.38. *Let* $\lambda > 0$ *and* $k \in (0,1)$. *Then, there exists at least one solution* u *of (1),(2) such that its null set* N(u) *satisfies the estimate*

$$N(u) \supset \{ x \in \Omega : d(x,\partial\Omega) \geq (\frac{\lambda_c}{\lambda})^{1/2} \} \tag{42}$$

with

$$\lambda_c = \frac{2[N+k(N-2)]}{(1+k)^2} .$$

Proof. First of all we remark that, by rescaling, it is easy to see that a function of the form $u(x) = C|x-x_0|^{2/(1+k)}$ is a solution of (1),(2) on $B_R(x_0)$ if and only if $\lambda = \lambda_{c,R} = \lambda_c R^{-2}$ and $C = R^{-2/(1+k)}$. Denote this solution by $u_{c,R}$ and, then, $u_{c,R}(x) = R^{-2/(1+k)}|x-x_0|^{2/(1+k)}$. Now for $\lambda > 0$ fixed, let $R_\lambda > 0$ be given by $R_\lambda^2 = \lambda_c/\lambda$,and let $x_0 \in \Omega$ such that $d(x_0,\partial\Omega) \geq R_\lambda$. On the ball $B_{R_\lambda}(x_0)$ consider the function $\underline{u}_\varepsilon^{R_\lambda}(x)$ as the minimal solution of (12),(13) and let $u_\varepsilon(x,R)$ be the function contructed as in (41) from $\underline{u}_\varepsilon^{R_\lambda}(x)$. By Corollary 2.36 $u_\varepsilon(x,R_\lambda)$ is now a supersolution of (12),(13) on Ω and then from the method of super and subsolutions we deduce that there exists, at least, one solution $\hat{u}_\varepsilon$ of (12),(13) on Ω and such that $0 \leq \hat{u}_\varepsilon(x) \leq u_\varepsilon(x,R_\lambda)$ a.e. $x \in \Omega$. Making $\varepsilon \downarrow 0$, by Theorem 2.32 we know that there exists a function $\underline{u}_0^{R_\lambda}(x)$ solution of (1),(2) on $B_{R_\lambda}(x_0)$ and such that $\underline{u}_\varepsilon^{R_\lambda} \downarrow \underline{u}_0^{R_\lambda}$ in $L^p(B_R(x_0))$, for every $1 \leq p < \infty$. Using the Theorem 2.31 , we can extract a subsequence, also denoted by $\hat{u}_\varepsilon$, such that $\hat{u}_\varepsilon \to \hat{u}_0$ weakly in $H^1(\Omega)$ with $\hat{u}_0$ solution of (1),(2) on Ω . Finally, by Proposition 2.33 we deduce that

$$0 \leq \hat{u}_0(x) \leq \underline{u}_0^{R_\lambda}(x) \leq R^{-2/(1+k)}|x-x_0|^{2/(1+k)} \text{ for a.e. } x \in B_{R_\lambda}(x_0),$$

and then the estimate (42) holds for the solution $\hat{u}_0$.□

Remark 2.21. As in Chapter 1, the estimate (42) leads implicitly to a condition between the size of Ω and the size of u (in this case given by $\|u\|_\infty = 1$); nevertheless, in contrast with the case of monotone equations, the proof of the above theorem shows that this estimate is not optimal in the sense that there can exist solutions u of (1),(2) with $N(u)$ strictly greater than the region of the right hand side of (42). (This happens, for instance, when k and N are such that $u_0^{R_\lambda} \neq u_{c,R}$). Finally, we remark that Hausdorff estimates and other regularity results for the free boundary $\mathcal{F}(u)$ are given in Phillips [2], when u is assumed to be a variational solution of (1),(2).□

2.4. NONISOTROPIC EQUATIONS.

Up to this point of the book, the existence and properties of the free boundary $\mathcal{F}(u)$ have been studied for *isotropic equations*, i.e, equations invariant by symmetry. This property allowed us to construct (local) super and subsolutions as radially symmetric functions $\eta(|x-x_0|)$, getting in this way estimates on the location and measure of the null set $N(u)$ which was optimal when the domain Ω was a ball. The main goal of this subsection is to consider several classes of nonisotropic equations and to show how local super and subsolutions can still be constructed under adequate balances between the nonlinearities.

First of all, we remark that some nonisotropic equations accept radially symmetric functions as local super and subsolutions and, in consequence, results on the free boundary similar to those given in Section 1.1 can be formulated easily for these equations. This is, for instance, the case of the general second order semilinear equation given by (64),(67) and (68) of Chapter 1. More generally, the main idea in the proof of Theorem 1.13 can also be applied to some quasilinear equations as, for instance,

$$-\Delta_p u + B(x,u,\nabla u) + f(u) = g \quad , \tag{1}$$

assumed that p and f satisfy the assumption (48) of Chapter 1, and that the term $B(x,u,\nabla u)$ is of the form

$$|B(x,u,\nabla u)| \leqslant \mu(|u|)|\nabla u|^{p-1} \tag{2}$$

where μ is an increasing real function, or (in the special case of $p=2$) for convection terms B such that

$$B(x,u,\nabla u) = \Sigma\, B_j(x)D^j\, b_j(u), \tag{3}$$

where $B_j \in L^\infty$ and $b_j \in C^1(\mathbb{R})$ for every $j=1,\dots,N$. Indeed, under the mentioned conditions it is easy to see that there exists a positive constant k such that the functions $\overline{u}(x:x_0) = \eta(|x-x_0|)$ are local supersolutions on $B_R(x_0)$ of equation (1) if $\eta = \eta(r)$ is taken as a solution of the Cauchy problem

$$-(|\eta'|^{p-2}\eta')' - \frac{k}{r}|\eta'|^{p-2}\eta' + f(\eta) = 0 \qquad (4)$$

$$\eta(0) = \eta'(0) = 0 \qquad (5)$$

(The existence of such a solution is assured by the assumption of f). It turns out that the conditions on $B(x,u,\nabla u)$ may be improved in different ways; a growing condition more general than (2) will be exhibited in Section 3.1, and, for convection terms like (3), the assumptions $p = 2$ and $b_j \in C^1$ will be removed in the next subsection.

Another previous remark is that some particular, but interesting, nonisotropic equations may be reformulated in an isotropic way by an ingenious trick. This is the case of the following quasilinear equations involving another pseudo-laplacian operator

$$-\sum_{i=1}^{N} \frac{\partial}{\partial x_i}\left(\left|\frac{\partial u}{\partial x_i}\right|^{p-2} \frac{\partial u}{\partial x_i}\right) + f(u) = g, \qquad (6)$$

where $p > 1$. Indeed, as was already remarked in Bamberger [1], some straightforward computations show that by introducing the new argument

$$s = |x|_{p'} = \left(\sum_{i=1}^{N} |x_i|^{p'}\right)^{1/p'}, \quad \frac{1}{p} + \frac{1}{p'} = 1,$$

solutions of the homogeneous equation associated to (6) (i.e. for $g \equiv 0$) can be constructed as solutions of the ordinary equation

$$-\frac{1}{s^{N-1}} \frac{d}{ds}\left(s^{N-1}\left|\frac{d\eta}{ds}\right|^{p-2} \frac{d\eta}{d\eta}\right) + f(\eta(s)) = 0. \qquad (7)$$

In consequence, all the results of Section 1.1 can be reformulated and proved for the equation (6) by working with balls $B^*_R(x_0)$ in the metric $|x|_{p'}$ instead of the euclidean one $|x|_2$, i.e, $B^*_R(x_0) = \{x \in \mathbb{R}^N : |x-x_0|_{p'} < R\}$.

We end the introduction to this section by pointing out that some nonisotropic equations in divergence form are treated in Subsection 2.4a, while the general case of nonisotropic equations (not in divergence form).

is developed in Subsection 2.4b where the special case of the Hamilton-Jacobi-Bellman equation is also considered.

2.4a. Equations in divergence form. On the diffusion-convection balance. The existence of the free boundary $\mathcal{F}(u)$ can still be derived from an adequate diffusion-absorption balance, as in previous sections, for non-isotropic equations such as

$$-\sum_{i=1}^{N} \frac{\partial}{\partial x_i} a_i\left(\frac{\partial u}{\partial x_i}\right) + f(u) = g \quad \text{in} \quad \Omega \tag{8}$$

$$u = h \quad \text{on} \quad \partial\Omega . \tag{9}$$

Here, the structural assumptions will be the following:

$$a_i \in C^{\infty}(\mathbb{R}) \cap C^1(\mathbb{R} - \{0\}),\ a_i(s) > 0 \ \text{ if } \ s \neq 0 \ ,\ a_i(-s) = -a_i(s), \forall i = 1,..,N \tag{10}$$

and

$$f \ \text{ is a nondecreasing continuous function, } f(0) = 0 \tag{11}$$

Existence, uniqueness and comparison results under these circunstances can be given in a variational setting as well as for g merely in L^1 (see Chapter 4). Concerning the existence of the free boundary $\mathcal{F}(u)$, it seems natural to ask for nonisotropic super and subsolutions, i.e,

$$\overline{u}(x:x_0) = \sum_{i=1}^{N} \eta_i(|x_i - x_{0,i}|) \tag{13}$$

if $x = (x_1, \ldots x_N)$. The keystone in this approach is the following simple lemma.

Lemma 2.39. *Assume that* a_j *and* f *are such that*

$$\int_{0^+} \frac{ds}{A_j^{-1}(F(s))} < +\infty \tag{14}$$

for every $1 \leq j \leq N$, *where*

$$A_j(r) = \int_0^r a_j'(s)sds \qquad \textit{and} \qquad F(r) = \int_0^r f(t)dt. \tag{15}$$

For $\tau > 0$ *let*

$$\psi_j(\tau) = \int_0^\tau \frac{ds}{A_j^{-1}\left(\frac{F(s)}{N}\right)} \tag{16}$$

and define $\eta_j(s) = \psi_j^{-1}(s)$. *Then, given* $x = (x_{0,1},,x_{0,N})$ *, the function*

$$\overline{u}(x) = \sum_{i=1}^{N} \eta_j(|x_j - x_{0,j}|) \tag{17}$$

satisfies that $\overline{u}(x_0) = 0$, $\nabla\overline{u}(x_0) = 0$

$$-\sum_{i=1}^{N} \frac{\partial}{\partial x_i} a_i\left(\frac{\partial\overline{u}}{x_i}\right) + f(\overline{u}) \geq 0. \tag{18}$$

Proof. By construction it is clear that $\eta_j(0) = \eta'(0)$ and

$$-a_j(\eta_j')' + \frac{1}{N} f(\eta_j) = 0$$

for every $1 \leq j \leq N$. Then, by remarking that any nondecreasing function f vanishing at the origin satisfies

$$f\left(\sum_{i=1}^{N} s_i\right) \geq (1/N) \sum_{i=1}^{N} f(s_i) \quad , \quad s_i \geq 0 ,$$

we obtain the conclusion. □

Using the above lemma we can repeat the same program of Chapter 1 in order to study the free boundary $F(u)$ for solutions of (8),(9). Indeed it suffices to argue on balls of $\mathbb{R}^N$ for the metric $|x|_\infty = \max |x_i|$ (i.e. on cubes) instead of on euclidean balls. As a sample, we give here only a similar version of Theorem 1.9. Statements concerning boundary estimates and solutions with compact support are left to the reader.

Theorem 2.40. *Assume that (14) holds for every* $1 \leq j \leq N$ *and let* $u \in W^{1,1}(\Omega) \cap L^{\infty}(\Omega)$ *be a solution of (8),(9). Then we have the following estimate for the null set* $N(u)$,

$$N(u) \supset \{ x \in N(g) \cup N(h|_{\partial\Omega}) : d_{\infty}(x, S(g) \cup S(h|_{\partial\Omega})) \geq L\} \tag{19}$$

where $d_{\infty}(x,y) = \max_i |x_i - y_i|$ *,and* L *is a certain positive constant depending in an increasing way on* $\|u\|_{\infty}$.

Proof. Let $x_0 \in N(g) \cup N(h|_{\partial\Omega})$ and let $R = d_{\infty}(x_0, S(g) \cup S(h|_{\partial\Omega}))$. On the ball $B_R^{\infty}(x_0) = \{ x \in \mathbb{R}^N : d_{\infty}(x_0,x) < R\}$, the function $\bar{u}(x)$ given by (17) satisfies (18) and, on the other hand, if $M = \|u\|_{\infty}$, we have

$$u \leq M \leq \bar{u} \qquad \text{on} \qquad \partial B_R^{\infty} ,$$

if we choose R such that

$$M \leq \sum_{j=1}^{N} \eta_j(R) .$$

In particular, taking $R = L$, with

$$L \geq \max_j \{ \psi_j(\tfrac{M}{N})\}, \tag{20}$$

then $\bar{u}$ is a supersolution on $B_R^{\infty}(x_0)$ and, by comparison, $u \leq \bar{u}$ on $B_R^{\infty}(x_0)$. Analogously, a local subsolution vanishing at $x = x_0$ may be constructed for such a point x_0 and the conclusion follows.□

Remark 2.22. It is also possible to study the obstacle problem (or general multivalued equations) associated to equation (8). (See e.g. Diaz-Herrero [2]).□

As we have already pointed out, very often the nonisotropic character of the equation is due to the presence of first order (or convection) terms. In the rest of this subsection we shall illustrate how an adequate *balance between the diffusion and convection* terms may be the reason of the occurrence of the free boundary $\mathcal{F}(u)$. To explain this we consider the following model problem

$$-\sum_{i=1}^{N} \frac{\partial}{\partial x_i} a_i\left(\frac{\partial u}{\partial x_i}\right) + \sum_{i=1}^{N} \frac{\partial}{\partial x_i} b_i(u) = g(x) \quad \text{in} \quad \Omega \tag{21}$$

$$u = h \quad \text{on} \quad \partial\Omega , \tag{22}$$

where a_i satisfy (10) and b_i are continuous real functions.

The behaviour of solutions of (21),(22) may differ sharply in the directions x_i , $i = 1,\dots,N$, according to the nature of the functions a_i and b_i. In this way, *unidirectional phenomena* may occur, as stated in the following result (we assume, for simplicity, that there is comparison of solutions for (21) ; some sufficient conditions for this are given in Chapter 4).

<u>Proposition 2.41.</u> *Let* Ω *be a bounded domain of* $\mathbb{R}^N$. *Let* $g \in L^\infty(\Omega)$ $h \in W^{1,p}(\Omega) \cap L^\infty(\Omega)$ *be nonnegative functions such that*

$$g = 0 \text{ and } h = 0 \text{ a.e. on } \bar{\Omega} \cap (-\infty,R] \times \mathbb{R}^{N-1} \tag{23}$$

Assume that

$$b_1(s)s \geq 0 \tag{24}$$

and that

$$\int_{0^+} \frac{ds}{a_1^{-1}(b_1(s))} < +\infty . \tag{25}$$

Then, if $u \in W^{1,p}(\Omega) \cap L^\infty(\Omega)$ *is a solution of* (21),(22), *there exists* $\underline{R} \leq R$ *such that*

$$u(x) = 0 \quad \text{a.e.} \quad x \in \bar{\Omega} \cap (-\infty,\underline{R}\,] \times \mathbb{R}^{N-1}. \tag{27}$$

<u>Proof</u>. Due to assumption (25), for $r > 0$ we can define $\psi : [0,\infty) \to [0,\infty)$ by

$$\psi(r) = \int_0^r \frac{ds}{a_1^{-1}(b_1(s))} . \tag{28}$$

Now, take $\eta = \psi^{-1}$. As in Lemma 1.3, we have that for every $\tau \geq 0$ the function $\eta((s-\tau)^+)$ is a solution of the Cauchy problem

$$- \eta' + a_1^{-1}(b_1(\eta)) = 0 \tag{29}$$

$$\eta(0) = 0. \tag{30}$$

Let $\tau_0 > 0$ such that $\eta(\tau_0) = M$, where $M = \| u \|_\infty$ (i.e. $\tau_0 = \Psi(M)$). Then if we define $\Omega_1^R = \{x \in \Omega : x_1 \leq R\}$ and $\overline{u}(x) = \eta((x_1-\tau_0)^+)$, for $x = (x_1,\ldots,x_N) \in \Omega_1^R$, we have that

$$\overline{u} \geq u \quad \text{on} \quad \partial\Omega_1^R$$

and

$$- \sum_{i=1}^N \frac{\partial}{\partial x_i} a_i\left(\frac{\partial \overline{u}}{\partial x_i}\right) + \sum_{i=1}^N \frac{\partial}{\partial x_i} b_i(\overline{u}) = -a_1(\eta')' + b_1(\eta)' = 0 \geq g \quad \text{in } \Omega_1^R$$

Therefore, by the comparison principle $0 \leq u \leq \overline{u}$ on Ω_1^R and so $u(x) = 0$ for a.e. $x \in \Omega_1^R$ such that $x_1 \leq \underline{R}$, $\underline{R} = R - \tau_0$. □

Remark 2.23. It is clear that similar results can be obtained depending on the sign of u and of the function $b_1(s)s$, assumed that (25) (or/and the integral condition at 0^-) is satisfied. So, for instance, if $b(s)s \leq 0$ the conclusion of Proposition 2.41. holds for nonpositive solutions, while for nonnegative solutions we deduce that

$$u(x) = 0 \quad \text{for a.e.} \quad x \in \Omega , \quad x_1 \geq \overline{R}$$

for some $\overline{R} \geq R$, where R is such that

$$g(x)=0 \text{ and } h(x)=0 \text{ for a.e. } x \in \Omega , \; x_1 \geq R. \; \square$$

A very special consequence of Proposition 2.40 and the above remark is the following.

Corollary 2.42. *Let* Ω *be a bounded domain of* $\mathbb{R}^2$. *Let* $g \in L^\infty(\Omega)$ *and* $h \in W^{1,p}(\Omega) \cap L^\infty(\Omega)$ *be nonnegative functions such that*

$$g = 0 \quad \text{and } h = 0 \quad \text{a.e. on} \quad \overline{\Omega} \cap (-\infty, R_1] \times (-\infty, R_2]$$

Let $u \in W^{1,p}(\Omega) \cap L^\infty(\Omega)$ *be a nonnegative solution of the problem*

$$-\Delta u + \frac{\partial}{\partial x_1} u^\lambda + \frac{\partial}{\partial x_2} u^\mu = g \quad \text{in} \quad \Omega \tag{31}$$

$$u = h \qquad \text{on} \quad \partial\Omega \tag{32}$$

where λ *and* μ *are assumed such that* $1/2 < \lambda < 1$, $1/2 < \mu < 1$. *Then there exists* R_1 *and* R_2, $R_1 \leqslant R_1$ *and* $\underline{R}_2 \leqslant R_2$, *such that*

$$u(x) = 0 \quad \text{a.e. on} \quad \overline{\Omega} \cap (-\infty, R_1] \times (-\infty, R_2]\ . \square \tag{33}$$

Remark 2.24. It is clear that the same arguments can be applied to equations more general than (21):for instance these with x-dependent terms, involving monotone absorption terms, and so on. On the other hand, the regularity of g and h may also be weakened. We also note that, again, the existence of the boundary $\mathcal{F}(u)$ is obtained from a balance between the nonlinearities (hypotheses (25)) as well as a balance between the sizes of u and the null set of the sata $N(g) \cup N(h)$ (which is implicitly included in (27) or (33)). As in some other cases, condition (25) is also necessary for the existence of $\mathcal{F}(u)$. $\square$

2.4b. Fully nonlinear equations. Optimal strategy for the Hamilton-Jacobi-Bellman equation.

In this subsection we study second order, nonlinear equations of the general form

$$F(x, u, Du, D^2u) = 0 \qquad \text{in} \qquad \Omega\ , \tag{34}$$

$$u = h \qquad \text{on} \qquad \partial\Omega\ , \tag{35}$$

where F is a real function on the set $\Omega \times \mathbb{R} \times \mathbb{R}^N \times \mathbb{R}^{N\times N}$, where $\mathbb{R}^{N\times N}$ denotes the $N(N+1)/2$ dimensional space of real $N \times N$ symmetric matrices. Here Du and D^2u represent the gradient and the Hessian of u.

In studying the formation of the free boundary $F(u)$ for solutions of (34),(35), the notion of ellipticity will play an important role. More precisely, according to the considerations made at the Introduction, it seems natural that the existence of $F(u)$ should be related to the loss of the ellipticity of the equation. In other words, we shall be able to find such a free boundary for degenerate elliptic equations.

When $F(x,r,p_i,z_{ij})$ is differentiable with respect to z_{ij} the operator F is called elliptic in a subset U, $U \subset \Omega \times \mathbb{R} \times \mathbb{R}^N \times \mathbb{R}^{N\times N}$, if the matrix $[F_{ij}(\gamma)]$ given by

$$F_{ij}(\gamma) = \frac{\partial F}{\partial z_{ij}}(\gamma) \quad , \quad i,j = 1,\dots,N$$

is positive for $\gamma = (x,r,p_i,z_{ij}) \in U$. If we let $\lambda(\gamma)$, $\Lambda(\lambda)$ denote, respectively, the minimum and maximum eigenvalues of $[F_{ij}(\gamma)]$, we call F uniformly elliptic (resp. strictly elliptic) in U, if Λ/λ (resp. $1/\lambda$) is bounded in U, equivalently, if

$$\lambda|\xi|^2 \leqslant F_{ij}\, \xi_i \xi_j \leqslant \Lambda|\xi| \tag{36}$$

with $\Lambda/\lambda \leqslant \mu$ (resp. $1/\lambda \leqslant \mu$) for some $\mu \in (0,\infty)$ (resp. $1/\lambda \leqslant \mu$) for all $\xi \in \mathbb{R}^N$, $(x,r,p_i z_{ij}) \in U$. In (36) and throughout this subsection we adopt the standard summation convention in which repeated indices indicate summation from 1 to N. Finally the operator F is called degenerate elliptic in U if $[F_{ij}(\gamma)]$ is a semidefinite positive matrix, i.e., $\lambda(\gamma) \geqslant 0$, for all $\gamma \in U$.

Motivated by some important examples, it is useful to extend the above notions to the case in which F is not differentiable but, for instance, merely Lipschitz continuous with respect to z_{ij}. In this last case $[F_{ij}]$ exists for almost all $z_{ij} \in \mathbb{R}^{N\times N}$ and the definitions may be understood by replacing the expression "for all $\gamma \in U$" by "for $\gamma \in U$ where $[F_{ij}(\gamma)]$ exists." For a general F not differentiable, the ellipticity notions are expressed in the following terms : F is elliptic (resp.

degenerate elliptic) in U if

$$F(x,r,p_i,z_{ij} + m_{ij}) - F(x,r,p_i,z_{ij}) > 0 \quad (\text{resp.} \geq 0) \tag{37}$$

on U, where $m_{ij} \in \mathbb{R}^{N\times N}$, $m_{ij} \geq 0$. Analogously, F is uniformly elliptic (resp. strictly elliptic) in U if there exists a constant μ and positive functions λ, Λ on U such that

$$\lambda(x)\mathrm{tr}\, m_{ij} \leq F(x,r,p_i,z_{ij}+m_{ij}) - F(x,r,p_i,z_{ij}) \leq \Lambda(x)\mathrm{tr}\, m_{ij} \tag{38}$$

with $\Lambda/\lambda \leq \mu$ (resp. $1/\lambda \leq \mu$), $m_{ij} \geq 0$ and $\mathrm{tr}\, m_{ij}$ the trace of m_{ij}. (Note that the uniform ellipticity of F on U implies that F is Lipschitz continuous with respect to z_{ij}).

It is clear that the class of quasilinear equations can be reformulated in terms of equation (34). Among the relevant particular cases of equation (34) that are not in quasilinear form we mention the Monge-Ampere equation

$$\det D^2u - f(x,u,Du) = 0 \tag{39}$$

and the Hamilton-Jacobi-Bellman equation

$$\inf_{\nu\in V} \{ A_\nu u(x) + g_\nu(x) \} = 0 , \tag{40}$$

where A_ν denotes a family of quasilinear partial differential operators depending on a parameter ν which belongs to a set V. About both equations we only comment, for the moment, that (39) appears in Differential Geometry (T.Aubin [1]) and that if, for instance, $U = \{u\}$, with $u \in C^2(\Omega)$, then (39) is elliptic only for uniformly convex functions u. With respect to equation (40) we point out that it appears in the study of optimal cost in stochastic control problems (Krilov [1]), being there A_ν linear operators. (The quasilinear choice of A_ν is, essentially, motivated the work of Trudinger [1]). The elliptic character of the operator given in (40) depends on the nature of A_ν. For instance, if A_ν are linear operators like

$$A_\nu u = L_\nu u = a_\nu^{ij}(x)\, D_{ij}u + b_\nu^i(x) D_i u + c_\nu(x)u$$

then the operator F given by (40) is Lipschitz continuous with respect to z_{ij}, and thus it is elliptic (resp. uniformly elliptic) if

$$\lambda(x)|\xi|^2 \leqslant a_\nu^{ij}(x)\xi_i\xi_j \leqslant \Lambda(x)|\xi|^2$$

with $\lambda,\Lambda > 0$ (resp. $\Lambda/\lambda \in L^\infty(\Omega)$). Finally, we also recall that another particular non quasilinear equation, associated to some nonlinear parabolic diffusion problems (Benilan-Ha [1], G.Diaz-J.I.Diaz [1]) is given by

$$\phi(-\Delta u) + f(x,u,Du) = 0 \tag{41}$$

where ϕ is a continuous nondecreasing real function. Note that the operator given by (41) is, in general, degenerate elliptic in $\mathcal{U} = \mathbb{R} \times \mathbb{R}^N \times \mathbb{R}^{N\times N}$ and that, in fact it is not uniformly elliptic if, for instance, $\phi \in C^1(\mathbb{R})$ and $\phi'(0) = 0$.

Existence, uniqueness and regularity theorems for the problem (34),(35) are given by different authors (see for instance, the exposition given in Gilbarg-Trudinger[1]). Here, for the sake of simplicity of notation, we shall only deal with classical solutions $u \in C^2(\Omega) \cap C^0(\overline{\Omega})$ of this problem; however, a weaker notion of solution u with $u \in W^{2,N}_{loc}(\Omega)$ may also be considered for our purposes. On the other hand, some comparison results are summarized in Chapter 4.

After this long preamble we turn to the study of the free boundary $\mathcal{F}(u)$. To do this we shall give first a direct consequence of the following comparison result (see Remark 4.61) which allows us (as in Section 2.1) to infer many results directly from the corresponding theorems for quasilinear equations.

Proposition 2.43. *For* $k = 1,2$, *let* $F_k \in C^\circ(\Omega \times \mathbb{R} \times \mathbb{R}^N \times \mathbb{R}^{N\times N})$ *and* $u_k \in C^2(\Omega) \cap C^\circ(\overline{\Omega})$ *be such that*

$$F_1[u_1] \geqslant 0 \geqslant F_2[u_2] \tag{42}$$

where $F_k[u] = F_k(x,u,Du,D^2u)$ *for every* $u \in C^2(\Omega)$. *Assume that* F_2 *is an elliptic operator (eventually degenerated) such that*

$$F_2(x,r,p_i,z_{ij}) > F_2(x,r+s,p_i,z_{ij}) \tag{43}$$

for every $s \geq 0$. *Finally, suppose that*

$$F_1[u_1] \leq F_2[u_1] \tag{44}$$

Then $u_1 \leq u_2$ *on* $\partial\Omega$ *implies* $u_1 \leq u_2$ *in* Ω . □

Remark 2.25. We notice that assumption (43) may be replaced by other suitable hypotheses on the dependence of F_2 with respect to z_{ij}. On the other hand, it is clear that (as in Theorem 2.1) the condition (44) is usually obtained through structural assumptions on F_1 and F_2 , verified at least on a subset $U \subset \Omega \times \mathbb{R} \times \mathbb{R}^N \times \mathbb{R}^{N\times N}$, with $(x,u_i,Du_i,D^2u_i) \in U$ for $i = 1$ and 2. □

The following particular example shows how the existence of the free boundary $F(u)$ can be derived from the results of Chapter 1. For the sake simplicity of the notation we shall rewrite equation (34) in terms of

$$\tilde{F}(x,u,Du,D^2u) + g(x) = 0 \quad \text{in} \quad \Omega \tag{45}$$

being $g(x) = F(x,0,0,0)$ and $\tilde{F} = F - g$.

Theorem 2.44. *Let* $u \in C^2(\Omega) \cap C^\circ(\overline{\Omega})$, $u \geq 0$ *in* Ω , *satisfying (34), (35) and suppose that*

$$\tilde{F}(x,u,Du,D^2u) \leq - \phi(-\Delta_p u) - f(u) \quad \text{in} \quad \Omega \tag{46}$$

for some $p > 1$ *and some continuous functions* ϕ *and* f *such that*

ϕ *is odd, strictly increasing function and* $\phi(0) = 0$, (47)

f *is nondecreasing and* $f(0) = 0$. (48)

Assume that

$$\int_{0^+} \frac{ds}{\Phi(s)^{1/p}} < +\infty \quad , \tag{49}$$

where $\Phi(s) = \int_0^t \phi^{-1}(f(t))dt$. *Then we have the following estimate for the null set* $N(u)$.

$$N(u) \supset \{x \in N(g) \; : \; d(x,S(g)) \geq L\} \tag{50}$$

for some $L > 0$.

Proof. If the function ϕ in (46) is the identity, the conclusion follows directly from Proposition 2.43 (applied to $F_1 = F$ and $F_2[u]=\Delta_p u-f(u)+g(x)$) and Theorem 1.9. In the general case, let $x \in N(g)$ and $R = d(x_0,S(g))$. By hypotheses (47) ,(48) and (49) we can apply Theorem 1.5 to find a radially symmetric function $\overline{u}(x) = \eta(|x-x_0|)$, verifying $\overline{u}(x_0) = 0$ and

$$\phi(-\Delta_p\overline{u}) + f(\overline{u}) \geq 0 \qquad \text{in} \quad B_R(x_0).$$

Moreover, choosing R large enough h (recall the proof of Theorem 1.5), $\overline{u}$ satisfies $\overline{u} \geq M$ with M any bound of $\sup|u|$ on any compact subset K of $N(g)$. Then, applying again Proposition 2.43 on the subset $\tilde{\Omega} = \Omega \cap B(x_0,R)$, to $F_1 = \tilde{F}$, $F_2[v] = -\phi(-\Delta_p v) - f(v)$ $\forall\, v \in C^2(\tilde{\Omega})$, $u_1 = u$ and $u_2 = \overline{u}$, we conclude that $u(x) \leq \overline{u}(x)$ in $\tilde{\Omega}$ and the estimate (50) follows. □

Remark 2.25. Obviously, other choices of the operator in the right-hand side of (46) lead, in an analogous way, to similar results. This is useful, for instance, when $F(x,u,Du,D^2u)$ depends adequately on Du (recall the comments in the introduction of this section as well as Proposition 2.41). The obstacle problem associated to the operator $F[u]$ may also be treated in the same way (see G.Diaz [2]for a result in this direction). On the other hand, we note that Theorem 2.44 may be applied to Hamilton-Jacobi-Bellman equations of the form

$$\inf_{\nu\in V} \{ L_\nu u(x) - u^q + g_\nu(x)\} = 0 \tag{51}$$

where, for instance, $L_\nu u$ is a second order linear operator like the one given in (64) of Section 2.1 and $0 < q < 1$ (using in this occasion Theorem 1.13). Finally, we point out that the study of the free boundary $\mathcal{F}(u)$ for solutions of the Monge-Ampere equation (39) is carried out in G.Diaz-J.I.Diaz[1] by using similar ideas but with much more involved techniques.□

Remark 2.27. The optimality of assumption (49) is examined in G.Diaz [4], where a strong maximum principle and other qualitative properties are established for solutions of equation (34).□

Now we shall center our attention in the Hamilton-Jacobi-Bellman equation (40) . More precisely, we shall restrict ourselves to the case in which $A_\nu u$ are second order linear operators with constant coefficients and $\nu \in V = \{1,2,...,n\}$. The problem under consideration can be stated as

$$\sup_{1\leq m\leq n} \{ - L_m u(x) - g_m(x)\} = 0 \qquad \text{a.e.} \quad \text{in} \quad \Omega \tag{52}$$

$$u = 0 \qquad \text{on} \quad \partial\Omega , \tag{53}$$

with

$$L_m v = \sum_{i,j=1}^{N} a_{ij}^m \frac{\partial^2 v}{\partial x_i \partial x_j} + \sum_{i=1}^{N} b_i^m \frac{\partial v}{\partial x_i} - c^m v , \tag{54}$$

where we assume that for every $1 \leq m \leq n$ we have

$$|a_{ij}^m| \leq C_0 , \quad |b_i^m| \leq C_0 \quad \text{and} \quad c^m \geq \alpha > 0. \tag{55}$$

Related to equation (52) the following question naturally arises: for a specific k , when does the equality

$$\sup_{1\leq m\leq n} \{-L_m u(x) - g_m(x)\} = -L_k u(x) - g_k(x) \tag{56}$$

hold on a subset of Ω ?.This question has an important interpretation in term of stochastical control and was answered in Friedman-Lions [1]. Note also that, again, this is related to the existence of the free boundary $F(-L_k u - g_k)$ defined by the boundary of the set where $-L_k u = g_k$.

Before explaining the results of the work mentioned above we recall that the existence and uniqueness of weak solutions of (52),(53) were given in Evans-Friedman [1],Evans-Lions [1] and P.L. Lions [3] under ellipticity hypothesis of the type

$$\sum_{i,j=1}^{N} a_{ij}^{m} \xi_i \xi_j \geq \lambda|\xi|^2 \quad , \quad \text{for every} \quad 1 \leq m \leq n \tag{57}$$

for some constant $\lambda > 0$. Finally, the existence and uniqueness of classical solutions was obtained in Evans [3](see also Trudinger [1]) for $g_m \in C^2(\Omega)$.

A sufficient condition to have the equality in (56) is the following

<u>Theorem 2.45.</u> *Assume that* (57) *holds. For* $\mu \geq 0$ *given, suppose that there exists a subset* $N_\mu^k \subset \Omega$ *where*

$$- L_k g_m(x) + L_m g_k(x) \geq \mu > 0 \quad \textit{for all} \quad m \neq k \tag{58}$$

Then, if u *is the solution of* (52),(53) *, we have that*

$$0 = \sup_{1 \leq m \leq n} \{- L_m u(x) - g_m(x)\} = - L_k u(x) - g_k(x) \tag{59}$$

for every $x \in N_\mu^k$ *such that*

$$d(x, \partial N_\mu^k) \geq \left[\frac{2M_k(\Lambda + C_0 N)}{\mu}\right]^{1/2} , \tag{60}$$

where $M_k = \|- L_k u + g_k\|_{L^\infty(\Omega)}$ *and* $\Lambda = \sup_m \left(\sum_{i=1}^{N} a_{ii}^m\right)$,(i.e. $\Lambda \leq NC_0$).

<u>Proof</u>. Without loss of generality we may take $k = 1$ and $g_1 = 0$, (if $g_1 \neq 0$ we consider $u - u_1$, where $- Lu_1 = g_1$ in Ω). So, by (59)

$$- L_1 g_m(x) \geqslant \mu > 0 \quad \text{in} \quad N^1_\mu \ , \text{ for } 1 < m \leqslant n. \tag{61}$$

Suppose that there exists $x_0 \in N^1_\mu$ with $d(x_0, \partial N^1_\mu) \geqslant R$ (R to be determined) where

$$- L_1 u(x_0) < 0 \ , \tag{62}$$

and we wish to derive a contradiction. First of all, we recall a useful approximation argument: for each $0 < \varepsilon < 1$, we consider a convex function $\phi_\varepsilon \in C^\infty$ such that

$$\phi_\varepsilon(t) = 0 \text{ if } t \leqslant 0 \ , \ \phi'_\varepsilon(t) = 1 \text{ if } t \geqslant \varepsilon \quad \text{and} \quad 0 \leqslant \phi'_\varepsilon(r) \leqslant 1.$$

Let $F_\varepsilon(t_1, t_2) = t_1 + \phi_\varepsilon(t_2 - t_1)$ and

$$F^n_\varepsilon(t_1, \ldots, t_n) = F_\varepsilon(t_1, F^{n-1}_\varepsilon(t_2, \ldots, t_n)).$$

Obviously , F^k_ε is convex, increasing in t_i and satisfies

$$0 \leqslant \frac{\partial F^k_\varepsilon}{\partial t_i} \leqslant 1 \quad , \quad \sum \frac{\partial F^k_\varepsilon}{\partial t_i} = 1$$

Moreover , if u_ε is the classic solution of

$$\begin{aligned} &F^n_\varepsilon(-L_1 u^\varepsilon, -L_2 u^\varepsilon - g_2, \ldots, -L_n u^\varepsilon - g_n) = 0 && \text{in} \quad \Omega \\ &u_\varepsilon = 0 && \text{on} \quad \partial\Omega \end{aligned} \tag{63}$$

then it is known (Evans[3]) that

$$u^\varepsilon \to u \quad \text{and} \quad L_1 u^\varepsilon \to L_1 u \tag{64}$$

uniformly in compact subsets. Now, from (62) and (64) we get

$$- L_1 u^\varepsilon(x_0) \leqslant - \gamma < 0 \tag{65}$$

for some small ε and for some γ independent of ε. Let us define the auxiliary function

$$z^{\varepsilon}(x) = - L_1 u^{\varepsilon}(x) + C|x - x_0|^2 \tag{66}$$

where $x, x_0 \in N^1_\mu$ and C is a positive constant to be chosen. As $u^{\varepsilon} \in C^2(\Omega)$ we can take R large enough so that $z^{\varepsilon} \geqslant 0$ on $\partial B_R(x_0)$. More precisely by (64) it is clear that there exists a positive constant K_1 (independent of ε) such that

$$\| - L_1 u^{\varepsilon}\|_{L^\infty(\Omega)} \leqslant M_1. \tag{67}$$

Then, it is enough to take

$$R \geqslant \left(\frac{M_1}{C}\right)^{1/2}. \tag{68}$$

Let $y_0 \in \overline{B}_R(x_0)$ be such that

$$z^{\varepsilon}(y_0) = \min_{B_R(x_0)} z^{\varepsilon}(x)$$

Since $z(x_0) < 0$, also $z(y_0) < 0$, and therefore $y_0 \in B_R(x_0)$ and

$$- L_1 u^{\varepsilon}(y_0) \leqslant - L_1 u^{\varepsilon}(x_0) \leqslant - \gamma . \tag{69}$$

Rewriting equation (63) as

$$0 = -L_1 u^{\varepsilon} + \phi_{\varepsilon}(F^{n-1}(-L_2 u^{\varepsilon} - g_2, \ldots, -L_n u^{\varepsilon} - g_n)) + L_1 u^{\varepsilon})$$

and choosing $\varepsilon < \gamma$, as $\phi_{\varepsilon}(s) \leqslant s$, we have

$$F^{n-1}_{\varepsilon}(-L_2 u^{\varepsilon} - g_2, \ldots, -L_n u^{\varepsilon} - g_n) + L_1 u^{\varepsilon} \geqslant \varepsilon \quad \text{at } y_0.$$

But noting that

$$\frac{\partial F^n_{\varepsilon}}{\partial t_1} = 1 - \phi'_{\varepsilon}(F^{n-1}_{\varepsilon}(t_2, \ldots, t_n) - t_1) = 0 \quad \text{if} \quad F^{n-1}_{\varepsilon}(t_2, \ldots, t_n) - t_1 \geqslant \varepsilon ,$$

it follows that

$$\frac{\partial F_\varepsilon^n}{\partial t_1}(-L_1u^\varepsilon(y_0), F_\varepsilon^{n-1}(-L_2u^\varepsilon(y_0) - g_2(y_0), \ldots)) = 0. \tag{70}$$

On the other hand, applying $-L_1$ to (63) we obtain

$$\sum_{m=1}^{n} \frac{\partial F_\varepsilon^n}{\partial t_m}(\cdot)\,[-L_m(-L_1u^\varepsilon) + L_1g_m] \geqslant 0, \tag{71}$$

and so, using (70), we have that

$$\sum_{m=2}^{n} \frac{\partial F_\varepsilon^n}{\partial t_m}(\cdot)\,[-L_mz^\varepsilon - G_m] \geqslant 0 \quad \text{at} \quad y_0\,, \tag{72}$$

where

$$G_m(x) = -L_m(C|x - x_0|^2) - L_1g_m(x)\;,\; x \in B_R(x_0)\;. \tag{73}$$

But since y_0 is a minimum point of z^ε, this implies that

$$-L_mz^\varepsilon(y_0) \leqslant c^mz^\varepsilon(y_0) < 0\;.$$

Them, if we take $C > 0$ such that

$$G_m(y_0) > 0, \tag{74}$$

the left-hand side of (72) is strictly negative and we arrive at a contradiction. To assure (74) we note that $\Sigma a_{ij}\xi_i\xi_j \leqslant \Sigma a_{ii}|\xi|^2$ and then

$$-L_m(C|x - x_0|^2) \geqslant -2C\Lambda - 2C\,C_0N + \alpha C|x - x_0|^2.$$

Then, using the hypothesis (61), (74) holds if C is such that

$$C < \frac{\mu}{2(\Lambda + C_0N)}\;. \tag{75}$$

Replacing this C in R we obtain the estimate (60). □

Remark 2.28. To motivate the condition (59), note that if the assertion of Theorem 2.45 holds and if $x \in N_\mu^k$ and satisfies (60), then the function $v_m = - L_m u - g_m$ satisfies

$$- L_k v_m = - L_k(-L_m u) + L_k g_m = - L_m(-L_k u) + L_k g_m \qquad (76)$$

Since also $v_m \leqslant 0$, the right-hand side of (76) cannot be "too positive". In Theorem 2.45 we assume that this right-hand side is uniformly negative.□

Remark 2.29. Theorem 2.45 is still true for $\Omega = \mathbb{R}^N$, the set of parameters $V = \mathbb{N}$ and conditions of ellipticity more general than (58) (Friedman-Lions [1]) On the other hand, the bound M_k can be estimated only in terms of C_0, α and the norm $W^{2,\infty}$ of g_m (see e.g. G.Diaz [3]).□

Remark 2.30. For two operators, Theorem 2.45 asserts that

$$\text{if } \; -L_1 g_2 + L_2 g_1 \geqslant \mu > 0 \; \text{ in } \; N_\mu^1 \text{, then } \; -L_1 u = g_1 \text{ on a subset of } N_\mu^1 .$$

In the special case of $- L_1 w \equiv w$ and $g_1 \equiv 0$ this gives the same estimate obtained in Section 2.2 for the obstacle problem.

$$u \leqslant 0 \quad , \quad - L_2 u - g_2 \leqslant 0 \qquad u(-L_2 u - g_2) = 0 \quad \text{a.e. in } \Omega .$$

Note that in Section 2.2 the proof of such estimates extends to L_2 with variable coefficients.□

2.5. OTHER BOUNDARY-VALUE PROBLEMS.

Most of the results in the above Sections remain true if we replace the Dirichlet condition by other different boundary conditions. To illustrate this, and for the sake of simplicity in the notation, we shall consider such semilinear problems as the following

$$- \Delta u + f(u) = g(x) \qquad \text{in} \qquad \Omega \tag{1}$$

$$\frac{\partial u}{\partial n} + b(x,u) = e(x) \qquad \text{on} \qquad \partial\Omega \tag{2}$$

where $\vec{n}$ is the unit outward vector to $\partial\Omega$. The existence and properties of the free boundary $\mathcal{F}(u)$ for solutions of (1),(2) under the fundamental hypothesis on f

$$\int_{0^+} \frac{ds}{F(s)^{1/2}} < +\infty \quad , \quad (F(t) = \int_0^t f(s)ds) \tag{3}$$

will be discussed in subsection 2.5a. Another natural question is related to the behaviour of solutions of problems like (1),(2) when (3) is not necessarily satisfied. It turns out that, even in this case, a free boundary may occur, now defined by $\mathcal{F}(u|_{\partial\Omega})$, i.e. associated to the trace of u on $\partial\Omega$. This peculiar behaviour depends on the nature of the nonlinear boundary term $b(x,u)$ and appears when, in fact, $b(x,u)$ is a multivalued function, i.e, a maximal monotone graph of $\mathbb{R}^2$, for x fixed. This is the case of the thin (boundary) obstacle problem, also called the Signorini problem, which may be stated in the following terms: given $\psi \in H^{1/2}(\partial\Omega)$, and the convex set $K = \{v \in H^1(\Omega) : v \geq \psi \text{ in } \partial\Omega\}$, find $u \in K$ minimizing on K the functional

$$J(u) = \frac{1}{2} \int_\Omega |\nabla u|^2 dx + \frac{c}{2} \int_\Omega u^2 dx - \int_\Omega g\, u\, dx + \int_{\partial\Omega} e\, u\, d\sigma$$

where $c \geq 0$, $g \in H^{-1}(\Omega)$ and $e \in H^{-1/2}(\partial\Omega)$. Under regularity assumptions (e.g. $g \in L^2(\Omega)$, $\psi \in H^2(\Omega)$ and $e \in H^1(\Omega)$) it is well known (Brezis[5]) that the solution u is characterized by, $u \in H^2(\Omega)$ and

$$- \Delta u + c\, u = g \qquad \text{in} \qquad \Omega \tag{4}$$

$$\frac{\partial u}{\partial n} + \beta(u - \psi) \ni e \quad \text{on} \quad \partial\Omega , \tag{5}$$

where β is the maximal monotone graph of $\mathbb{R}^2$ given by

$$\beta(r) = 0 \text{ if } r > 0,\ \beta(0) = (-\infty, 0] \text{ and } \beta(r) = \emptyset \text{ (the empty set) if } r < 0. \tag{6}$$

This problem is considered in subsection 2.5b, where necessary and sufficient conditions are given in order to assure the formation of the coincidence set $N([u - \psi]|_{\partial\Omega}) = \{x \in \partial\Omega : u = \psi\}$.

2.5a. Nonlinear equations with other boundary conditions.

We consider the problem (1),(2) under the structural assumptions:

f is a continuous nondecreasing function, $f(0) = 0$,

and $b(x,u)$ is measurable in x and

$$b(x,u) \text{ is a continuous nondecreasing function, for } x \in \partial\Omega \text{ fixed}, b(x,0)=0. \tag{7}$$

Existence, comparison and uniqueness results for (1),(2) are commented in Chapter 4. Here we shall deal only with the study of the free boundary $F(u)$. First of all, we remark that "interior estimates" on the set $N(g)$ may be obtained as in Theorem 1.9, assumed that condition (3) holds. More precisely, we have that if $M \geq \|u\|_{L^\infty(K)}$, for any compact set K, $K \subset\subset N(g)$, then

$$N(u) \supset \{x \in N(g) : d(x, S(g)) \geq L\} \tag{8}$$

with $L = \psi_{1/N}(M)$, ($\psi_{1/N}$ given by (33) of Section 1.1). We point out that such a quantity M may be explained in many different ways under the above circunstances (see e.g. Ladyzenskaya-Ural'tseva [1], Theorem 2.1 Chap. 10 , Gilbarg-Trudinger [1], Section 17.9 , and Brezis [5]).

To improve estimate (8) on points near $\partial\Omega$, we introduce the following notation: given a smooth curve Γ in $\mathbb{R}^N$ and $x_0 \in \mathbb{R}^N$, we define

$$\theta(x_0, \Gamma) = \inf \{ \cos(\vec{n}(x) , \overrightarrow{x - x_0}) : x \in \Gamma \} , \tag{9}$$

where $(\vec{n(x)},\overrightarrow{x - x_0})$ represents the angle between the two vectors. It is clear that the value of $\theta(x_0,\Gamma)$ depends essentially on the "geometry" of Γ. For instance, if $\Gamma = \partial\Omega$ and Ω is an open convex bounded set, it is easy to see that $\theta(x_0,\Gamma) > 0$ for $x_0 \in \Omega$. We have

Theorem 2.46 . *Let* $u \in H^2(\Omega) \cap L^\infty(\Omega)$ *solution of (1), (2) with* $e \in L^\infty(\Omega)$. *Assume (3) and let* Ω, *and* g *be such that*

$$\theta(x_0,\partial\Omega \cap \overline{N(g)}) \geqslant 0 \qquad \forall x_0 \in N(g). \tag{10}$$

Let $\Gamma = \partial N(g) \cap S(e|_{\partial\Omega})$. *Then*

$$N(u) \supset \{x \in N(g) : d(x,S(g)) \geqslant L_1 \quad \text{and} \quad d(x,\Gamma) \geqslant L_2(x)\} \tag{11}$$

where $L_1 = \psi_{1/N}(M)$, $M \geqslant \|u\|_\infty$, *and*

$$L_2(x) = \psi_{1/N}\left[F^{-1}\left(\frac{N\|e\|_\infty^2}{2\,\theta(x,\Gamma)}\right)\right] \tag{12}$$

Proof.- By the comparison results it is enough to construct a super and a subsolution vanishing at x_0, where x_0 belongs to the set given by the right hand side of (11). More precisely, for $x_0 \in N(g)$, let $R = \min\{d(x_0,S(g)), d(x_0,\Gamma)\}$ and consider $\tilde{\Omega} = N(g) \cap B_R(x_0)$. Then, by comparison, we deduce that $u \leqslant \overline{u}$ on $\tilde{\Omega}$, for any $\overline{u}(x)$ satisfying

$$-\Delta\overline{u} + f(\overline{u}) \geqslant 0 \qquad \text{in} \quad \tilde{\Omega} \tag{13}$$

$$\overline{u} \geqslant M \qquad \text{in} \quad \partial\tilde{\Omega} - \partial\Omega \tag{14}$$

$$\frac{\partial\overline{u}}{\partial n} \geqslant \|e\|_{L^\infty(\partial\Omega)} \qquad \text{on} \quad \Gamma \tag{15}$$

$$\frac{\partial\overline{u}}{\partial n} \geqslant 0 \qquad \text{on} \quad \partial\tilde{\Omega} \cap (\partial\Omega - \Gamma). \tag{16}$$

Defining $\overline{u}(x) = \eta(|x - x_0| : 1/N)$, with $\eta(r : 1/N) = \psi_{1/N}^{-1}(r)$, we know that (13) holds and (14) is satisfied if $d(x_0,S(g)) \geqslant L_1$. On the other hand, recalling that

$$\eta'(r) = (\tfrac{2}{N})^{1/2}\, F(\eta(r))^{1/2}$$

(see (38) of Section 1.1) then

$$\frac{\partial \bar{u}}{\partial n}(x) = \sum_{i=1}^{N} \frac{\partial \bar{u}}{\partial x_i}\, n_i(x) = (\tfrac{2}{N})^{1/2} F(\eta(|x-x_0|))^{1/2} \cos(\vec{n}(x), \overrightarrow{x-x_0}) \geq$$

$$\geq (\tfrac{2}{N})^{1/2}\, F(\eta(|x-x_0|))^{1/2} \cdot \theta(x_0, \partial\Omega \cap \partial N(g)).$$

Thus, (16) is derived from assumption (10), and (15) is verified if $d(x_0,\Gamma) \geq L_2(x_0)$. The local subsolution is constructed by analogy. □

In some cases the condition $d(x,\Gamma) \geq L_2(x)$ in (11) may be substituted by another one, easier to verify. This can be shown by means of the strong maximum principle. So, we shall not need any hypothesis on the geometry of $\partial\Omega$ (or Γ) but only to suppose that $\partial\Omega$ satisfies an interior sphere condition.

Theorem 2.47 . *Let* u *be solution of (1),(2), with* $g = 0$ *on* Ω *and* $e(x) \geq 0$ *on* $\partial\Omega$, $e(x) \in C^\circ(\partial\Omega)$. *Assume that* $b(x,u) = b(u)$ *and that* b *is strictly increasing. Then* $0 \leq u(x) < b^{-1}(\|e\|_\infty)$ *in* Ω *and* $\frac{\partial u}{\partial n} > e - \|e\|_\infty$ *in* $\partial\Omega$. *In consequence, if (3) holds, in the estimate (11) we can substitute* $L_2(x)$ *by* L_2 *with*

$$L_2 = \psi_{1/N}(b^{-1}(\|e\|_\infty)). \tag{17}$$

Proof. From the comparison results we deduce that $u \geq 0$ and so that $\Delta u \geq 0$ in Ω. Assume that $u \in C^\circ(\bar{\Omega})$ (otherwise we argue by approximation). If u takes a maximum in $\bar{\Omega}$ at x_0 and $u(x_0) \geq b^{-1}(\|e\|_\infty)$, then by the strong maximum principle (see Gilberg-Trudinger [1]or the comments at the end of subsection 1.2a) , we deduce that $x_0 \in \partial\Omega$ and $\frac{\partial u}{\partial n}(x_0) > 0$, and so the boundary condition (2) cannot be satisfied. This proves that $u(x) < b^{-1}(\|e\|_\infty)$ and so that $\frac{\partial u}{\partial n} > e - \|e\|_\infty$ in $\partial\Omega$. Finally, the rest of the statement is obtained by remarking that now we can replace condition (15) by $b^{-1}(\|e\|_\infty) \leq \bar{u}(x)$ on Γ. □

Remark 2.32 . The assumptions of the above theorem are fulfilled in the special case of the following third boundary value problem , of great interest in the study of chemical reactions (see Aris [1]):

$$- \Delta u + f(u) = 0 \quad \text{in} \quad \Omega \tag{18}$$

$$\frac{\partial u}{\partial \nu} + \mu(u-1) = 0 \quad \text{on} \quad \partial\Omega , \tag{19}$$

where μ is a given positive number (so, $0 \leqslant u(x) < 1$ on Ω). We point out that a very detailed study of the free boundary $F(u)$ for solutions of this problem was made in Friedman-Phillips [1] when Ω is a convex domain of $\mathbb{R}^2$ and f is as in subsection 2.1a. By imbedding the problem in the family of elliptic problems

$$- \Delta u_\lambda + \lambda f(u_\lambda) = 0 \quad \text{in} \quad \Omega$$

$$\frac{\partial u_\lambda}{\partial n} + \mu\sqrt{\lambda}\,(u_\lambda - 1) = 0 \quad \text{on} \quad \partial\Omega ,$$

sharper estimates than that given above are then obtained in a way similar to that of Theorem 1.54. Moreover, if f satisfies the uniqueness condition (12) of Section 2.1, then it is shown that there exists a $\lambda_* > 0$ such that $N(u_\lambda)$ is a closed convex domain with $C^{1,\beta}$ boundary for any $\lambda > \lambda_*$, $N(u_{\lambda*})$ consists of a single point, and $N(u_\lambda) = \emptyset$ if $\lambda < \lambda_*$.□

Remark 2.33 . From Chapter 1 it is clear that the above theorem holds for equations more general than (1), when in (2) the term $\frac{\partial u}{\partial n}$ is replaced by the corresponding co-normal derivative. So the absorption term f may also depends on x and then the results may be applied to some systems (see Remark 1.8 and Diaz-Hernandez [1]). It is also possible to consider the quasilinear problem

$$- \Delta_p u + f(u) = g(x) \quad \text{in} \quad \Omega \tag{20}$$

$$|\nabla u|^{p-2} \nabla u \cdot n + b(x,u) = e(x) \quad \text{on} \quad \partial\Omega . \tag{21}$$

On the other hand, we remark that if Ω is unbounded, it is also possible to construct global supersolutions with compact support, as in subsection 1.1d, assumed, for instance, that Ω is the complementary of some convex set (Diaz [ ]). No geometrical assumptions on $\partial\Omega$ are needed if $\partial\Omega$ is bounded or $e(x) \equiv 0$ and b satisfies the additional condition $|b(x,r)| \geq C[F(r)]^{1/2}$ if $|r| < \varepsilon$, for some $C > 0$ and $\varepsilon > 0$ (Diaz[2]). □

2.5b. The Signorini problem.

An already classical problem, arising in several different contexts (see Duvaut-Lions[1], Baiocchi- Capelo [1], Kinderlehrer-Stampacchia[2], Friedman [3], etc) consists in the obtaining of $u \in H^1(\Omega)$, solution of the following variational inequality

$$u \in K_\psi = \{v \in H^1(\Omega) : v \geq \psi \text{ on } \partial\Omega\} \tag{22}$$

$$\int_\Omega \nabla u \cdot \nabla(v-u)dx + c\int_\Omega u(v-u)dx \geq \int_\Omega g(v-u)dx + \int_{\partial\Omega} e(v-u)d\sigma \quad \forall v \in K_\psi \tag{23}$$

where $c \geq 0$, $\psi \in H^{1/2}(\partial\Omega)$, $e \in H^{-1/2}(\partial\Omega)$ and $g \in H^{-1}(\Omega)$ are given (note the abuse of the notation in (23)). The existence and uniqueness for this problem are well known, for which some additional assumptions are needed in the semi-coercive case $c \equiv 0$ (for instance if $\Psi \equiv 0$ it is enough to have

$$\int_\Omega g\, dx + \int_{\partial\Omega} e d\sigma > 0 , \tag{24}$$

see the above references). By the regularity results (Brezis [5]), it is known that, under the additional assumptions

$$\psi \in H^{3/2}(\partial\Omega) , \quad e \in H^{1/2}(\partial\Omega) \quad \text{and} \quad g \in L^2(\Omega) , \tag{25}$$

the solution u belongs to $H^2(\Omega)$ and satisfies the complementary formulation

$$- \Delta u + cu = f \qquad \text{in} \qquad \Omega \tag{26}$$

$$\frac{\partial u}{\partial n} + \beta(u-\psi) \ni e \quad \text{on} \quad \partial\Omega \tag{27}$$

where β is the maximal monotone graph of $\mathbb{R}^2$ given by (6) . Let us call $P(f,\psi,e)$ the problem (26),(27) or (22),(23). Here we are interested in finding conditions on the data for which the solutions give rise to a free boundary originated, in this case, from the coincidence set $I_\psi = \{x \in \partial\Omega : u = \psi\}$ (i.e, $I_\psi = N((u-\psi)|_{\partial\Omega})$). For the sake of simplicity, we shall always assume the regularity condition (25).

As a previous remark, we note that the formulation in (26),(27) may be simplified in different ways. For our purposes, the following, trivial, lemma will be useful.

Lemma 2.48. *Let* u *be the solution of* $P(f,\psi,e)$, *and let* u *be such that*

$$-\Delta u_0 + cu_0 = g \quad \text{in} \quad \Omega \tag{28}$$

$$u_0 = \psi \quad \text{on} \quad \partial\Omega \tag{29}$$

Then, the function $\tilde{u}$ *given by* $\tilde{u} = u - u_0$ *satisfies* $P(0,0,\tilde{e})$ *with* $\tilde{e} = e - \frac{\partial u_0}{\partial n}$, *and conversely.* □

Remark 2.34 . The function $\tilde{e}$ can be made explicit by using the Green function associated to (28),(29). Indeed, if we assume that the boundary obstacle ψ in fact is the trace on $\partial\Omega$ of a function $\psi \in H^2(\Omega)$ (for formulation (26),(27)) or $\psi \in H^1(\Omega)$ (for (22),(23)). Then

$$u_0(x) = \psi(x) + \int_\Omega g^*(\xi)\, G(x,\xi)d\xi \quad , \text{with} \quad , g^* = g + \Delta\psi - c\psi$$

where $G(y,x)$ is the Green function for the domain Ω associated to (28) (29) (see e,g, Friedman[2],Stakgold [1],Roach [1]). In particular,

$$\tilde{e}(x) = e(x) - \frac{\partial\psi}{\partial n}(x) - \int_\Omega g^*(\xi)\, \frac{\partial G}{\partial n_x}(x,\xi)d\xi \quad , \quad x \in \partial\Omega. \tag{30}$$

We recall that, by the weak and strong maximum principles, the functions $G(x,\xi)$ and $\frac{\partial G}{\partial n_x}(x,\xi)$ are respectively positive and negative functions in Ω. □

Concerning the existence of the coincidence set I_ψ, we shall start by giving a necessary condition.

Theorem 2.49. *Let us assume* I_ψ *regular,* $|I_\psi| > 0$. *Then* $\tilde{e} \leq 0$ *on* I_ψ.

Proof. By Lemma 2 and the assumption, we can suppose, equivalently that the solution $\tilde{u}$ of $P(0,0,\tilde{e})$ is such that $\tilde{u} = 0$ on I_ψ. From the formulation (22),(23) for $\psi = 0$ (i.e. $K_\psi = K_0$) we deduce that for every $v \in K_0 \cap H^2(\Omega)$ the function $w = v - u \in H^2(\Omega)$ and satisfies

$$- \int_\Omega u\, \Delta w\, dx + c \int_\Omega u\, w\, dx + \int_{\partial\Omega} u \frac{\partial w}{\partial n} d\sigma \geq \int_{\partial\Omega} \tilde{e}\, w\, d\sigma. \tag{31}$$

Now, let $\theta \in C^2(\partial\Omega)$ be such that $\theta \geq 0$ on I_ψ and $\theta = 0$ on $\partial\Omega - I_\psi$, and consider $w_0 \in H^2(\Omega)$ such that $-\Delta w_0 + c w_0 = 0$ in Ω and $w_0 = \theta$ on $\partial\Omega$ (note that $w_0(x) = \int_{\partial\Omega} \theta(\xi) \frac{\partial G}{\partial n_x}(x,\xi) d\xi$). Taking $w = w_0$ in (31) (this is possible because $w_0 + u \in K_0 \cap H^2(\Omega)$), we deduce that

$$0 \geq \int_{\partial\Omega} \tilde{e}\, w_0 d\sigma = \int_{I_\psi} \tilde{e}\, \theta\, d\sigma .$$

Since θ is arbitrary, it follows that $\tilde{e} \leq 0$ on I_ψ. □

In contrast with the above result, the nonnegativity of $\tilde{e}$ on a part Γ of $\partial\Omega$ is not enough to assure that $I_\psi \cap \Gamma \neq \emptyset$.

Counterexample 2.1 . Given $R > 0$, define $\Omega = \{(x,y) \in \mathbb{R}^2 : y > 0, x^2 + y^2 < R^2, y - \sqrt{3}x < 0$ (see Figure 12). Inspired in Shamir [1], define $u(x,y) = \mathrm{Re}(z^{3/2})$, $z = x + iy$, i.e,

$$u(x,y) = \rho^{3/2} \cos \frac{3\theta}{2} \quad , \quad x = \rho \cos \theta\ , \quad y = \rho \operatorname{sen} \theta\ . \tag{32}$$

It is not difficult to check that $u \in H^2(\Omega)$ and $\Delta u = 0$ a.e. on Ω. Moreover, $u \geq 0$ on $\partial\Omega$ and if we take $\Gamma = \{(x,0) : 0 < x < R\}$, we have that $u|_\Gamma > 0$ but $\frac{\partial u}{\partial n} = -\beta(u) + e = e = 0$ on Γ. Note also that taking $\Omega = \{(x,y) : y > 0\}$ the function u given in (32) satisfies $P(0,0,0)$ in Ω. So the equality in the condition $\tilde{e} = 0$ leads to an indetermination because now $u = 0$ on $\{(x,0) : x < 0\}$ and at the same time $u > 0$ on $\{(x,0) : x > 0\}$. □

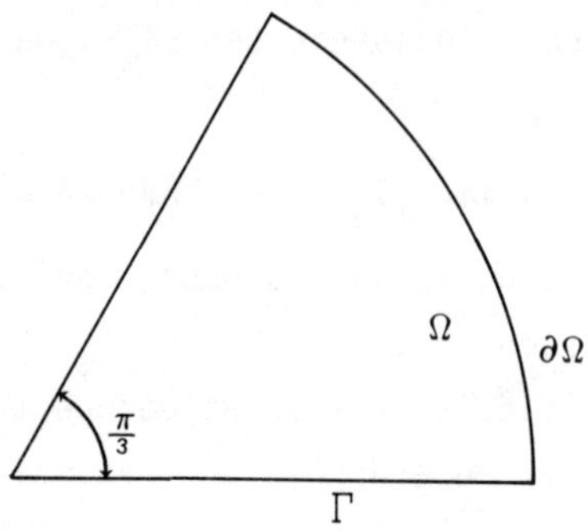

Figure 12

The following result shows that the necessary condition is "almost" sufficient for the formulation of the coincidence set.

Theorem 2.50. *Let* Ω *be a convex open bounded set of* $\mathbb{R}^N$ *and let* $u \in H^2(\Omega) \cap L^\infty(\Omega)$ *solution of* (26),(27). *Assume* ψ, e, *and* g *such that*

$$\tilde{e}(x) \leqslant -\varepsilon \qquad on \qquad \Gamma_\varepsilon \subset \partial\Omega \tag{33}$$

for some $\varepsilon > 0$ ($\tilde{e}$ *given in Lemma 2.48; see* (30)). *Then we have the estimate*

$$I_\psi \supset \{x \in \Gamma_\varepsilon : d(x, \partial\Omega - \Gamma_\varepsilon) \geqslant R\} \tag{34}$$

with R *given by*

$$R = \left[\frac{2MN}{\varepsilon(N-1)H}\right]^{1/2} \quad \text{if } H > 0 \quad \text{and} \quad R = \frac{2MN}{\varepsilon} \quad \text{if } H = 0, \tag{35}$$

where H *is the mean curvature of* $\partial\Omega$ *and* $\|u\|_\infty \leqslant M$.

Proof. Again, by Lemma 2, it suffices to show the estimate (34) for $\tilde{u}$ solution of $P(0,0,\tilde{e})$ (remark that then $I_\psi = \{x \in \partial\Omega : \tilde{u}(x) = 0\}$). Now, let $x_0 \in \Gamma_\varepsilon$, and such that $d(x_0, \partial\Omega - \Gamma_\varepsilon) = R$. Let $\tilde{\Omega} = \Omega \cap B(x_0, R)$ and define $\partial_1\tilde{\Omega} = \partial\tilde{\Omega} \cap \partial\Omega$ and $\partial_2\tilde{\Omega} = \partial\tilde{\Omega} - \partial\Omega$. For $C > 0$, to be chosen later, we shall construct $U \in H^2(\tilde{\Omega})$ such that $U \geqslant 0$ and

$$\Delta U \leqslant -C \quad \text{in } \tilde{\Omega}\text{, and} \quad U = 0 \quad \text{and} \quad \frac{\partial U}{\partial n} = \geqslant \varepsilon \quad \text{in } \partial_1\tilde{\Omega}\ .$$

To do that, Let V_δ^- be a tubular semineighbourhood of Γ_ε defined by the usual parametric representation $x = x(\omega,t) = \omega + t\vec{n}(\omega)$, $\omega \in \Gamma_\varepsilon, t \in (-\delta,0)\,\delta \leqslant R$, where $\vec{n}(\omega)$ is the outward normal unit vector to $\partial\Omega$ at ω, and $\delta > 0$ is such that $V_\delta^- \subset \tilde{\Omega}$. Taking $U(x) = U(t\vec{n}) = \phi(t)$, and recalling the expression of the Laplacian operator on V_δ^+ (see Theorem 1.19), the construction of such a U is reduced to the following problem:

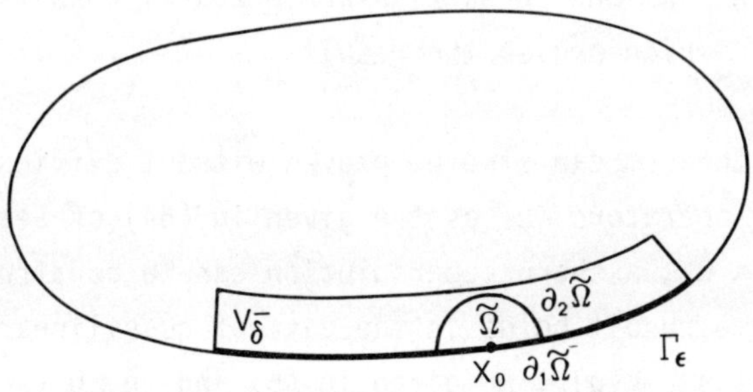

Figure 13

Find $\phi(t) \geqslant 0$ solution of the linear Cauchy Problem

$$\phi''(t) + (N-1)H\phi'(t) \leqslant -C$$

$$\phi(0) = 0 \quad , \quad \phi'(0) \geqslant -\varepsilon\ .$$

Among the multiple choices, we shall take $\phi(t) = -\varepsilon t$ and, then, $C = (N-1)H\varepsilon$, if $H \neq 0$ and $\phi(t) = -\varepsilon t(\frac{t}{R} - 1)$ and $C = \frac{\varepsilon}{R_0}$ if $H = 0$ (note that in fact $\phi(t) = -\varepsilon t(\frac{t}{2R} - 1)$ and $C = \frac{\varepsilon}{2R_0}$ is also avaible for every $H \geqslant 0$).
Now we introduce in $\tilde{\Omega}$ the auxiliary function

$$\bar{u}(x) = U(x) + \frac{C}{2N}|x - x_0|^2,$$

where C is taken as mentioned above. We have

$$-\Delta\bar{u} + c\bar{u} = -\Delta U + cU - C + \frac{cC}{2N}|x - x_0|^2 \geqslant 0 \quad \text{in} \quad \tilde{\Omega}\ .$$

Moreover, on $\partial_1\tilde{\Omega}$ it is clear that $\overline{u} \geq 0$ and that

$$\frac{\partial \overline{u}}{\partial n}(x) = \frac{\partial U}{\partial n}(x) + \frac{C}{N}|x - x_0|\cos(\vec{n}(x),\overrightarrow{x - x_0}) \geq -\varepsilon ,$$

because of the convexity of Ω . On the other hand, in $\partial_2\tilde{\Omega}$,

$$\overline{u}(x) \geq \frac{C}{2N} R^2 \geq M \geq u(x)$$

if $R \geq 2MN/C^{-1}$. In conclusion, by the comparison theorems (see Chapter 4) we deduce that $u \leq \overline{u}$ on $\overline{\tilde{\Omega}}$ and in particular $0 \leq u(x) \leq \frac{C}{2N}|x - x_0|^2$ for $x \in \partial\Omega \cap B(x_0,R)$, which proves the result.□

Remark 2.35 . The above theorem can also be proved without difficulty for general second order linear operators L as the given in (64) of Section 1.1 We also note that if $c > 0$, a sharp supersolution can be constructed (see Remark 2.9). The same result holds in the case of quasilinear problems as (20),(21) with $b(x,u) = \beta(u)$, β given in (6) and $e = 0$ (see Diaz-Jimenez [1]). Note that, in this last case, the simplification made by Lemma 2.48 is not available. Also, easy modifications in the proof of Theorem 2.50 lead to the same kind of result for mixed boundary conditions of the type

$$\frac{\partial u}{\partial n} + \beta(u-\psi) \ni e \quad \text{on} \quad \partial_1\Omega \;, u = h \quad \text{on} \quad \partial_2\Omega \;,\; \partial\Omega = \partial_1\Omega \cup \partial_2\Omega \qquad (36)$$

Finally, the coincidence set for the thin (interior) obstacle problem (Friedman [3]p. 105) may also be considered.□

Remark 2.36 . Results giving estimates on the noncoincidence set $\partial\Omega - I\psi$ can be found from Theorem 2.49 . The behaviour of u near the coincidence set is also studied in Caffarelli [2] and Kinderlehrer [2]where some kind of non-degeneracy inequality is given as a step to prove the regularity $C^{1,\alpha}$ of the solution.□

The multivaluedness at $r = 0$ of the maximal monotone graph $\beta(r)$ given in (5) (and more precisely, the fact that $D(\beta) = [0,\infty) \subsetneq \mathbb{R}$) seems to be a necessary condition for the formation of the coincidence set. Indeed, consider the problem

$$- \Delta u + cu = 0 \quad \text{in} \quad \Omega \tag{37}$$

$$\frac{\partial u}{\partial n} + b(u) = e \quad \text{on} \quad \partial\Omega \tag{38}$$

where $c \geqslant 0$ and b is a nondecreasing continuous function with $b(0) = 0$. Assume that $e(x) \leqslant 0$ on $\partial\Omega$ with $e < -\varepsilon$ in some $\Gamma_\varepsilon \subset \partial\Omega$. By the comparison results $u \leqslant 0$ in $\overline{\Omega}$ and from the strong maximum principle $u < 0$ in Ω. So if $u(x_0) = 0$ for some $x_0 \in \partial\Omega$ then $\frac{\partial u}{\partial n}(x_0) > 0$ and therefore u cannot vanish on Γ_ε.

Remark 2.37. Introducing the pseudo-differential operator $A : H^{1/2}(\partial\Omega) \to H^{-1/2}(\partial\Omega)$ given by $Aw = \frac{\partial v}{\partial n}$, where v is the solution of $- \Delta v + cv = 0$ and $v = w$ on $\partial\Omega$, it is not difficult to show that (37),(38) may be equivalently formulated in terms of

$$Aw + b(w) = e$$

where w is the trace on $\partial\Omega$ of the solution of (37),(38) (Brezis [5], Perriot [1], Diaz-Jimenez [2]). The above considerations show that, in constrast with the case in which A is a local operator (e.g. $A = - \Delta$ plus boundary conditions), a free boundary $F(w)$ is formed only when $b = \beta$ is multivalued (problem $P(0,0,e)$) and not in other cases, even if b is not Lipschitz continuous at the origin. □

2.6 BIBLIOGRAPHICAL NOTES

Section 2.1. As indicated, problem (6), (7) appears in exothermic chemical reactions (Aris [1]). The results of Subsection 2.1a are taken from Friedman-Phillips [1]. The greater part of the results of Section 1.4 are proved, in the articles mentioned, for this general formulation. The exterior problem is discussed by Kawohl [5].

The treatment of the nonlinear system (18), made in Subsection 2.1b, follows Brezis-Lieb [1]. We refer the reader to this article for concrete problems of mathematical physics formulated in these terms.

The results of Subsection 2.1c are taken from Aronson-Crandall-

Peletier [1]. After this important work several generalizations and variants are available in the literature (see e.g. Dal Passo [1], De Motloni-Schiaffino-Tesei [1], Bertsch-Rostamian [1], Langlais-Phillips [1] and Peletier-Tesei [1], [2]).

Subsection 2.1d follows Peletier-Serrin [1]. Note the strong difference between the behaviour of solutions for the one-dimensional equation and for the case of any dimension.

Section 2.2. As we have already indicated in Section 1.5, it seems that the study of the free boundary $F(u)$ has its origins in the study of the support of solutions of second order variational inequalities (a general question in this sense was already posed by J.L. Lions [2]).

Theorems 2.17 and 2.18 are taken from Diaz [4] and generalize previous results due to Bensoussan-Brezis-Friedman [1] (see also the approach made by Nagai [1] and Yamada [1], [2] by using Bessel functions: Remark 2.9).

For unbounded domains, the compactness of the support of the obstacle problem was first proved in Brezis [7] and then generalized in Diaz [1], [2] in Redheffer [2],and finally in Benilan-Brezis-Crandall [1], from which Theorem 2.19 is taken. A similar result also holds for the quasilinear equation (Diaz-Herrero [1], [2]). The construction of super and subsolutions with compact support for the obstacle problem has been applied to the study of some special variational inequalities such as: the rectangular dam problem (Shimborski [1]); a free boundary problem in potential theory (Kinderlehrer-Stampacchia [1]); and, specially, subsonic flows (Brezis-Stampacchia [2], [3], Brezis-Duvaut [1], Brezis [8], Diaz [4], Diaz-Dou [1], Shimborsky [2], [3], Hummel [1], etc.).

We mention that a similar treatment of the free boundary associated with quasi-variational inequalities is also possible (see Bensoussan-Brezis-Friedman [1]). The support of the solution of a system of quasi-variational inequalities has been studied by Bensoussan-Friedman [1].

The rearrangement result given in Theorem 2.22, under this formulation seems to be new. It generalizes a previous result due to Bandle-Mossino [1], where the obstacle problem is considered. In this article, obstacles $\psi \in H^2(\Omega) \cap H_0^1(\Omega)$ are considered. See also Maderna-Salsa [2] for a related result as well as for the capacitary variational inequality.

With respect to zero order reactions we point out that the corresponding equation has already been treated by Diaz-Hernandez [1], Bandle-Sperb-Stakgold [1] and, more systematically, by Frank-Wendt [1], who carefully studied the dependence of the null set with respect to a parameter λ.

The study of solutions of the obstacle problems near the free boundary and other properties (Hausdorff estimates, flatness condition etc.) was first established by Caffarelli [1], [3] (see also Alt-Phillips [1] and Friedman-Phillips [1]). The starshapedness of the coincidence set has been proved by Kawohl [1], [4] (see also Sakaguchi [1]). The convexity of the coincidence set was first shown for N = 2 by Friedman-Phillips [1] and, in any dimensions, by Kawohl [6]. More recently, Diaz-Kawohl [1] proved that u is a log concave function (and, in particular, all the level sets $\{u \geqq c\}$ are convex) for suitable obstacles ψ. More geometrical properties and references can be found in Kinderlehrer [1], Kinderlehrer-Stampacchia [2] and Friedman [3].

The stability result contained in Theorem 2.25 is due to Rodrigues [1] and adapts an argument of Alt-Caffarelli [1]. Other regularity results contained in Rodrigues [1] refer to an estimate of the variation in the Lebesgue measure of the coincidence sets of solutions of different obstacle problems (improving a previous result of Caffarelli [4]) as well as to the stability of the free boundary under small changes of the domain (the Lewy conjecture).

We also mention here the numerical approach of the free boundary for the obstacle problem made in Brezzi-Caffarelli [1]. They approximate (estimating the difference) the continuous free boundary by discrete free boundaries generated by the solution of the finite element approximation of the problem.

<u>Section 2.3</u> Lemma 2.25 and Theorem 2.27 are due to Phillips [2]. The behaviour near the free boundary, Hausdorff measure estimates and other properties are given by Phillips [2] and Alt-Phillips [1]. These works extend the results of Alt-Caffarelli [1] (see also Friedman [3]) for a minimum problem that, roughly speaking, corresponds to q = -1; (the associated exterior problem is treated in Tepper [1], [2]). The regularity given in Theorem 2.27 was also shown by Giaquinta-Giusti [1].

Theorems 2.28, 2.29 and Lemma 2.34 follow Brauner-Nicolaenko [2], and Theorems 2.31 and 2.32 Brauner-Nicolaenko [3]. The proof of Proposition 2.33 is due to Brauner (personal communication) and adapts an argument of Brauner-Eckhaus-Garbey-Van Harten [1]. Theorem 2.35 collects several results from Brauner-Nicolaenko [2] and Misiti-Guyot [1]. Finally, Theorem 2.35 is a

slight improvement of a previous result of Brauner-Nicolaenko [2].

Other related references are: Mignot-Puel [1] and Conrad-Issard-Rouch-Brauner-Nicolaenko [1] for recent studies of nonlinear eigenvalue problems (including the singular equation (1)); Perry [1] and Luning-Perry [1], [2] for the case of homogeneous Dirichlet conditions (which appear in negative order chemical reactions and non-Newtonian fluid flows, respectively); and Crandall-Rabinowitz-Tartar [1] and Bouillet-Gomes [1] for the singular equation with the perturbed term with reverse sign.

Section 2.4. Local supersolutions like that given in (13) have already been used by Evans-Knerr [1] (see also Vazquez [1]). Radially symmetric supersolutions for equation (6) are exhibited in Diaz-Herrero [1], [2].

Unidirectional phenomena have been systematically studied by Diaz-Veron [1] for first-order quasilinear equations. The diffusion-convection balance (25) is also of interest in the study of the uniqueness of solutions associated with different boundary conditions (Carrillo [1], Carrillo-Chipot [2]). We also remark that the dam problem may be formulated in terms of equation (21) by taking $a_i(s) = s$, b_1 the maximal monotone graph given in (67) of Section 2.2 and $b_i \equiv 0$ if $i > 1$ (see, e.g. Alt [1], Carrillo-Chipot [1], Friedman [3] and the references therein). Note that, in this case, the balance (25) holds. For a study of the diffusion-convection balance for a quasilinear parabolic equation, see Diaz-Kersner [1].

Section 2.5. Theorem 2.44 in this general formulation seems to be new. A related work, containing also a strong maximum principle, is that by G. Diaz [4]. The existence of the free boundary for the Monge-Ampere equation (39) is studied by G. Diaz-J.I. Diaz [2]. We also mention here the treatment of the obstacle problem for some fully nonlinear operators given by G. Diaz [1], [2].

The study of the optimal strategy for the Hamilton-Jacobi-Bellman equation, Theorem 2.45, follows Friedman-Lions[1] (see also G. Diaz [3] for the estimate of some of the involved bounds).

Section 2.5. Theorem 2.46 follows Diaz [4] and Diaz-Hernandez [1]. The study of general (eventually multivalued) boundary conditions associated with the obstacle problem is due to Nagai [1], Yamada [1], [2] and Diaz [4]. The case of unbounded domains was considered under a different hypothesis by Diaz [2] and by Diaz [1], [2], [4] and Redheffer [2] for the obstacle problem. A careful treatment of the free boundary for the third boundary value problem for semilinear equations is given by Friedman-Phillips [↑].

The results of Subsection 2.5b for the Signorini problem are taken from Diaz-Jimenez [1]. Previous results on the existence and location of the coincidence set are those by Friedman [1] and Diaz [3]. A different qualitative study of the coincidence set is due to Lewy [1] and Athanasopoulos [1].

As a final comment, in the opinion of the author it would be interesting to extend the results of Subsection 2.1c to the any-dimensional case, and also to obtain some geometrical properties of the null set N(u) such as, for instance, the convexity or an isoperimetric inequality in the case of fully nonlinear equations.

3 Existence and location of the free boundary by means of energy methods

In previous chapters the existence and location of the free boundary $F(u)$ were obtained by using, in a fundamental way, the comparison principle. The main goal of this Chapter is to exhibit an alternative method to derive this a kind of results; the energy method. This method is based on the general idea of finding some suitable energy function of the solution and proving that it satisfies some ordinary differential inequality which leads to the vanishing of the solution on some adequate subset. In contrast with the comparison principle, energy methods are available for equations of higher order and even without any monotonicity assumption on the nonlinear terms of the equation.

In Section 3.1, an energy method is introduced for the study of sufficient conditions for the existence of the free boundary $F(u)$ for a very general class of second order quasilinear equations. The case of quasilinear equations of any order is treated in Section 3.2 by using other energy functions and subsets. Both methods are also available for the limiting problems given by the obstacle problem and the case of some singular equations.

3.1. SECOND ORDER QUASILINEAR EQUATIONS

3.1a. The main result.

In this section we shall study the null set $N(u)$ for local weak solutions of the equation

$$- \operatorname{div} A(x,u,\nabla u) + B(x,u,\nabla u) + C(x,u) = 0. \tag{1}$$

This study will give us interior estimates for some global nonhomogeneous boundary value problems such as, for instance, the following Dirichlet problem;

$$- \operatorname{div} A(x,u,\nabla u) + B(x,u,\nabla u) + C(x,u) = g(x) \quad \text{in} \quad \Omega \tag{2}$$

$$u = h \quad \text{on} \quad \partial\Omega. \tag{3}$$

The structural assumptions that we will adopt in this section are the following : for some open region G of $\mathbb{R}^N$, $A(x,r,\xi)$ is a vector function defined on $G \times \mathbb{R} \times \mathbb{R}^N$, $B(x,r,\xi)$ is a real function defined on the same set and $C(x,r)$ is a real function defined on $G \times \mathbb{R}$. All these are Caratheodory functions (i.e. measurables on x and continuous in other arguments) and they satisfy:

$$|A(x,r,\xi)| \leqslant C_1|\xi|^{p-1} \quad \text{for some} \quad p > 1 \tag{4}$$

$$A(x,r,\xi)\xi \geqslant C_2|\xi|^p \tag{5}$$

$$|B(x,r,\xi)| \leqslant C_3|r|^\alpha|\xi|^\beta \quad \text{for some} \quad \alpha,\beta > 0, \tag{6}$$

$$C(x,r) \geqslant C_4|r|^{q+1} \quad \text{for some} \quad q \geqslant 0 \tag{7}$$

where C_i, $i = 1,4$ are positive constants ($C_3 \geqslant 0$). We remark that all the assumptions are fulfilled in the model equation

$$-\Delta_p u + \lambda|u|^{q-1} u = 0 ,$$

where Δ_p is the pseudo-Laplacian operator defined in the Introduction. We also remark that no monotonicity assumption is made either on the dependence of A with respect to ξ or on that of C with respect to r.

The results that we shall give here are (as in above chapters) independent of the existence, regularity and uniqueness theory. Indeed, we shall work with locally weak solutions in the following sense:

Definition 3.1. *A measurable function* $u \in L^1_{loc}(G)$ *is a weak solution of (1) if*

i) $\nabla u \in L^p_{loc}(G)$

ii) $B(\cdot,u,\nabla u) \in L^1_{loc}(G)$

iii) $C(\cdot,u) \in L^1_{loc}(G)$

and for any $\phi \in C_0^\infty(G)$ *the following equality holds*

$$\int_\Omega \{A(x,u,\nabla u)\nabla\phi + B(x,u,\nabla u)\phi + C(x,u)\phi\}dx = 0. \tag{8}$$

To state our main result, for every $x_0 \in G$, we introduce the diffusion and absorption energy functions , defined on the ball $B_\rho(x_0)$ by

$$E(\rho) = \int_{B_\rho(x_0)} A(x,u,\nabla u)\cdot\nabla u \, dx, \tag{9}$$

and

$$b(\rho) = \int_{B_\rho(x_0)} |u|^q \, dx. \tag{10}$$

We have

Theorem 3.1. *Assume the structural conditions (4), (5), (6) and (7) under the following circumstances:*

$$0 \leq q < p - 1. \tag{11}$$

If $C_3 \neq 0$ *, we also assume that*

$$0 \leq \beta < p \quad , \quad \alpha = [q - \beta(q + 1)]/p \tag{12}$$

and

$$C_3 < C_4 \text{ if } \beta = 0 \text{ , } C_3 < C_2 \text{ if } \beta = p \text{ and } C_3 < C_4\left(\frac{p}{p-\beta}\right)^{\frac{(p-\beta)}{p}}\left(C_2\frac{p}{\beta}\right)^{\frac{\beta}{p}} \text{ if } 0 < \beta < q+1. \tag{13}$$

Let u *be a weak solution of (1),* $x_0 \in G$ *and* $\rho_0 = d(x_0, \partial G)$ *. Then there exists a positive constant* C *(depending in an increasing way on* $E(\rho_0)$ *and* $b(\rho_0)$*) such that* $u(x) = 0$ *a.e. in* $B_\rho(x_0)$ *for every* $\rho \in [0,\rho_0]$ *such that*

$$\rho \leq \rho_0 - C. \tag{14}$$

Remark 3.1. The above result has a nature similar to that of Theorem I.9. Indeed: on the one hand conditions (12) and (13) are superfluous if $B(x,r,\xi) \equiv 0$ (i.e. $C_3 = 0$). On the other hand, (14) makes sense only if $\rho_0 \geq C$ and so it can be compared with the condition $L \geq \psi_{1/N}(M)$ in that theorem. Nevertheless, the conclusion of Theorem 3.1 is stated in terms of the energies $E(\rho_0)$ and $b(\rho_0)$, i.e. using a local information on u. Such a information holds for many equations in which it is possible to have global "a priori" estimates on the energies, and so the constant C can be taken uniformly, for every x_0. □

The main ingredients in the proof of Theorem 3.1 are the next two auxiliary lemmas. The first one is like a local "complementary formula". (Now we shall write B_ρ to represent $B_{\rho_0}(x_0)$ and $S_\rho = \partial B_\rho$).

Lemma 3.2. *Under the hypotheses of Theorem 3.1, we have that* $A(\cdot,u,\nabla u)\cdot\nabla u$, $|u|^q$, $|A(\cdot,\nabla u)|u$ *and* $B(\cdot,u,\nabla u)u$ *belong to* $L^1(B_{\rho_0})$. *Moreoever, for almost all* $\rho \in (0,\rho_0)$, *we have*

$$\int_{B_\rho} A(x,u,\nabla u)\cdot\nabla u\,dx + C_4\int_{B_\rho} |u|^{q+1}dx + \int_{B_\rho} B(x,u,\nabla u)u\,dx \leqslant \int_{S_\rho} A(x,u,\nabla u)\cdot\vec{\nu}\,u\,ds \tag{15}$$

where $\vec{\nu} = \vec{\nu}(s)$ *is the outward normal vector at* $x \in S_\rho(x_0)$.

The second auxiliary lemma is the key-stone of the proof of Theorem 3.1. Its proof, quite technical, is given in subsection 3.1b.

Lemma 3.3. *Let* D *be a bounded open set of* $\mathbb{R}^N$, $N \geqslant 1$, *with a* C^1 *boundary* ∂D. *Assume*

$$0 \leqslant q \leqslant p - 1 < \infty.$$

Then there exists a constant C *depending on* q,p *and* D *such that for any* $v \in W^{1,p}(D)$ *we have*

$$\|v\|_{L^p(\partial D)} \leqslant C(\|\nabla v\|_{L^p(D)} + \|v\|_{L^{q+1}(D)})^\theta \|v\|^{1-\theta}_{L^{q+1}(D)} \tag{16}$$

where

$$\theta = \frac{N(p-1-q)+q+1}{N(p-1-q)+(q+1)p}. \tag{17}$$

If in particular $D = B_\rho(x_0)$ $(\partial D = S_\rho(x_0))$ *then (16) can be improved by the estimate*

$$\|v\|_{L^p(S_\rho)} \leqslant C(\|\nabla v\|_{L^p(B_\rho)} + \rho^{-\delta}\|v\|_{L^{q+1}(B_\rho)})^\theta \|v\|^{1-\theta}_{L^{q+1}(B_\rho)} \tag{18}$$

where

$$\delta = \frac{N(p-1-q)+(q+1)p}{p(q+1)} \tag{19}$$

and $C = C(N,p,q)$.

To go ahead, we assume , for a moment, the above lemmas already proved.
Proof of Theorem 3.1.First step. If u is a local weak solution of (1) then

$$E(\rho) + C_4 b(\rho) + \int_{B_\rho} B(x,u,\nabla u)u\,dx \geq C_5(E(\rho) + b(\rho)) \tag{20}$$

where $C_5 = C_5(C_2, C_4, p, q, \beta) > 0$. This can be obtained in the following way. By using Young's inequality , for any $\varepsilon > 0$ and $\tau > 1$ we have

$$C_3 |u|^{\alpha+1} |\nabla u|^{\beta} \leq \frac{\varepsilon C_3}{\tau} |u|^{\tau(\alpha+1)} + \frac{(\tau-1)}{\tau} C_3\, \varepsilon^{\frac{-1}{\tau-1}} |\nabla u|^{\frac{\beta\tau}{\tau-1}}. \tag{21}$$

If we choose $\tau = \frac{q+1}{\alpha+1}$, then $\frac{\beta\tau}{\tau-1} = p$ thanks to assumption (12). So, using (6) we have

$$B(x,u,\nabla u)u \leq \varepsilon\, C_3 \frac{p-\beta}{p} |u|^{q+1} + \frac{\beta C_3}{p} \varepsilon^{-\frac{(p-\beta)}{\beta}} |\nabla u|^{p}.$$

Therefore, by assumption (5) and the definitions of $E(\rho)$ and $b(\rho)$ we have

$$\left|\int_{B_\rho} B(x,u,\nabla u)u\,dx\right| \leq \varepsilon C_3 \left(\frac{p-\beta}{p}\right) b(\rho) + \frac{\beta C_3}{C_2 p} \varepsilon^{-\frac{(p-\beta)}{\beta}} E(\rho).$$

Since C_3 satisfies (13), it is possible to find $\varepsilon > 0$ depending on p, β, C_2, C_4 such that

$$\varepsilon\, C_3 \left(\frac{p-\beta}{p}\right) < C_4 \quad \text{and} \quad \frac{\beta C_3}{C_2 p} \varepsilon^{-\frac{(p-\beta)}{\beta}} < 1.$$

In this case, it is enough to take

$$C_5 = \min \left\{C_4 - \varepsilon C_3 \left(\frac{p-\beta}{p}\right) , 1 - \frac{\beta C_3}{C_2 p} \varepsilon^{-\frac{(p-\beta)}{\beta}} \right\}$$

in order to have (20).
End of the proof. By Lemma 3.2 , structural assumptions (5) and (7),and the first part of the expression (20),we have

$$C_5(E(\rho) + b(\rho)) \leq \int_{S_\rho} A(x,u,\nabla u)\cdot \vec{\nu} u\,ds \tag{22}$$

but, by (4) and Hölder

$$\int_{S_\rho} A(x,u,\nabla u)\cdot\vec{\nu}\ u\ ds \leqslant C_1(\int_{S_\rho} |\nabla u|^{p-1}|u|dx) \leqslant C_1(\int_{S_\rho} |\nabla u|^p)^{(p-1)/p}(\int_{S_\rho} |u|^p)^{1/p}. \tag{23}$$

On the other hand, by using spherical coordinates (ω,r) with center x_0 we have

$$E(\rho) = \int_0^\rho \int_{S^{N-1}} A(r\omega,u,\nabla u)\cdot\nabla u\ r^{N-1}\ d\omega\ dr.$$

Hence, E is almost everywhere differentiable and

$$\frac{dE}{d\rho}(\rho) = \int_{S^{N-1}} A(r\omega,u,\nabla u)\cdot\nabla u\ \rho^{N-1}\ d\omega\ dr = \int_{S_\rho} A(r\omega,u,\nabla u)\cdot\nabla u\ d\omega$$

which, by (5), implies

$$\frac{dE}{d\rho}(\rho) \geqslant C_2 \int_{S_\rho} |\nabla u|^p\ ds. \tag{24}$$

So, by (22),(23),(24) and the Lemma 3.3. (expression (18)) we have

$$E(\rho) + b(\rho) \leqslant K(\frac{dE}{d\rho})^{\frac{p-1}{p}}(E(\rho)^{\frac{\theta}{p}}\ b(\rho)^{\frac{1-\theta}{q+1}} + \rho^{-\delta\theta}b(\rho)^{\frac{1}{q+1}})$$

for some constant K. But, since

$$b(\rho)^{\frac{1}{q+1}} \leqslant b(\rho_0)^{\theta(\frac{1}{q+1}-\frac{1}{p})}\ b(\rho)^{\frac{\theta}{p}+\frac{1-\theta}{q+1}},$$

by Young's inequality we get

$$E(\rho) + b(\rho) \leqslant K_1\rho^{-\delta\theta}(\frac{dE}{d\rho})^{\frac{p-1}{p}}(E(\rho) + b(\rho))^{\omega} \tag{25}$$

where

$$K_1 = 2K[\max(1,b(\rho_0)^{\theta(\frac{1}{q+1}-\frac{1}{p})})\max(\rho_0^{\delta\theta},1)], \text{ and } \omega = \frac{\theta}{p} + \frac{1-\theta}{q+1}.$$

Hence we deduce that E satisfies the differential inequality

$$K_1\rho^{-\delta\theta}\left(\frac{dE}{d\rho}\right)^{\frac{p-1}{p}} \geqslant (E(\rho)+b(\rho))^{1-\omega} \geqslant E(\rho)^{1-\omega}$$

i.e.

$$K_2\rho^{-\frac{\delta\theta p}{p-1}} \frac{dE}{d\rho}(\rho) \geqslant E(\rho)^{\frac{(1-\omega)p}{p-1}} \tag{26}$$

if $K_2 = K_1^{\frac{p}{p-1}}$. Remark that $\frac{1}{p} < \omega < \frac{1}{q+1}$ thanks to (11), and so $0 < \frac{(1-\omega)p}{p-1} < 1$. Integrating (25) on (ρ,ρ_0), we get

$$\frac{K_2}{\left(1-\frac{(1-\omega)p}{p-1}\right)}\left\{E(\rho_0)^{1-\frac{(1-\omega)p}{p-1}} - E(\rho)^{1-\frac{(1-\omega)p}{p-1}}\right\} \geqslant$$

$$\geqslant \frac{1}{\left(1+\frac{\delta\theta p}{p-1}\right)}\left\{\rho_0^{1+\frac{\delta\theta p}{p-1}} - \rho^{1+\frac{\delta\theta p}{p-1}}\right\}$$

and, hence, if $\rho_1 < \rho_0$ is such that

$$\rho_1^{1+\frac{\delta\theta p}{p-1}} = \rho_0^{1+\frac{\delta\theta p}{p-1}} - \frac{K_2}{\left(1-\frac{(1-\omega)p}{p-1}\right)} E(\rho_0)^{1-\frac{(1-\omega)p}{p-1}}$$

then $E(\rho_1) = 0$, and so $E(\rho) = 0$ for $\rho \leqslant \rho_1$. This implies, by (25) that $b(\rho) = 0$, which means $u(x) = 0$ a.e. in B_ρ for $\rho \leqslant \rho_1$. □

Remark 3.2. In fact, if we compute the exponents in the proof of Theorem 3.1, we can improve (14), getting

$$\rho_1^\nu = \rho_0^\nu - C\,\frac{E^\gamma(\rho_0)}{p-q-1}\max(1,\rho_0^{\nu-1})\max(1,b^\eta(\rho_0))$$

where $C = C(C_1,C_2,C_3,C_4,N,p,q,\beta)$ and

$$\gamma = \frac{p-q-1}{N(p-1-q)+p(q+1)}, \quad \nu = \frac{p(q+1)+N(p-1-q)}{(p-1)q}$$

$$\eta = \frac{p-1-q}{(p-1)(q+1)} - \frac{q+1-p}{N(p-1-q)+p(q+1)}. \ \square$$

Proof of Lemma 3.1. From (4) $A(\cdot,u,\nabla u)\cdot\nabla u \in L^1(B_{\rho_0}(x_0))$. As $\nabla u \in L^p(B_\rho(x_0))$ and $u \in L^q(B_{\rho_0}(x_0))$ we deduce from Poincare's inequality and the Sobolev imbedding theorem that $u \in L^{p^*}(B_{\rho_0}(x_0))$, $\frac{1}{p^*} = \frac{1}{p} - \frac{1}{N}$, if $p < N$, or $u \in L^r(B_{\rho_0}(x_0))$, $\forall r < \infty$, if $p \geq N$. Hence $u \in L^{q+1}(B_{\rho_0}(x_0))$. From (6) $|B(x,u,\nabla u)u| \leq C_3|u|^{\alpha+1}|\nabla u|^\beta$. As $|u|^{\alpha+1} \in L^{(q+1)/(\alpha+1)}(B_{\rho_0}(x_0))$ and $|Du|^\beta \in L^{p/\beta}(B_{\rho_0}(x_0))$ and as $\beta/p + (\alpha+1)/(q+1) = 1$, $B(\cdot,u,\nabla u)u \in L^1(B_{\rho_0}(x_0))$. As $A(\cdot,u,\nabla u) \in L^{p/(p-1)}(B_{\rho_0}(x_0))$ and $u \in L^{p^*}(B_{\rho_0}(x_0))$ if $p < N$ (or $u \in L^r(B_{\rho_0}(x_0))$ if $p \geq N$) and as $\frac{1}{p^*} + \frac{(p+1)}{p} = 1 - 1/N < 1$, $|A(\cdot,u,\nabla u)|u \in L^1(B_{\rho_0}(x_0))$ and $\int_{S_\rho(x_0)} A(x,u,\nabla u)\cdot\vec{\nu}\, u\, ds$ exists for almost all $\rho \in (0,\rho_0)$.

Now we define for $m \in \mathbb{N}$, $T_m(u) = \text{sign}(u)\min(m,|u|)$ and for $n \in \mathbb{N}$ and $\rho \in (0,\rho_0)$, we consider the sequence of functions $\psi_n : [0,\rho_0] \to \mathbb{R}^+$ such that

$$\psi_n(r) = 1 \quad \text{if} \quad r \in [0,\rho - \tfrac{1}{n}] ,$$

$$\psi_n(r) = 0 \quad \text{if} \quad r \in [\rho,\rho_0] ,$$

$$\psi_n(r) = -n(\rho-r) \quad \text{if} \quad r \in [\rho - \tfrac{1}{n}, \rho].$$

From a result of Stampacchia [ ], $\phi_{n,m}(x) = T_m(u(x))\,\psi_n(|x-x_0|)$, belongs to $W_0^{1,p}(B_{\rho_0}(x_0))$, so it is an admissible test function, and we have

$$\int_{B_{\rho_0}(x_0)} |A(x,u,\nabla u)\cdot\nabla\phi_{n,m} + B(x,u,\nabla u)\phi_{n,m} + C(x,u)\phi_{n,m}|dx = 0 . \tag{27}$$

But $\int_{B_{\rho_0}(x_0)} A(x,u,\nabla u)\cdot\nabla\phi_{n,m}dx = \int_{B_\rho(x_0)} \psi_n A(x,u,\nabla u)\cdot \nabla T_m(u) + \dots$ $+ T_m(u) A(x,u,\nabla u)\cdot\nabla\psi_n\, dx$. We deduce from Lebesgue's theorem , as m goes to infinity , that

$$\int_{B_{\rho_0}(x_0)} [\psi_n A(x,u,\nabla u)\cdot\nabla u + B(x,u,\nabla u)u + C_4|u|^{q+1}]dx \leq -\int_{B_{\rho_0}(x_0)} u\, A(x,u,\nabla u)\cdot\nabla\psi_n dx \tag{28}$$

But $-\int_{B_{\rho_0}(x_0)} u\, A(x,u,\nabla u)\cdot\nabla\psi_n\, dx = n \int_{\rho - \frac{1}{n} < |x-x_0| < \rho} u\, A(x,u,\nabla u)\cdot \frac{x - x_0}{|x-x_0|} dx$. Using spherical coordinates (r,ω) with center x_0 we have

$$n \int_{\rho-\frac{1}{n}<|x-x_0|<\rho} u\, A(x,u,\nabla u)\cdot \frac{x-x_0}{|x-x_0|}\, dx = n \int_{\rho-\frac{1}{n}}^{\rho} \int_{S^{N-1}} u\, A(x,u,\nabla u)\cdot\vec{\nu}\, r^{N-1}\, d\omega\, dr$$

where $x = r\omega$. From Lebesgue's differentiation theorem and the fact that $r \to \int_{S^{N-1}} u\, A(r\omega,u,\nabla u)\cdot \vec{\nu}\, r^{N-1}\, d\omega \in L^1(0,\rho_0)$, we deduce that, for almost all $\rho \in (0,\rho_0)$,

$$\lim_{n\to\infty} n \int_{\rho-\frac{1}{n}}^{\rho} \int_{S^{N-1}} u\, A(r\omega,u,\nabla u)\cdot \vec{\nu}\, r^{N-1} d\omega = \int_{S_\rho(x_0)} u\, A(x,u,\nabla u)\cdot\vec{\nu}\, ds. \tag{29}$$

Going to the limit $(n\to\infty)$ in (28) we deduce (15).□

<u>Remark 3.3.</u> We can relax the hypotheses (12) and (13) assuming that $u \in L^\infty_{loc}(G)$. This is the case if, for instance,

$$|C(x,r)| \leqslant C_6|r|^{p-1} + C_7 \tag{30}$$

(see Serrin [1]). From the proof of Lemma 3.3 it is easy to see that we just have to suppose $\alpha \geqslant 0$, $0 \leqslant \beta \leqslant p$ and C_3 small enough. We also remark that for bounded functions the following estimate for the diffusion energy function holds:

$$E(\rho_0) \leqslant \int_{B\rho(x_0)} |\nabla u|^p\, dx \leqslant C\, \rho^N$$

for some adequate C, depending only on $\sup_{B\rho} |u|$ and the structural conditions (see Ladyzhenskaya-Ural'tseva [1] p. 247).□

As an application of Theorem 3.1 we have the following global result, stated, for simplicity, for the equation (2) on $\Omega = \mathbb{R}^N$.

<u>Theorem 3.4.</u> *Under the same assumptions as in Theorem 3.1, let* $u \in W^{1,p}(\mathbb{R}^N) \cap L^{q+1}(\mathbb{R}^N)$ *be any weak solution of*

$$-\operatorname{div} A(x,u,\nabla u) + B(x,u,\nabla u) + C(x,u) = g(x) \quad \textit{in} \quad \mathbb{R}^N \tag{31}$$

where $g \in L^{(q+1)/q}(\mathbb{R}^N)$ *and the support of* g *is in* $B_{\rho_0}(0)$ *for some*

$\rho_0 > 0$. *Then there exists* $\rho_1 > \rho_0$, ρ_1 *depending only on* $\|g\|_{L^{(q+1)/q}(\mathbb{R}^N)}$ *and the structural constants, such that the support of* u *is in* $B_{\rho_1}(0)$. *Moreover, if we assume that* $p < N$ *or* $C(x,r)$ *satisfies* (30), *the result remains true, if we just suppose that* $u \in L^{q+1}(\mathbb{R}^N)$, $\nabla u \in L^p(\mathbb{R}^N)$ *and* $f \in L^{p/(p-1)}(\mathbb{R}^N)$.

Proof. Since u is a weak solution of (31), we have

$$\int_{\mathbb{R}^N} A(x,u,\nabla u)\cdot\nabla\phi dx + \int_{\mathbb{R}^N} B(x,u,\nabla u)\phi dx + \int_{\mathbb{R}^N} C(x,u)\phi dx = \int_{\mathbb{R}^N} g(x)\phi dx, \quad (32)$$

for any $\phi \in C_0^\infty(\mathbb{R}^N)$. Using the same truncation method as in Lemma 3.2, we have for any $\zeta \in C_0^\infty(\mathbb{R}^N)$, $\zeta \geqslant 0$,

$$\int_{\mathbb{R}^N} \{\zeta A(x,u,\nabla u)\cdot\nabla u + uA(x,u,\nabla u)\cdot\nabla\zeta + B(x,u,\nabla u)u\zeta + C(x,u)u\zeta\}dx = \int_{\mathbb{R}^N} gu\zeta\, dx. \quad (33)$$

If we take $\zeta = \zeta_n$ such that $0 \leqslant \zeta_n \leqslant 1$, $\zeta_n(x) = 1$ if $|x| \leqslant n$, $\zeta_n(x) = 0$ if $|x| \geqslant n+1$, $\|\nabla\zeta_n\|_\infty \leqslant 2$, then

$$\left|\int_{\mathbb{R}^N} u\, A(x,u,\nabla u)\cdot\nabla\zeta_n\, dx\right| \leqslant 2\, C_1\Big(\int_{\{n<|x|<n+1\}}|u|^p\, dx\Big)^{1/p} \Big(\int_{\{n<|x|<n+1\}}|\nabla u|^p\, dx\Big)^{(p-1)/p}. \quad (34)$$

If $n \to \infty$ we deduce (as in the first step of the proof of Theorem 3.1)

$$C_5 \int_{\mathbb{R}^N} \{A(x,u,\nabla u)\cdot\nabla u + |u|^{q+1}\}dx \leqslant \int_{\mathbb{R}^N} g\, u\, dx. \quad (35)$$

From Young's inequality, $\int g\, u\, dx \leqslant \varepsilon \int |u|^{q+1}dx + C_\varepsilon \int |g|^{(q+1)/q}dx$. But if $\varepsilon < C_5$ we deduce that

$$\int_{\mathbb{R}^N} A(x,u,\nabla u)\cdot\nabla u\, dx + \int_{\mathbb{R}^N}|u|^{q+1}dx \leqslant K \int_{\mathbb{R}^N} |g|^{(q+1)/q}\, dx \quad (36)$$

for some structural constant K. Hence $E(\infty)$ and $b(\infty)$ remain bounded independently of u and, for any $r > 1$ and $x_0 \in \mathbb{R}^N$,

$$C \min_{(1/p)<\tau\leqslant 1} \Big\{ \frac{E^\gamma(r)}{\tau p - 1} \max(1,r^{\nu-1})\max(b^\mu(r),b^\eta(r)\Big\} \leqslant K\, r^{\nu-1}, \quad (37)$$

where C depends on the structural constants and $\|g\|_{L^{(q+1)/q}(\mathbb{R}^N)}$. If we apply Theorem 3.1 in $B_r(x_0)$, where $|x_0| = \gamma_0 + r$, we deduce that

$$S(u) \subset B_{\rho_1}(0) \quad , \rho_1 = \rho_0 + \max(1,K) \quad , \quad S(u) = \text{support of } u.$$

If we suppose $p < N$, then $\nabla u \in L^p(\mathbb{R}^N)$ implies $u \in L^{p^*}(\mathbb{R}^N)$ with $\frac{1}{p^*} = \frac{1}{p} - \frac{1}{N}$. Hence $u \in L^{q+1}(\mathbb{R}^N) \cap L^p(\mathbb{R}^N)$, so $u \in W^{1,p}(\mathbb{R}^N)$ and we can go to the limit in (34) and get (36). If $C(x,r)$ satisfies (30) then $u \in L^\infty(\mathbb{R}^N - B_{2\rho_0}(0))$ (see Remark 3.3) and $u \in L^p(\mathbb{R}^N)$ from interpolation results.□

As in the previous chapter, the <u>obstacle problem</u> can be treated as the limiting case $q = 0$. From the proof of Theorem 3.1 we see that the hypothesis of continuity on $r \to C(x,r)$ can be relaxed. So we can consider variational inequalities in a weak sense such as the following:

$$u \geq 0 \quad , \quad -\operatorname{div} A(x,u,\nabla u) + B(x,u,\nabla u) \geq g(x) \quad \text{in} \quad G. \tag{38}$$

Now the solutions must satisfy $u \in L^1(G)$, $\nabla u \in L^p(G)$, $B(\cdot,u,\nabla u) \in L^1(G)$ and

$$\int_G \{A(x,u,\nabla u)\cdot\nabla\phi + B(x,u,\nabla u)\phi\}\, dx \geq \int_G g(x)\phi\, dx \tag{39}$$

for any $\phi \in C_0^\infty(G)$, $\phi \geq 0$. Finally, if for some $\varepsilon > 0$, we have

$$g(x) \leq -\varepsilon \quad \text{a.e.} \quad \text{in} \quad G,$$

we can apply Theorem 3.1 provided C_3 is small enough.

<u>Remark 3.4</u>. A careful review of the proof of Theorem 3.1 shows that the exponent q can also be taken satisfying $-1 < q < 0$. In this way, the energy method is also available for some <u>singular equations</u> (see Section 2.3). We remark that the possibility of taking negative values of q in interpolation inequalities (see inequality (41)) was already pointed out in Nirenberg [1].□

3.1b. Proof of the interpolation-trace Lemma.

For the sake of simplicity we restrict ourselves to $v \in C^1(\overline{G})$, since $C^1(\overline{G})$ is dense in $W^{1,p}(G)$. The proof of inequality (16) is divided into four steps (see Bidaut-Veron [2] for a similar result).

First step. From a well-known result (see e.g. Ladyzhenskaya-Uralseva[1] p. 45) for any $\varepsilon > 0$ there exists $C_\varepsilon > 0$, such that, for any $v \in C^1(\overline{G})$, the following holds:

$$\|v\|_{L^{q+1}(G)} \leq \varepsilon \|\nabla v\|_{L^p(G)} + C_\varepsilon \|v\|_{L^{q+1}(G)} . \tag{40}$$

If we set $C_2 = \max(1 + \varepsilon , C_\varepsilon |G|^{1-\frac{1}{q+1}})$, we get

$$\|v\|_{W^{1,p}(G)} \leq C_2(\|\nabla v\|_{L^p(G)} + \|v\|_{L^{q+1}(G)}). \tag{41}$$

Second step. We start from the elementary trace result (see e.g. Adams[1]): there exists $C_3 > 0$ such that for any $u \in C^1(\overline{G})$ we have

$$\|u\|_{L^1(\partial G)} \leq C_3 \|u\|_{W^{1,1}(G)} , \tag{42}$$

and for $p > 1$ we apply (42) to $u = v|v|^{(p-1)}$, $v \in C^1(\overline{G})$, so

$$\int_{\partial G} |v|^p \, d\sigma \leq C_3\{p \int_G |v|^{(p-1)} |\nabla v| dx + \int_G |v|^p dx\} .$$

Since $\int_G |v|^{(p-1)} |\nabla v| dx \leq \|\nabla v\|_{L^p(G)} \|v\|^{(p-1)}_{L^p(G)}$, we get

$$\int_{\partial G} |v|^p \, d\sigma \leq C_3\{p \|\nabla v\|_{L^p(G)} \|v\|^{(p-1)}_{L^p(G)} + \|v\|^p_{L^p(G)}\} ,$$

which implies

$$\|v\|_{L^p(\partial G)} \leq (p\, C_3)^{1/p} \|v\|^{1/p}_{W^{1,p}(G)} \|v\|^{(p-1)/p}_{L^p(G)} . \tag{43}$$

Third step. Set $0 \leq q \leq p-1 < \infty$, we claim that there exists a constant $C_4 > 0$ such that, for any $v \in C^1(\overline{G})$, we have

$$\|v\|_{L^p(G)} \leqslant C_4\|v\|_{W^{1,p}(G)}^{(p\theta-1)/(p-1)} \|v\|_{L^{q+1}(G)}^{p(1-\theta)/(p-1)} . \tag{44}$$

Case 1: assume $p < N$. From Sobolev's inequality, we have $\|v\|_{L^\tau(G)} \leqslant C\|v\|_{W^{1,p}(G)}$ with $\frac{1}{\tau} = \frac{1}{p} - \frac{1}{N}$. Moreover,

$$\|v\|_{L^p(G)} \leqslant \|v\|_{L^\tau(G)}^{1-\lambda} \|v\|_{L^{q+1}(G)}^{\lambda} \tag{45}$$

where $\frac{1}{p} = \frac{\lambda}{q+1} + \frac{1-\lambda}{\tau}$, that is, $\lambda = \frac{p(q+1)}{N(p-1-q)+p(q+1)}$. Hence, from Sobolev's inequality

$$\|v\|_{L^p(G)} \leqslant C^{1-\lambda}\|v\|_{W^{1,p}(G)}^{1-\lambda} \|v\|_{L^{q+1}(G)}^{\lambda} , \tag{46}$$

and $1 - \lambda = \frac{N(p-1-q)}{N(p-1-q)+p(q+1)} = \frac{p\theta-1}{(p-1)}$, $\lambda = \frac{p(1-\theta)}{(p - 1)}$.

Case 2: assume $p \geqslant N \geqslant 1$. We set $\alpha = (N+1)/2$, $\rho = 2p/(N+1)$, $\beta = (q+1)(N+1)/2p$ and $\alpha^* = \alpha N/(N-\alpha)$ ($\alpha^* = \infty$ if $N = 1$). From Hölder's interpolating inequality we have

$$\|u\|_{L^\alpha(G)} \leqslant \|u\|_{L^{\alpha^*}(G)}^{1-\lambda} \|u\|_{L^\beta(G)}^{\lambda} , \tag{47}$$

where $\frac{1}{\alpha} = \frac{1-\lambda}{\alpha^*} + \frac{\lambda}{\beta}$ ((47) is valid even if $0 < \beta < 1$ with a simple change of function). From Sobolev's inequality we get

$$\|u\|_{L^\alpha(G)} \leqslant C_5\|u\|_{W^{1,\alpha}(G)}^{1-\lambda} \|u\|_{L^\beta(G)} . \tag{48}$$

Now we set $u = v|v|^{\rho-1}$ and we have

$$\|u\|_{L^\alpha(G)} = \|v\|_{L^{\alpha\rho}(G)}^{\rho} = \|v\|_{L^p(G)}^{\rho} ,$$

$$\|u\|_{L^\beta(G)} = \|v\|_{L^{\beta\rho}(G)} = \|v\|_{L^{q+1}(G)}^{\rho} ,$$

$$\|u\|_{W^{1,\alpha}(G)} = \|v\|_{L^p(G)}^{\rho} + \left(\int_G (\rho|v|^{\rho-1}|\nabla v|)^\alpha dx\right)^{1/\alpha} \quad \text{and}$$

$\int_G(|v|^{\rho-1}|\nabla v|)^{\alpha}dx \leq (\int_G |v|^{\alpha\rho}dx)^{1-1/\rho}(\int_G|\nabla v|^{\alpha\rho}dx)^{1/\rho}$, which yields $\|u\|_{W^{1,\alpha}(G)} \leq \rho\|v\|^{\rho-1}_{L^p(G)}\|v\|_{W^{1,p}(G)}$ and (48) becomes

$$\|v\|^{\rho}_{L^p(G)} \leq C_6\rho^{1-\lambda}\|v\|^{(\rho-1)(1-\lambda)}_{L^p(G)}\|v\|^{1-\lambda}_{W^{1,p}(G)}\|v\|^{\lambda\rho}_{L^{q+1}(G)}. \qquad (50)$$

If we compute the exponents we get

$$\frac{1-\lambda}{\lambda\rho+1-\lambda} = \frac{N(p-1-q)}{p(q+1)+N(p-1-q)} = \frac{p\theta-1}{(p-1)} \quad \text{and}$$

$$\frac{\lambda\rho}{\lambda\rho+1-\lambda} = \frac{p(q+1)}{N(p-1-q)+p(q+1)} = \frac{p(1-\theta)}{(p-1)} \quad ,$$

which is (44).

<u>Step 4</u>: <u>end of the proof</u>. We use (43) and (44) and get

$$\|v\|^{p}_{L^p(\partial G)} \leq C_7\|v\|^{1/p}_{W^{1,p}(G)}\|v\|^{(p\theta-1)/p}_{W^{1,p}(G)}\|v\|^{1-\theta}_{L^{q+1}(G)}$$

where $\theta = \dfrac{N(p-1-q)+q+1}{N(p-1-q)+p(q+1)}$; using (41) yields finally

$$\|v\|_{L^p(\partial G)} \leq C(\|\nabla v\|_{L^p(G)} + \|v\|_{L^{q+1}(G)})^{\theta}\|v\|^{1-\theta}_{L^{q+1}(G)}.$$

To end the proof of Lemma 3.3, we consider the particular case of $D = B_\rho(x_0)$. For the sake of simplicity, we suppose $x_0 = 0$ and we perform the following change of variable : $x = \rho y$, $x \in B_\rho(0)$, $y \in B_1(0)$. If $u \in W^{1,p}(B_\rho(0))$, the function v defined by $v(y) = u(x)$ belongs to $W^{1,p}(B_1(0))$ and from (16) we have

$$\|v\|_{L^p(S_1(0))} \leq C(\|\nabla v\|_{L^p(B_1(0))} + \|v\|_{L^{q+1}(B_1(0))})^{\theta}\|v\|^{1-\theta}_{L^{q+1}(B_1(0))}. \qquad (51)$$

But $\nabla v(y) = \rho\,\nabla u(x)$, $\|v\|_{L^{q+1}(B_1(0))} = \rho^{-\frac{N}{q+1}}\|u\|_{L^{q+1}(B_\rho(0))}$

$$\| v \|_{L^{q+1}(B_1(0))} = \rho^{1-\frac{N}{q+1}} \| \nabla u \|_{L^p(B_1(0))}$$

$$\| v \|_{L^p(S_1(0))} = \rho^{-\frac{N-1}{p}} \| u \|_{L^p(S_\rho(0))} .$$

As $1 - \frac{N}{p} + \frac{N-1}{\theta p} - \frac{(1-\theta)}{\theta} \frac{N}{q+1} = 0$ and

$$-\frac{N}{q+1} + \frac{N-1}{\theta p} - \frac{1-\theta}{\theta} \frac{N}{q+1} = -\frac{N(p-1-q)+(q+1)p}{p(q+1)}$$

we get (18).□

3.2. QUASILINEAR ELLIPTIC EQUATIONS OF ARBITRARY ORDER.

In this section we shall give a different energy method for the study of the null set of the solution of quasilinear elliptic equations of arbitrary order. For the sake of clarity of the exposition we shall only consider the case of equations on the whole space $\Omega = \mathbb{R}^N$, although it is also possible to consider the case in which Ω is an open regular set of $\mathbb{R}^N$ (see Remark 3.7). Our formulation is the following : Given $1 < p < \infty$, $m \geq 1$ and $g \in W^{-m,p'}(\mathbb{R}^N)$, we consider the equations

$$(-1)^m \sum_{|\alpha|=m} D^\alpha(|D^\alpha u|^{p-2} D^\alpha u) + f(u) = g \quad \text{in} \quad \mathbb{R}^N \tag{1}$$

where

$$f \in C^0(\mathbb{R}) \quad , \quad sf(s) \geq |s|^{q+1} \quad \text{for all} \quad s \in \mathbb{R} . \tag{2}$$

The main result is

Theorem 3.5. *Let* $g \in W^{-m,p'}(\mathbb{R}^N)$. *Assume* (2) *and*

$$0 < q < p - 1. \tag{3}$$

Then, at least there exists one solution $u \in W^{m,p}(\mathbb{R}^N)$, *with* $f(u) \in L^1(\mathbb{R}^N)$, *satisfying (1) in the sense of distributions and such that the support of* u *is compact.*

Remark 3.5. Note that f is not assumed to be nondecreasing and so the uniqueness of the solutions of (1) is not assured in general. □

As in the above section, the proof will be carried out by proving that some "energy functions" satisfy some suitable ordinary differential inequalities.

Proof of Theorem 3.5. The wanted solution u of (1) with compact support will be obtained as the zero extension to $\mathbb{R}^N$ of a function u defined on a open ball G containing the support of f and such that $u \in W_0^{m,p}(G)$ and u satisfies the equation (1) on G (as well as homogeneous Dirichlet conditions on ∂G). In order to show that such a $\tilde{u}$ obtained in this way belongs to $W^{m,p}(\mathbb{R}^N)$ and satisfies (1), it suffices to show that if G is big enough, the support of $\tilde{u}$ is strictly contained in G. Thus , we need to estimate the support of $\tilde{u}$ independently of G. To clarify the notation we shall identify $\tilde{u}$ and u. By the existence results (see Brezis-Browder [1,2]or Chapter 4) we know that $u \in W_0^{m,p}(G)$, $f(u) \in L^1(G)$, $uf(u) \in L^1(G)$ and u is a solution of the equation (1) on G. So we have

$$\sum_{|\alpha| = m} \int_G |D^\alpha u|^{p-2} D^\alpha u \cdot D^\alpha v \, dx + \langle f(u),v\rangle = \langle g,v\rangle \qquad (4)$$

for every $v \in W_0^{m,p}(G)$, where the brackets represent the duality action between $W^{-m,p'}(G)$ and $W_0^{m,p}(G)$. Set $x = (x_1,\ldots,x_N)$, $y = (x_1,\ldots,x_{N-1})$, $t = x_N$. We can trivially suppose that the half-space $\{t > 0\}$ intersects G and does not intersect the support of g.

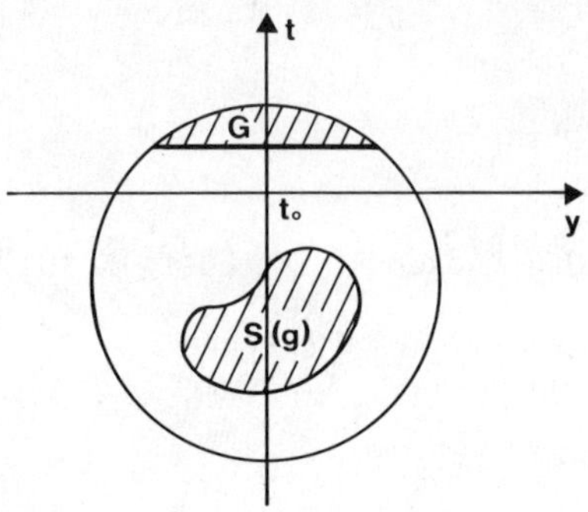

Figure 14.

For any $t_0 \geq 0$ the function

$$w(x) = \begin{cases} (t-t_0)^m u(x) & \text{if} \quad t > t_0,\ x \in G \\ 0 & \text{if} \quad t \leq t_0,\ x \in G \end{cases} \tag{5}$$

belongs to $W_0^{m,p}(G)$. (This is a simple but important remark for the present proof). Since $wf(u) \geq 0$ (recall that $sf(s) \geq 0$), by a result of Brezis-Browder [2](see also Remark 4.12) we have

$$<f(u),w> = \int_G f(u)w\,dx = \int_{\substack{x\in G \\ t>t_0}} (t-t_0)^m\, u\, f(u)dx.$$

Setting $v = w$ in (4) we obtain

$$\sum_{|\alpha|=m} \int_{\substack{x\in G \\ t>t_0}} |D^\alpha u|^{p-2}\, D^\alpha u \cdot D^\alpha((t-t_0)^m u)dx + \int_{\substack{x\in G \\ t>t_0}} (t-t_0)^m u f(u)dx = 0. \tag{6}$$

We proceed to compute $D^\alpha(t-t_0)^m u)$. Setting $D_t = \partial/\partial t$, we have, with an obvious notation and for some constants a_{ijm},

$$D^\alpha = D_t^j\, D_y^\beta \quad , \quad |\beta| = |\alpha| - j \quad , \quad \sum_{|\alpha|=m} = \sum_{j=0}^m \sum_{|\beta|=m-j} \tag{7}$$

$$D^\alpha((t-t_0)^m u) = (t-t_0)^m D^\alpha u + \sum_{i=1}^j a_{ijm}\,(t-t_0)^{m-i}\, D_t^{j-i}\, D_y^\beta\, u. \tag{8}$$

Using the notation

$$|D^j u(x)|^p = \sum_{|\alpha|=j} |D^\alpha u(x)|^p \ , \quad j = 1,\ldots,m$$

we introduce the "energy function"

$$I_s(t_0) = \int_{\substack{x\in G \\ t>t_0}} (t-t_0)^s |D^m u|^p\, dx + \int_{\substack{x\in G \\ t>t_0}} (t-t_0)^s\, |u|^{q+1}\, dx.$$

Noting that

$$|D_t^{j-i}\, D_y^\beta\, u(x)| \leq |D^{m-i}\, u(x)|$$

(because $|\beta|+j = m$), some elementary computations with (2) (6),(7) and (8) give (we drop $x \in G$ in the integrals):

$$I_m(t_0) \leqslant C_m \sum_{i=1}^{m} \int_{t>t_0} (t-t_0)^{m-i} |D^m u|^{p-1} |D^{m-i} u| dx \leqslant$$

$$\leqslant C_m \sum_{i=1}^{m} [\int_{t>t_0} (t-t_0)^{m-i} |D^m u|^p dx]^{1/p'} [\int_{t>t_0} (t-t_0)^{m-i} |D^{m-i} u|^p]^{1/p},$$

where we have applied Holder's inequality . To proceed we need the following lemma:

Lemma 3.6. (Weighted interpolation inequality). *Set* $\mathbb{R}^N_+ = \{x \in \mathbb{R}^N : x_N > 0\}$, $t = x_N$. *Let* m, j, k *be integers* , $m \geqslant 1$, $0 \leqslant j < m$, $k \geqslant 0$. *Let* $1 < p < \infty$ *and* $0 < q \leqslant p - 1$. *Then*

$$[\int_{\mathbb{R}^N_+} t^k |D^j u|^p dx]^{1/p} \leqslant C [\int_{\mathbb{R}^N_+} t^k |D^m u|^p dx]^{\theta/p} [\int_{\mathbb{R}^N_+} t^k |u|^{q+1}]^{(1-\theta)/(q+1)} \tag{9}$$

where θ *is given by*

$$\frac{1}{p} = \frac{j}{N+k} + \theta \left(\frac{1}{p} - \frac{m}{N+k}\right) + (1 - \theta) \frac{1}{(q+1)} \tag{10}$$

and the constant C *depends only on* N,m,j,p,q *and* k.

Proof of Theorem 3.5. (continuation). Since the zero extension operator maps $W_0^{m,p}(G)$ into $W^{m,p}(\mathbb{R}^N)$ and commutes with D^α, $|\alpha| \leqslant m$, we apply the Lemma 3.6 on half-spaces and thus the constants are independent of G. So we obtain

$$I_m(t_0) \leqslant C \sum_{i=1}^{m} [\int_{t>t_0} (t-t_0)^{m-i} |D^m u|^p dx]^{1/p'+\theta_i/p} [\int_{t>t_0} (t-t_0)^{m-i} |u|^{q-1} dx]^{\frac{(1-\theta_i)}{(q+1)}},$$

$$\frac{1}{p} = \frac{m - i}{N+m-i} + \theta_i \left(\frac{1}{p} - \frac{m}{N+m-1}\right) + (1 - \theta_i) \frac{1}{r} . \tag{11}$$

We set $\lambda_i = 1/p' + \theta_i/p + (1-\theta_i)/(q+1)$. Applying the inequality $A^a B^b \leqslant C(A+B)^{a+b}$ and computing explicitly λ_i through (11) we obtain

$$I_m(t_0) \leqslant C \sum_{i=1}^{m} I_{m-i}(t_0)^{\lambda_i} \tag{12}$$

$$\lambda_i = 1 + i/(N+m-1+\sigma(q+1)) \quad \text{with} \quad \sigma = pm/(p-q-1) \tag{13}$$

and the constant C of (10) depends only on N,m,p,q. Note that $\sigma > 0$ and $\lambda_i > 1$. Since $I_s' = -sI_{s-1}$, (12) is an ordinary differential inequality for which we shall prove later (see Lemma 3.8) that necessarily $I_m(t)$ must be with compact support $[0,a]$,

$$a \leqslant C\, I_0(0)^{1/(N+\sigma(q+1))} \quad , \quad \sigma = pm/(p-q-1) \ ,$$

where the constant C depends only on N,m,p,q. So, supp u is included in the half-space $\{t \leqslant a\}$. Again by the result of Brezis-Browder [2],

$$< f(u),u >_G = \int_G uf(u)dx.$$

Since the integrals defining $I_0(0)$ are performed on a subset of G, setting $v = u$ in (4) and using (2) and Young's inequality, we obtain

$$I_0(0) \leqslant C(p,q)\| f\|^{p'}_{W^{-m,p'}(\mathbb{R}^N)} . \tag{14}$$

Considering half-spaces orthogonal to the coordinate axes, it is clear that supp u is bounded independently of G. □

In order to give the proof of Lemma 3.6, we first recall the Nirenberg's interpolation inequalities

Lemma 3.7. (Nirenberg [1]). *Let* $u \in L^{q+1}(\mathbb{R}^N_+)$ *and* $D^m u \in L^p(\mathbb{R}^N_+)$, $0 < q$, $1 < p < \infty$. *Then for* $0 \leqslant j < m$ *we have*

$$\| D^j u\|_q \leqslant C\| D^m u\|^{\theta}_p \| u\|^{1-\theta}_{q+1}$$

where

$$\frac{1}{q} = \frac{j}{N} + \theta\left(\frac{1}{p} - \frac{m}{N}\right) + (1-\theta)\frac{1}{(q+1)}$$

for all θ *in the interval* $j/m \leqslant \theta \leqslant 1$ *and the constant* C *depends only on* N,m,j,p,q *and* θ.

Proof of Lemma 3.6. Since we suppose k integer, a very simple proof can be given using a device of Adams [1]. We argue by induction on k. For $k = 0$ it is true by the former lemma. Suppose it is true for k. Consider

$$\Pi = \{(x,z) \in \mathbb{R}^{N+1} : x \in \mathbb{R}^N_+ \ , 0 < z < t\} \ , \quad u^*(x,z) = u(x) \quad \text{in } \Pi$$

The domain Π can be mapped onto a product domain by a nonsingular linear transformation for which the lemma remains true (see Nirenberg [1]). Therefore, by the induction hypothesis (9),(10) hold with u, $\mathbb{R}^N_+$, and N replaced by u^* , Π and $N+1$. So in (10) we have $N + k + 1$ instead of $N + k$. We conclude the proof by taking into account that

$$\int_\Pi t^k |u^*|^{q+1} \, dx\, dz = \int_{\mathbb{R}^N} t^{k+1} |u|^{q+1} dx$$

and the analogous relations for the derivatives.□

Now we turn to the consideration of ordinary differential inequalities as given in (12) for the proof if Theorem 3.5. So we are interested in the study of the behaviour of nonoscillatory solutions of the general differential inequality

$$|z(t)|^{\lambda_0} \leq B \sum_{i=1}^{m} |z^{(i)}(t)|^{\lambda_i} \quad \text{for all} \quad t \geq 0 \tag{15}$$

where $B,\lambda_0,\ldots,\lambda_m$ are arbitrary positive numbers. We associate to (15) a new set of exponents μ_i defined as follows:

$$\frac{1}{\mu} = \frac{1}{\lambda_0} + \max_i \frac{m}{i} \left(\frac{1}{\lambda_i} - \frac{1}{\lambda_0} \right) \tag{16}$$

$$\frac{1}{\mu_i} = \frac{i}{m}\frac{1}{\mu} + \frac{m-i}{m}\frac{1}{\lambda_0} \ , \quad i = 0,1,\ldots,m. \tag{17}$$

Therefore $\mu_m = \mu \leq \lambda_m$,$\mu_0 = \lambda_0$, and $\mu_i > 0$. The exponent μ is the greatest number such that the μ_i defined by (17) satisfy

$$\mu_i \leqslant \lambda_i \quad \text{for} \quad i = 0,1,\ldots,m.$$

We also note that

$$\lambda_0 \lesseqqgtr \mu \text{ if and only if } \lambda_0 \lesseqqgtr \min\{\lambda_1,\ldots,\lambda_m\}.$$

Lemma 3.8. *Assume* $z \in C^m(\mathbb{R}_+) \cap L^\infty(\mathbb{R}_+)$, $m \geqslant 1$, $z^{(m)}$ *is a monotone function in* $\overline{\mathbb{R}}_+$ *and* z *satisfy (15). Then*

$$|z(t)|^{\lambda_0} \leqslant K|z^{(m)}(t)|^{\mu} \quad \text{for all} \quad t \geqslant 0 \tag{18}$$

where μ *is given by (16) and the constant* K *depends only on* $m,\lambda_0,\lambda_1,\ldots\lambda_m$, B *and only on* $|z^m(0)|$ *if* $\frac{m-1}{m}\lambda_i < \lambda_0$ *for* $i = 1.,,,m-1$. *Finally, if* $\lambda_0 < \mu$ *then* z *has compact support* $[0,a]$ *and*

$$a \leqslant C|z^{(m)}(0)|^{1/(\tau-m)} \quad , \quad \tau = m\mu/(\mu-\lambda_0) \tag{19}$$

where C *is a positive constant depending only on* m,λ_0,μ *and* K.

Proof. The hypotheses imply that for $0 \leqslant i \leqslant m$, $z^{(i)}(t) \to 0$ as $t \to \infty$ and that $z'(t)$ is monotone and has constant sign in $\overline{\mathbb{R}}_+$. (If not, z would be unbounded). Therefore,

$$|z^{(i)}(t)| = \| z^{(i)} \|_{L^\infty(t,\infty)} .$$

Thus by L^∞ interpolation (for half-lines)

$$|z^{(i)}(t)| \leqslant C|z^{(m)}(t)|^{i/m}\, |z(t)|^{(m-1)/m} . \tag{20}$$

Since $\mu_i \leqslant \lambda_i$, from (20)

$$|z^{(i)}(t)|^{\lambda_i} \leqslant |z^{(i)}(0)|^{\lambda_i-\mu_i}\, |z^{(i)}(t)|^{\mu_i} \leqslant K_i|z^{(i)}(t)|^{\mu_i} \tag{21}$$

where K_i depends only on $m,i,\lambda_0,\lambda_1,\ldots,\lambda_m$, $|z^m(0)|$ and $|z(0)|$. Moreover

$$|z^{(i)}(t)|^{\mu_i} \leq C\,\frac{1}{\varepsilon}\,|z^{(m)}(t)|^{\frac{i}{m}\mu_i}\;\varepsilon|z(t)|^{\frac{m-i}{m}\mu_i} \leq$$

$$\leq C\,[\,(1/\varepsilon)^{q_i}|z^m(t)|^{\mu} + \varepsilon^{q_i'}\,|z^m(t)|^{\lambda_0}\,], \tag{22}$$

where we have used Young's inequality and the definition of μ_i, setting

$$\frac{1}{q_i} = \frac{i}{m}\,\frac{\mu_i}{\mu} \quad , \quad \frac{1}{q_i'} = \frac{m-i}{m}\,\frac{\mu_i}{\lambda_0}\,.$$

Inserting (22) and (21) in (15) and choosing ε small enough we obtain (18). Note that setting $t=0$ in (15) and (20) we obtain

$$|z(0)|^{\lambda_0} \leq C \sum_{i=1}^{m} |z^{(m)}(0)|^{\frac{i}{m}\lambda_i}\;|z(0)|^{\frac{m-i}{m}\lambda_i}$$

which implies (e.g., by Young's inequality) that $|z(0)|$ can be bounded in terms of $|z^{(m)}(0)|$ if $\frac{m-i}{m}\,\lambda_i < \lambda_0$ for $i = 1,\ldots,m-1$, in which case the constant K in (18) depends on $|z^m(0)|$ but not on $|z(0)|$.
Finally, from (20),(18),

$$|z^{(m-1)}(t)| \leq C|z^{(m)}(t)|^{\frac{m-1}{m}+\frac{1}{m}\frac{\mu}{\lambda_0}} = C\,|z^{(m)}(t)|^{\frac{\tau-m+1}{-m}},$$

with τ given in (19). Then, if $\lambda_0 < \mu$, integrating this first order differential inequality (for $z^{(m-1)} \neq 0$) we obtain that $z^{(m-1)}$ has compact support. The proof is completed by successive integrations between t and ∞ (recall that $z^{(i)}(t) \to 0$ as $t \to \infty$). □

Remark 3.6. The asymptotic behaviour of nonoscillatory solutions of general classes of higher order ordinary differential equations has been studied in Kiguradze [1], [2]and Canturia [1]. The proof of Lemma 3.6 is due to Bernis [3](see also Bernis [4]for additional results). □

The above treatment can also be applied to many other situations we shall mention in the following remarks.

Remark 3.7. By adapting Brezis-Browder's result it is also possible to consider the case of nonhomogeneous Dirichlet data. The proof of Theorem 3.5 yields the following result: " *Let* Ω *be an unbounded open set of* $\mathbb{R}^N$

with compact boundary. Set

$$h(s) = \sup_{|s|\leq|t|} |f(t)|.$$

Let $\Phi \in W^{m,p}(\Omega)$ *such that* $\Phi h(\Phi) \in L^1(\Omega)$. *Assume* (2) *and* (3). *Then there exists* u *such that* $u - \Phi \in W_0^{m,p}(\Omega)$, $f(u) \in L^1(\Omega)$, (1) *holds on* $D'(\Omega)$ *and* u *has compact support*". It is also possible to consider equations involving intermediate derivatives and variable coefficients such as

$$\sum_{|\alpha|\leq m} (-1)^{|\alpha|} D^\alpha [a_\alpha(x)|D^\alpha u|^{p-2} D^\alpha u] + f(u) = g$$

(see Bernis [3]). A different proof of Theorem 3.5 and the above results valid only for p = 2 was given in Bernis [2].□

Remark 3.8. When f is assumed to be non-decreasing, the solutions of (1) realize the minimum of the functional

$$\operatorname*{Min}_{W^{m,p}(\mathbb{R}^N)} \frac{1}{p} \int_{\mathbb{R}^N} \sum_{|p|=m} |D^\alpha u|^p dx + \int_{\mathbb{R}^N} F(u)dx - \int_{\mathbb{R}^N} gudx \qquad (F'=f) \qquad (22)$$

(note that, in this case, there is a unique solution of (1)). This is the situation if, for instance,

$$f(r) = |r|^{q-1}r \quad , \quad q > 0 . \quad (F(r) = \frac{1}{q+1} |r|^{q+1}). \qquad (23)$$

The minimization problem (22) corresponding to the case q = 0, leads to the multivalued equation

$$(-1)^m \sum_{|\alpha|=m} D^\alpha(|D^\alpha u|^{p-2} D^\alpha u) + \operatorname{sign} u \ni g. \qquad (24)$$

Equation (24) appears in several different contexts and has been largely considered in the particular case of the one-dimensional semilinear fourth-order equation N = 1, p = 2 and m = 4 (Berkovitz-Pollard [1], Redheffer [1], Hestenes-Redheffer [1], Bernis [1], [2] and Brunousky-Mallet-Paret [1]), or in the same case but for spherically symmetric solutions in $\mathbb{R}^N$ (Bidaut-Veron [1],[2]).□

Remark 3.9. A careful review of the proof of lemmas 3.6. and 3.8, as well as of the proof of Theorem 3.5, shows that it is also possible to consider the case in which $-1 < q < 0$. That is: there exists at least one $u \in W^{m,p}(\mathbb{R}^N) \cap L^{q+1}(\mathbb{R}^N)$ minimizing the functional

$$\underset{W^{m,p}(\mathbb{R}^N)\cap L^{q+1}(\mathbb{R}^N)}{\mathrm{Mim}} \quad \frac{1}{p} \int_{\mathbb{R}^N} \sum_{|\alpha|=m} |D^\alpha u|^p dx + \frac{1}{q+1} \int_{\mathbb{R}^N} |u|^{q+1} dx - \int_{\mathbb{R}^N} gu$$

and with compact support in $\mathbb{R}^N$, assumed $-1 < q < p - 1$. (See Bernis [2] and [3]).□

Remark 3.10. The above energy method also gives information on the behaviour of the solution of (1), even if the assumption $q < p - 1$ does not hold (Bernis [4]).□

3.3. BIBLIOGRAPHICAL NOTES

Section 3.1. The first author to introduce an energy method to study the existence of the free boundary $\mathcal{F}(u)$ was Antoncev [1], [2], who applied a *global* energy method for a class of second order degenerate parabolic equations. Later, a *local* energy method was used by Diaz-Veron [2], [3] (the results given in Subsection 3.1a) to study general quasilinear elliptic equations such as (1), as well as its parabolic version. We mention that in Antoncev [2] a second order elliptic equation is also considered, but its formulation is completely different from equation (1) and the energy function is also different from that given in (9).

Section 3.2. The existence of the free boundary $\mathcal{F}(u)$ for problems of order 4 and in dimension 1 was first established when proving the compactness of the support of the solution (see Berkovitz-Pollard [1], Redheffer [1], and Hestenes-Redheffer [1], Bernis [1], [2] and Brunovsky-Mallet-Paret [1]). The case of radially symmetric solutions was considered by Bidaut-Veron [1], [2].

The compactness of the support for nonlinear equations of arbitrary order in dimension N was first proved by Bernis [2], using an energy method. Further improvements to this method were given in Bernis [3], [4]. The exposition made here follows Bernis [3].

We remark that the results of this chapter also apply to the obstacle problem as well as to equations with a singular term such as, for instance, that studied in Section 2.3.

Energy methods have also been applied to elliptic-parabolic systems by Antoncev [3].

4 The general theory for second order nonlinear elliptic equations: a particular overview

The main objective of this chapter is to review some existence, uniqueness, comparison and regularity results for a class of second order nonlinear elliptic equations which contains the quasilinear model equation of Chapter 1, as well as a large part of the variants considered in Chapters 2 and 3. The main aim of this survey is to complete the information on the problems considered in previous chapters and it is far from being an exhaustive survey of the theory of second order elliptic equations. Thus, results are sometimes discussed in relation to equations that we are already using, rather than in terms of their most general application.

Section 4.1 deals with the case in which the solutions are sought in an energy space associated, in a natural way, with the equation. We first collect some results in the spirit of the calculus of variations, giving later their general or abstract formulation, which leads to the consideration of monotone operators and their generalizations. Some regularity results are reviewed, emphasizing the L^{∞}-estimates, of special interest in earlier chapters. This is also the case of some comparison results that are given here. We also comment briefly on some existence methods via the existence of super and subsolutions.

Finally, Section 4.2 is devoted to the case in which the right-hand term of the equation is not in the dual of the energy space. We first consider the case of semilinear equations in L^1 and other spaces, and follow with some abstract results which, formulated in terms of accretive operators, allow the consideration of quasilinear equations.

4.1. SOLUTIONS IN THE ENERGY SPACE

4.1a. Some first existence results via minimization of functionals

In this section we shall review some first existence results for the Dirichlet problem

$$- \operatorname{div} A(x,u,\nabla u) + f(x,u) = g(x) \quad \text{in} \quad \Omega \tag{1}$$

$$u = h \quad \text{on} \quad \partial\Omega \tag{2}$$

The question of the existence of solutions of partial differential equations with prescribed values on the boundary of the region was the twentieth of the prophetic problems included by Hilbert in his famous list of the most important mathematical questions of his time. After Hilbert's selection in 1900, this subject has burst into flower throughout our century and has been the starting point of many new branches of Mathematics. Many deep answers to this question have been given by numerous authors by developing the Calculus of Variations and have also answered other related crucial problems in Hilbert's list : the twenty-third and the nineteenth. In this context, the main idea to solve (1),(2) is to understand the equation as the Euler-Lagrange equation associated with the minimization ,in a set of admissible functions, of the integral

$$J(u) = \int_\Omega F(x,u,\nabla u)\, dx \tag{3}$$

for some suitable function F. Indeed, roughly speaking, if u belongs to a set K_h of smooth enough functions defined on $\overline{\Omega}$, satisfying $u = h$ on $\partial\Omega$, and if u minimizes J on K_h, then

$$J(u) \leqslant J(u + t\zeta)$$

for all $t \in \mathbb{R}$ and every $\zeta \in K_0$, i.e, $\zeta = 0$ on $\partial\Omega$. So, the function $\Phi(t) = J(u + t\zeta)$ is such that $\Phi(0) \leqslant \Phi(t)$ $\forall t \in \mathbb{R}$, and Φ has a minimum at $t = 0$, whence $\Phi'(0) = 0$ which, by differentiation, gives that the first variation of J at u vanishes, i.e.

$$\int_\Omega \Big(\sum_{i=1}^{N} \frac{\partial F}{\partial p_i} \frac{\partial \zeta}{\partial x_i} + \frac{\partial F}{\partial u} \zeta \Big) dx = \partial J(u)(\zeta) = 0 \tag{4}$$

for every $\zeta \in K_0$. That is, the function u can be understood as a weak solution of the Euler-Lagrange equation

$$-\sum_{i=1}^{N} \frac{\partial}{\partial x_i}\left(\frac{\partial F}{\partial p_i}(x,u,\nabla u)\right) + \frac{\partial F}{\partial u}(x,u,\nabla u) = 0. \tag{5}$$

Before other remarks, we note that, obviously, equation (1) coincides with (5) if we set

$$A(x,u,p) = \left(\frac{\partial F}{\partial p_1}(x,u,p),\ldots,\frac{\partial F}{\partial p_N}(x,u,p)\right) \tag{6}$$

$$f(x,u) - g(x) = \frac{\partial F}{\partial u}(x,u,p). \tag{7}$$

In this way, equation (1) is more general than (5) since the A's for a variational problem would satisfy (6) and hence, if they are regular, must satisfy

$$\frac{\partial}{\partial p_i} A_j(x,u,p) = \frac{\partial}{\partial p_j} A_i(x,u,p)$$

if $A(x,u,p) = (A_i(x,u,p))$, $i = 1,\ldots,N$.

We have found weak solutions of (5) as <u>stationary points</u> of J (i.e. functions for which the first variation of J is zero) and then a natural question is to study which of these points are extremals of J. If, for instance, J is of class C^2 and $u \in C^1(\overline{\Omega})$ minimizes J among all functions of class C^1 and with prescribed boundary value, then

$$\frac{d^2}{dt^2} J(u + t\zeta)\Big|_{t=0} \geqslant 0$$

for all $\zeta \in C_0^1(\Omega)$. From here, it is easy to see that, for each $x_0 \in \Omega$, u must satisfy

$$\sum_{i,j=1}^{N} \frac{\partial^2 F}{\partial p_i \partial p_j}(x_0,u(x_0),\nabla u(x_0))\xi_i\xi_j \geqslant 0 \quad \forall \xi = (\xi_1,\ldots\xi_N) \in \mathbb{R}^N. \tag{8}$$

If (8) is satisfied with the strict inequality, the functional J is called <u>regular elliptic</u>. This is equivalent to the convexity of F(x,u,p) with respect to p as well as to the <u>strong ellipticity</u> of Euler-Lagrange equation (5) (see Morrey [1]). However, this condition is not sufficient, in general, to assure that a stationary point u is a minimum point (the answer is positive if F does not depend on u)

The existence of a minimum point for J can be proved by using the so called *direct methods in the Calculus of Variations*. Roughly speaking, the main idea is to show that:

i) the integral to be minimized is bounded from below (in the class of admissible functions), so that the infimum and, therefore, a minimizing sequence, exists;

ii) the functional $J(u)$ is lower-semicontinuous (l.s.c) with respect to a suitable notion of convergence in the class of admissible functions;

iii) a minimizing sequence converges with respect to an admissible function u.

Perhaps the simplest statement in this context is the Weierstrass theorem, which asserts that if X is a separable topological space, $J:X \to (-\infty,+\infty]$ is an (l.s.c) functional, $J \not\equiv +\infty$, and if K is a compact subset of X, then there exists a minimum point for J in K. In any case, direct methods were also used by Riemann, Hilbert, Lebesgue and, especially Tonelli. In the works of these authors, a crucial point was to introduce the adequate notion of convergence in the class of admissible functions. Tonelli was able to deal only with integral functions of the type

$$F(x,u,\xi) \geq C|\xi|^p - \psi(x) \quad , p > 1 , C > 0 , \psi \in L^1(\Omega) \qquad (9)$$

in the particular case of $N = 2$ by working in the class of absolutely continuous functions with the uniform convergence. Later, around 1930, Morrey completed this program by using some function classes of the type of the Sobolev spaces $W^{1,p}(\Omega)$ and proving, in particular, existence results such as the following :

<u>Theorem 4.1.</u> *Let Ω be an open bounded set with smooth boundary. Suppose that*

i) $F(x,u,\xi)$ is measurable in x for all (u,ξ), continuous in u and ξ for a.e. x and convex in ξ for a.e. x and all u.

ii) F *is coercive in the sense that, for instance,* F *satisfies* (9). *Finally, let* K *be a class of functions* $u \in W^{1,p}(\Omega)$ *which is closed with respect to the weak convergence in* $W^{1,p}(\Omega)$ *and in which the greatest lower bound of* $J(u)$ *is finite. Then* $J(u)$ *takes on its minimum in* K. □

The proof of the above theorem is based in the lower-semicontinuity with respect to the weak convergence in $W^{1,p}(\Omega)$ of J, which is derived from hypotheses i) and ii). Today, the literature on the lower-semicontinuity of functionals like J is very broad. For instance, it was proved in Ball[1]and Marcellini-Sbordone [1]that the convexity of $F(\cdot,\cdot,\xi)$ is a necessary condition for J to be l.s.c. (A general exposition is made, e.g, in Giaquinta [1]).

Rather than giving here a detailed proof of Theorem 1.1, we shall present a particular result, *of the same nature*, but including some other situations which have appeared in previous Chapters : the case of Ω unbounded and the case in which $F(x,u,\xi)$, for (x,ξ) fixed, is neither differentiable nor convex in u. As a model of the equation (1) we shall consider the particular equation

$$-\Delta_p u + f(x,u) = g(x) \quad \text{on} \quad \Omega, \tag{10}$$

where $p > 1$, Δ_p is the pseudo-Laplacian operator defined in the Introduction, and Ω is an open set of $\mathbb{R}^N$ not necessarily bounded. About the term $f(x,u)$ we shall assume that its primitive function

$$j(x,u) = \int_0^u f(x,r)dr$$

is such that

$$\left.\begin{array}{l} j:\Omega \times \mathbb{R} \to [0,+\infty) \text{ is measurable in } x \text{ and l.s.c on } u \text{ and} \\ \exists\, q > -1 \text{ and } C_1 \geq 0 \text{ such that } C_1|u|^{q+1} \leq j(x,u) \text{ a.e. } x \text{ and} \\ \forall u \in \mathbb{R}. \end{array}\right\} \tag{11}$$

So, we are interested in minimizing the functional

$$J(u) = \frac{1}{p}\int_\Omega |\nabla u|^p dx + \int_\Omega j(x,u(x))dx - \int_\Omega g\,u\,dx. \tag{12}$$

In order to study the Dirichlet problem (10),(2), the set of admissible functions on which J will be minimized depends on the boundedness or unboundedness of Ω, as well as on the values of p and q. For instance, when Ω is bounded, a choice of such a set including the boundary condition (2) is

$$K = \{u \in L^1(\Omega) : (u - h) \in W_0^{1,p}(\Omega) \text{ and } j(x,u) \in L^1(\Omega)\} .$$

This choice determines some natural assumptions on g and h in order to apply the direct methods. So, we will assume

$$g = g_0 + \sum_{i=1}^{N} \frac{\partial g_i}{\partial x_i} \quad , \quad g_0, g_i \in L^{p'}(\Omega). \tag{13}$$

That is, $g \in W^{-1,p'}(\Omega)$, the dual of $W_0^{1,p}(\Omega)$. To ensure that J is finite on K, it is enough to assume

$$h \in W^{1,p}(\Omega) \quad , \quad j(\cdot,h(\cdot)) \in L^1(\Omega). \tag{14}$$

The case of Ω unbounded is somewhat different because the condition $(u - h) \in L^p(\Omega)$ is too restrictive. Nevertheless, if $0 < q \leq p - 1$ then conditions $j(\cdot,u(\cdot)) \in L^1(\Omega)$ and (11) with $C_1 > 0$ show that $u \in L^{q+1}(\Omega)$ and we can still work with the topology of $W^{1,p}(\Omega)$ because this space coincides with $W^{1,p}(\Omega) \cap L^{q+1}(\Omega)$ and both topologies are equivalent. We shall refer later to the case $q > (p - 1)$.

Theorem 4.2. (a) *Assume Ω bounded,* $p > 1$, (11), (13) *and* (14). *Then there exists at least $u \in K$ minimizing J on K.* (b) *For Ω unbounded, the same conclusion is true if, in addition,* $C_1 > 0$, $q \leq p - 1$ *and* h *has compact support.*

Proof. We shall only prove part b). Slight modifications allow us to conclude part a). Let M be the infimum of J in K and $\{u_n\}$ a minimizing sequence. Then, by the assumptions on q and g, $\int_\Omega |\nabla u|^p\,dx$ is bounded and so, there is some constant C independent of n such that

$$\int_\Omega |\nabla u_n|^p dx \leq C \, , \quad \int_\Omega j(x,u_n(x))dx \leq C \quad \text{and} \quad \int_\Omega |u_n|^{q+1} dx \leq C.$$

Therefore, if $v_n = u_n - h$

$$\int_\Omega |\nabla v_n|^p \, dx \leqslant C \quad \text{and} \quad \int_\Omega |v_n|^{q+1} \, dx \leqslant C.$$

Let $\tilde{v}$ denote the zero extension of v outside Ω. It is well-known that the operator $v \to \tilde{v}$ is an isometric embedding from $W_0^{1,p}(\Omega)$ into $W^{1,p}(\mathbb{R}^N)$ (see Adams [1]). On the other hand, due to the assumption $q \leqslant p - 1$ the norm

$$|||v|||_{1,p,q} = ||v||_{q+1} + \sum_{i=1}^{N} ||\frac{\partial v}{\partial x_i}||_p \tag{15}$$

is equivalent to the usual norm of $W^{1,p}(\mathbb{R}^N)$,as we can deduce from the Nirenberg-Gagliardo inequalities (see e.g. Lemma 3.7). Since $W_0^{1,p}(\Omega)$ is reflexive, there is a subsequence (which we still label $\{v_n\}$) and a function $v \in W_0^{1,p}(\Omega)$ such that $\nabla v_n \to \nabla v$ weakly in L^p and $\tilde{v}_n \to \tilde{v}$ weakly in $W^{1,p}(\mathbb{R}^N)$. Given a ball $B \subset \mathbb{R}^N$, the Rellich's compact embedding theorem implies the existence of a new subsequence of $\{\tilde{v}_n\}$ which converges strongly in $L^p(B)$ to $\tilde{v}$. A new extraction gives a subsequence which converges a.e. to $\tilde{v}$ in B. Covering $\mathbb{R}^N$ by an increasing sequence of balls, by diagonal extraction we obtain a subsequence $\{\tilde{v}_n\}$ which converges a.e. to $\tilde{v}$ in $\mathbb{R}^N$ and therefore a.e. in Ω. In conclusion,

$$u_n(x) = v_n(x) + h(x) \to u(x) = v(x) + h(x) \quad \text{a.e. } x \text{ in } \Omega .$$

Finally, $(u-h) \in W_0^{1,p}(\Omega)$ and $j(x,u) \in L^1(\Omega)$ by the lower semi-continuity of $j(x,\cdot)$ and Fatou's lemma. Using the l.s.c. of j and the L^p-norm, the L^p-weak convergence of ∇u_n to ∇u and Fatou's lemma again, we obtain that $J(u) \leqslant M$ that implies $J(u) = M$.□

Remark 4.1. If $q > p - 1$ and Ω is unbounded, an existence theorem in $W^{1,p}(\Omega)$ cannot be obtained. Indeed, the natural set of admissible functions is

$$V^{1,q+1,p}(\Omega) = \{u \in L^{q+1}(\Omega) \quad , \quad \nabla u \in L^p(\Omega)\} \tag{16}$$

with the topology induced by the norm $|||\cdot|||_{1,p,q}$ given in (15). When $q > p - 1$ this norm is not equivalent to that of the space $W^{1,p}(\Omega)$ and the minimum of J must be looked for in $V^{1,q+1,p}(\Omega)$. Again, the boundary condition holds by asking $(u - h) \in V_0^{1,q+1,p}(\Omega)$, where $V_0^{1,q+1,p}(\Omega)$ is the closure of $C_0^\infty(\Omega)$ in $V^{1,q+1,p}(\Omega)$. The existence result in such a space is an easy modification of the above proof. $\square$

A careful revision of the above proof shows that much more general statements are possible. For instance, concerning the general formulation in equation (1), if we assume A independent of u and such that

$$A(x,\xi)\cdot\xi \geq C_1 |\xi|^p - \psi(x) \quad , \quad p > 1 , \ C_1 > 0 , \ \psi \in L^1(\Omega), \tag{17}$$

and $f(x,u)$ such that its primitive in u, $j(x,u)$, satisfies (11), then we can find $u \in W^{1,p}(\Omega)$ minimizing the functional J given by (3) for the choice

$$F(x,u,\xi) = F_0(x,\xi) + j(x,u) - g_0(x)u - g_i(x)\xi_i \tag{18}$$

where F_0 is some function which is assumed to exist, such that

$$A(x,\xi) = \left(\frac{\partial F_0}{\partial p_1}(x,\xi), \ldots , \frac{\partial F_0}{\partial p_N}(x,\xi)\right). \tag{19}$$

We also assume Ω bounded or $q \leq p - 1$, applying, otherwise, the considerations made in Remark 4.1. Nevertheless, in some important cases, the *coercivity* assumption (17) is not satisfied. This happens for the equation of (non parametric) surfaces with prescribed mean curvature

$$-\sum_{i=1}^{N} \frac{\partial}{\partial x_i} \frac{u_{x_i}}{\sqrt{1+|\nabla u|^2}} + f(x,u) = 0. \tag{20}$$

This is also the case of many other equations; for instance, for the equation

$$-\operatorname{div}\left(\frac{\operatorname{Ln}(1+|\nabla u|)}{|\nabla u|}\nabla u\right) + f(x,u) = 0$$

in which the corresponding F_0 satisfying (19) is given by $F_0(\xi) = \xi \operatorname{Ln}(1+\xi)$. More generally, if we assume that there exists a convex function $a(t)$, for $t \in \mathbb{R}^+$, such that

$$\frac{a(|\xi|)}{|\xi|} \to 0 \quad \text{if} \quad \xi \to 0 , \quad \xi \in \mathbb{R}^N \tag{21}$$

and

$$\sum_{i=1}^{N} A_i(x,u,\xi_i)\xi_i \geq a(|\xi|) , \tag{22}$$

then the *direct methods* can be applied on the Orlicz-Sobolev spaces $W^{1,a}(\Omega)$ instead of on the space $W^{1,p}(\Omega)$. To introduce such an approach, assume that in (22) we have equality. In this case, equation (1) is the Euler-Lagrange equation of the minimum of J given by

$$J(u) = \int_\Omega \{ \int_0^{|\nabla u|} \frac{a(s)}{s} ds + \int_0^u f(x,t)dt - \int_\Omega g\,u\,dx \} dx , \tag{23}$$

that is, (1) takes the form

$$- \operatorname{div} \left(\frac{a(|\nabla u|)}{|\nabla u|^2} \nabla u\right) + f(x,u) = g(x) , \tag{24}$$

which coincides with equation (10) for the particular choice $a(t) = t^p$. We shall not give here the existence theorem (similar to Theorems 4.1 and 4.2) on such spaces; however, to point out in which way such a result would be adapted to the new setting, we recall the main definitions: The Orlicz-space $L^a(\Omega)$ is defined as follows:

$$L^a(\Omega) = \{f : \Omega \to \mathbb{R} \text{ measurable: } \exists\lambda \text{ such that } \int_\Omega a(|\frac{f(x)}{\lambda}|)dx < \infty\} .$$

$L^a(\Omega)$ is a Banach space with the norm

$$\| f \| = \inf \{\lambda > 0 : \int_\Omega a \left(\frac{|f(x)|}{\lambda} \right)dx \leq 1\}.$$

Finally, the Orlicz-Sobolev space $W^{1,a}(\Omega)$ is defined by

$$W^{1,a}(\Omega) = \{u \in L^a(\Omega) : \frac{\partial u}{\partial x_i} \in L^a(\Omega) \text{ for every } i = 1,\dots,N \}$$

with its natural norm. Note that if $a(t) = t^p$ then $W^{1,a} = W^{1,p}$, the usual Sobolev space. Existence results in Orlicz-Sobolev spaces are due to many authors such as Vishik, Donaldson, Gossez, Gossez-Hess, Fougeres, Vaudene etc. (see e.g. Gossez [1] and its bibliography). An alternative to avoid the use of such spaces when assumption (17) fails can be found, for instance, in Attouch-Damlamian [1], but many others authors have also considered non-coercive problems (see Baiocchi et al. [1]). Note that in Remark 1.2 equation (24) is formulated in other different way. □

Remark 4.3. There are still other directions in which theorems 4.1 and 4.2 can be generalized. On the one hand, with slight changes in the proof it can be applied to higher order problems, working in the Sobolev spaces $W^{m,p}(\Omega)$ (see e.g Giaquinta [1]and Bernis [2]). On the other hand, assumption (11) is often only needed in a neighbourhood of the origin. This is the case of equation (10) with f satisfying (11) for some $q > 0$. (See Benilan-Brezis-Crandall[1] and Bidaut-Veron [2]). □

Now we return to the problem of solving the boundary value problem (1), (2). As we have pointed out, for the special convex set $K = K_h$, any stationary point on K of the functional J given by (3),(6) and (7) is a weak solution of (1) assumed J to be of class C^1. Nevertheless this last condition is quite strong, and it is enough to let J be differentiable in $W^{1,p}(\Omega)$. Any case, it is clear that some control on the growth of the functions $F_u(x,u,\xi)$ and $F_{p_i}(x,u,\xi)$ is needed in order to give a meaning to the integrals which appear in formula (4).

It is not difficult to show that if u is a stationary point of J, $u \in K \subset W^{1,p}(\Omega)$, and K satisfies that

$$u + t v \in K \quad \text{for every } v \in C_0^\infty(\Omega) \text{ and } t \in [-1,1], \tag{25}$$

then some sufficient conditions assuring the differentiability of J in $W^{1,p}(\Omega)$ are the following.

$$\left.\begin{aligned} &|F_u(x,u,\xi)| \leqslant \mu[\psi_1(x) + |u|^q + |\xi|^{\frac{pq}{q+1}}] , \\ &|F_{p_i}(x,u,\xi)| \leqslant \mu[\psi_2(x) + |u|^{\frac{(q+1)(p-1)}{p}} + |\xi|^{p-1}] \end{aligned}\right\} \quad (26\text{ a})$$

if $1 < p < N$, and

$$\left.\begin{aligned} |F_u(x,u,\xi)| &\leq \mu(|u|)[\psi_3(x) + |\xi|^p], \\ |F_{p_i}(x,u,\xi)| &\leq \mu(|u|)[\psi_2(x) + |\xi|^{p-1}] \end{aligned}\right\} \quad (26\text{ b})$$

if $p > N$, where $q + 1 = \frac{pN}{N-p}$ and q is an arbitrary number if $p = N$. Here, the functions ψ_1, ψ_2 and ψ_3 belong to the space $L^{(q+1)/q}(\Omega)$, $L^{p/(p-1)}(\Omega)$ and $L^1(\Omega)$ respectively and μ is a positive constant or some continuous non-negative function respectively. Moreover, under these assumptions, (4) holds for every $\zeta \in W_0^{1,p}(\Omega)$. (See Ladyzhenskaya-Uraltseva [1]).

In many important applications, the class of functions K where the minimum point of J is found does not satisfy (25) and it is assumed to be merely a closed convex set. A special example of this class of problems, called generally Variational Inequalities, is the *obstacle problem* considered in Section 2.2, which corresponds to the choice

$$K = \{v \in W^{1,p}(\Omega) : v = h \text{ on } \partial\Omega, \ v \geq \psi \text{ on } \Omega\}$$

for some given function ψ. (Many other examples can be found, for instance, in the books, Baiocchi-Capelo [1], Bensoussan-Lions[1],Duvaut-Lions [1], Elliot-Ockendon [1], Friedman [3] and Kinderlehrer - Stampacchia [2]. Now we only have that

$$J(u) \leq J(u + t(v - u))$$

for all $t \in [0,1]$ and $v \in K$. Assumed (26), this is equivalent to saying that u satisfies

$$\int_\Omega \Big[\sum_{i=1}^N \frac{\partial F}{\partial p_i}(x,u,\nabla u)\frac{\partial(v-u)}{\partial x_i} + \frac{\partial F}{\partial u}(x,u,\nabla u)(v-u)\Big]dx \geq 0 \quad \forall\, v \in K. \quad (27)$$

Nevertheless, in some situations the integrand $F(x,u,\xi)$ is not differentiable and another characterization of u is needed. For instance, this is the case of the functional J given by (12) when $q \in (-1,0]$. In that case, it is easy to see that any minimum point u of J on a convex set

K of $W^{1,p}(\Omega)$(for instance, that associated to the Dirichlet boundary conditions) satisfies

$$\int_\Omega |\nabla u|^{p-2} \nabla u \cdot \nabla(v-u)dx + \int_\Omega [j(x,v)-j(x,u)]dx \geq \langle g , v-u \rangle \tag{28}$$

for every $v \in K$.

To end this subsection, we remark that the differentiability of J has also been considered in the class of Orlicz-Sobolev spaces $W^{1,a}(\Omega)$ (see e.g. Fucik-Kufner [1] and Landes [1]for a different approach).

4.1b. <u>Some extensions of variational problems: Monotone operators and its generalizations</u>

It is easy to see that if A satisfies

$$|A_i(x,u,\xi)| \leq C_1(|u|^{p+1} + |\xi|^{p-1} + \psi_i(x)), \quad \psi_i \in L^{p'}(\Omega), \ C_1 > 0$$

(or more generally under(6),(7) and (26)),then the formal operator

$$Au = - \operatorname{div} A(x,u,\nabla u) \tag{29}$$

determines a continuous operator from $W^{1,p}(\Omega)$ into its dual $(W^{1,p}(\Omega))'$ by means of

$$\langle Au,\zeta \rangle = \int_\Omega A(x,u,\nabla u) \cdot \nabla \zeta \, dx$$

for every $u,\zeta \in W^{1,p}(\Omega)$. So the Dirichlet problem (1),(2) leads to the abstract formulation

$$Au + f(\cdot,u) = g \quad \text{in} \quad V' \tag{30}$$

$$u \in K \tag{31}$$

where V is the Banach space $V = W^{1,p}(\Omega)$ and K is the affine subspace

$$K = h + W_0^{1,p}(\Omega) = \{ u \in W^{1,p}(\Omega) \ : u = h \text{ on } \partial\Omega \} .$$

When we study the abstract problem (30),(31) we can also obtain existence results for the Dirichlet problem (1),(2) and even without the symmetry restriction assumed in subsection 4.1a. There exists a long bibliography about this type of problems. The abstract operator is usually assumed to satisfy some kind of monotonicity condition that already appears derived from the convexity of the functional F in (3), when (1) is the associated Euler-Lagrange equation.

Definition 4.1. *Let* V *be a Banach space of dual* V'. *An operator* $A : V \to V'$ *is called monotone if*

$$< Au - Av \ , u - v >_{V'V} \geq 0$$

for every $u, v \in V$.

It is easy to see that the operator A given by (29) is a monotone operator on the space $V = W_0^{1,p}(\Omega)$ if, for instance, $A_i = A_i(x,\xi)$, i.e., independent of u , and

$$\sum_{i=1}^{N} (A_i(x,\tilde{\xi}) - A_i(x,\xi))(\tilde{\xi}_i - \xi_i) \geq 0$$

a.e. $x \in \Omega$, for every $\xi, \tilde{\xi} \in \mathbb{R}^N$. In order to study more general situations, several classes of operators for which (30) (31) can be solved have been introduced by different authors, namely Leray-Lions, Browder and Brezis (see, e.g., the exposition by J.L. Lions [1] and its bibliography). Among these classes of operators, known as "operators of the Calculus of Variations", "semi monotones","of type M " and "pseudo-monotones", the last , introduced by Brezis [1], seems to be the most useful. Its exact definition abstracts the main properties required by A in order to solve the abstract equation (30).

Definition 4.2. *An operator* $A : V \to V'$ *is called pseudo-monotone if for every* u_j *such that* $u_j \to u$ *in* V *weakly and* $\limsup \langle Au_j, u_j - u \rangle \leq 0$ *then* $\liminf < Au_j, u_j - v > \geq < Au, u - v > \quad \forall v \in V$.

Important examples of such pseudo-monotone operators are provided by the bounded, hemicontinuous monotone operators and, more in general, by the operators of the Calculus of Variations (see J.L. Lions [1]and, for the case of unbounded domains, Browder [2]). In particular, the operator A given in (29) is an operator of the Calculus of Variations (and, thus , pseudomonotone) assuming Ω bounded $V = K = W_0^{1,p}(\Omega)$, (23) and

$$\sum_{i=1}^{N} (A_i(x,u,\tilde{\xi}) - A_i(x,u,\xi))(\tilde{\xi}_i - \xi_i) > 0 \tag{32}$$

In fact, it turns out that assumption (32) is a necessary condition for the operator A to be pseudomonotone (Boccardo-Dacorogna [1])

The following existence result is due to Brezis-Browder [2],

Theorem 4.3 *Let* Ω *be an arbitrary open subset of* $\mathbb{R}^N$. *Let* $f(x,u):\Omega\times\mathbb{R}\to\mathbb{R}$ *be a Caratheodory function (i.e. measurable on* x *and continuous on* u) *such that*

$$\forall s > 0 , \sup_{|u|\leq s} |f(x,u)| \leq \phi_s(x) \in L^1(\Omega) \tag{33}$$

$$f(x,u)u \geq 0 \quad \text{a.e.} \quad x \in \Omega \quad \text{and} \quad \forall u \in \mathbb{R}. \tag{34}$$

Let $A: W_0^{m,p}(\Omega) \to W^{-m,p'}(\Omega)$ *be a pseudomonotone operator which maps bounded sets into bounded sets and which is coercive in the sense that*

$$\lim_{||u||_{m,p}\to\infty} \frac{<Au,u>}{||u||_{m,p}} = +\infty . \tag{35}$$

Then, for every $g \in W^{-m,p'}(\Omega)$ *there exists a* $u \in W_0^{m,p}(\Omega)$ *such that* $f(x,u)$ *and* $f(x,u)u \in L^1(\Omega)$ *, and*

$$<Au,v> + \int_\Omega f(x,u)v\, dx = <g,v> \tag{36}$$

for every $v \in W_0^{m,p}(\Omega) \cap L^\infty(\Omega)$ *and* $v = u.$ □

A sufficient condition in order that (35) holds for the particular A given in (29) is

$$\sum_{i=1}^{N} A_i(x,u,\xi)\xi_i \geq C\,|\xi|^p - \psi(x), \quad \psi \in L^1(\Omega) \ , \ C > 0 \tag{38}$$

(compare this with (9)).

There is also an abstract formulation of variational inequalities such as for instance, (27) in terms of

$$< Au + f(x,u) - g, v-u >_{V'V} \geq 0 \qquad \forall v \in K \tag{39}$$

$$u \in K \tag{40}$$

where A is a pseudomonotone operator and K is a closed convex set of V (Brezis [1]). For the special case of A given by (29), f satisfying (33) and (34) , and $K = \{v \in W_0^{1,p}(\Omega) \cap L^\infty(\Omega) : v \geq \psi$ a.e. on $\Omega\}$, an existence result was given in Boccardo-Giachetti [1] (see also their references).

With respect to variational inequalities with non-differentiable terms, as (28),we also remark that again the second order differential operator may be substituted by another more general generating a pseudo-monotone operator in $V = W^{m,p}(\Omega)$, $m \geq 1$ (see J.L. Lions [1] and Bidaut-Veron [2]).

An important question associated with the above variational inequalities is to try to find an equation, and not only an inequation, characterizing the solution. To undertake this question we shall restrict ourselves to the inequation (28) and, as a first step , we reformulate it in the following way

$$\int_\Omega |\nabla u|^{p-2}\nabla u \cdot \nabla(v-u)dx + J_h(v-h) - J_h(u-h) \geq < g, v-u > \quad \forall v \in W_0^{1,p}(\Omega) + \{h\}, \tag{41}$$

$$u - h \in W_0^{1,p}(\Omega), \tag{42}$$

where the functional J_h is defined by

$$J_h(v) = J_0(v+h) \quad , \quad J_0(v) = \begin{cases} \int_\Omega j(x,v)dx & \text{if} \quad j(x,v) \in L^1(\Omega) \\ +\infty & \text{otherwise} . \end{cases} \tag{43}$$

We note that J_0 is a convex, l.s.c, functional on $L^p(\Omega)$ (Brezis [4]) and so the restriction of J_0 on $W_0^{1,p}(\Omega)$ has the same properties. The second step in this program is to recall the notion of *subdifferential* of a convex l.s.c functional on a Banach space. This leads to the consideration of eventually mutivalued operators, and we shall take advantage of this opportunity to present some abstract results of interest in existence theory.

Definition 4.3. *Let* V *be a Banach space of dual* V' *and let* J *be a convex, l.s.c. function of* V *on* $(-\infty,+\infty]$, $J \not\equiv +\infty$. *We call subdifferential of* J *the (eventually multivalued) operator* $\partial J : D(\partial J) \to P(V')$ *given by*

$$\partial J(u) = \{w \in V' : J(v) - J(u) \geq \langle w,v-u\rangle_{V'V} \quad , \forall v \in V\} ,$$

where

$$D(\partial J) = \{u \in V : \partial J(u) \neq \emptyset\} \quad , (\emptyset \equiv \text{the empty set}).$$

It is not difficult to see (e.g. Barbu [1]) that the set $\partial J(u)$ is in fact single-valued when J is Gateaux-differentiable at u. However, there are frequent examples in which ∂J is a multivalued operator. For instance, this is the case of $V = \mathbb{R}$ and $J = j$, as in the following two examples:

Example 4.1. Let j be given by

$$j(r) = 0 \quad \text{if} \quad r \leq 0 \quad , \quad j(r) = r \quad \text{if} \quad r > 0.$$

Then

$$\partial j(r) = \{0\} \text{ if } r < 0 \quad , \quad \partial j(0) = [0,1] \text{ and } \partial j(r) = \{1\}, \text{if } r > 0.$$

Example 4.2. Let $j : \mathbb{R} \to (-\infty,+\infty]$ be given by

$$j(r) = +\infty \text{ if } \quad r < 0 \quad , j(r) = 0 \quad \text{if} \quad r \geq 0.$$

Then

$$\partial j(r) = \emptyset \text{ (the empty set) if } r < 0, \partial j(0) = (-\infty,0] \text{ and } \partial j(r) = \{0\} \text{ if } r > 0.$$

As in the Gateaux-differentiable case, the operator ∂J has some kind of monotonicity property derived from the convexity of J. To explain this we need to extend the notion of monotone operator given in Definition

4.1. to the class of eventually multivalued operators.

Definition 4.4. *Let* V *be a Banach space of dual* V'. *An operator* $A : D(A) \subset V \to \mathcal{P}(V')$ *is called monotone if*

$$\langle v - \tilde{v} , u - \tilde{u} \rangle \geq 0 \quad \textit{for every } u,\tilde{u} \in D(A), v \in Au \textit{ and } \tilde{v} \in A\tilde{u}.$$

If, in addition, there is no monotone operator B *on* V *such that* $A \subset B$ *in the sense of graphs, then* A *is called maximal monotone on* V *(or, sometimes, maximal monotone graph of* $V \times V'$*).*

It turns out that any subdifferential operator is a maximal monotone operator. On the other hand, introducing the duality map $J:V \to \mathcal{P}(V')$ given by $Ju = \{v \in V' : ||v|| = ||u||\}$, or equivalently $J = \partial \frac{1}{2} ||\cdot||^2$, then, at least for reflexive Banach spaces V, it is possible to characterize the maximal monotone operators by means of all the monotone operators satisfying the range condition

$$R(J + \lambda A) = V' \quad , \quad \text{for every} \quad \lambda > 0. \tag{44}$$

Note that if $V = H$ is an Hilbert space, then J is the identity and (44) becomes $R(I + \lambda A) = H$. We shall go back to other range conditions in Section 4.2. (Concerning the theory of maximal monotone operators, we refer to the books Brezis [6], Browder[1]and Pascali-Sburlan [1]).

The special class of maximal monotone operators on $V = \mathbb{R}$ (or maximal monotone graphs of $\mathbb{R}^2$, m.m.g, in an abbreviated way) turns out to be very useful in the treatment of nonlinear PDE's with discontinuous nonlinearities , as well as for Variational Inequalities. It can be easily shown that any m.m.g. β of $\mathbb{R}^2$ is the subdifferential of some convex l.s.c. funtion $j : \mathbb{R} \to (-\infty,+\infty]$, $j \not\equiv +\infty$, and, in fact, the following characterization holds:

Proposition 4.4. *Let* β *be a graph of* $\mathbb{R}^2$. *Then:* β *is a* m.m.g *if and only if there exists a nondecreasing real function* b *such that*

$$\beta(r) = [b(r-),b(r+)] \quad \textit{if} \quad -\infty < b(r-) \leq b(r+) < +\infty$$

$$\beta(r) = (-\infty,b(r+)] \quad \textit{if} \quad -\infty = b(r-) < b(r+) < +\infty$$

and

$$\beta(r) = [b(r-),+\infty) \qquad -\infty < b(r-) < b(r+) = +\infty . \square$$

Given a maximal monotone operator A on a Banach space V, sometimes it is useful to assign to A some single valued operators (called sections of A). When V is reflexive the most important section is the so-called principal section of A, which is denoted by A° and is obtained in the following way: $\forall u \in D(A)$, $A^\circ u$ = element of the set Au of minimum norm (such a minimum element always exists because Au is a closed convex set, see Barbu [1]). In the particular case of β being a m.m.g of $\mathbb{R}^2$, there are still other sections (i.e. real eventually discontinuous functions) of some interest:

$$\beta^\circ(r) = \{s_0 \in \beta(r) : |s_0| \leqslant |s| \quad \text{for every} \quad s \in \beta(r)\}$$

$$\beta^+(r) = \{s_+ \in \beta(r) : s \leqslant s_+ \quad \text{for every} \quad s \in \beta(r)\}$$

$$\beta^-(r) = \{s_- \in \beta(r) : s_- \leqslant s \quad \text{for every} \quad s \in \beta(r)\}.$$

To be more clear on this point , we remark that if we use the notation $\beta = \partial j$ in Example 4.1, then $\beta^\circ(r) = \beta^+(r) = \beta^-(r) = \{0\}$ if $r < 0$, $\beta^\circ(r) = \beta^+(r) = \beta^-(r) = \{1\}$ if $r < 0$ but $\beta^\circ(0) = \beta^-(0) = \{0\}$ and $\beta^+(0)=\{1\}$.

Now we return to the question of characterizing the solution of the Variational Inequality (41),(42). If, for the sake of simplicity, we assume $h = 0$ then, using the notion of subdifferential and choosing $V = W_0^{1,p}(\Omega)$, expression (41) is equivalent to

$$-\Delta_p u + \partial J_0(u) \ni g \quad \text{in} \quad V' = W^{-1,p'}(\Omega). \tag{45}$$

Thus, (45) leads to the study of ∂J_0 as an operator from $W_0^{1,p}(\Omega)$ into $W^{-1,p'}(\Omega)$. For different reasons (regularity results, study of the associated Cauchy Problem and so on),it is very interesting to know how regular any element of $\partial J_0(u)$ is, and, in particular, if they are functions and not only distributions of V' when $g \in W^{-1,p'}(\Omega)$.The study of such operators is due to Brezis [4] when j is independent on x and Ω is bounded, Grun-Rehome [1]for Ω bounded, and more generally, to Bidaut-Veron [2],[3] who also considers the case $h \neq 0$ (where ∂J_0 needs to be replaced in (38) by $\partial J_h(u-h)$) and Ω unbounded. These works also deal with the m-order space $W^{m,p}(\Omega)$. The following result collects a particular version of the above mentioned results

Theorem 4.5. *Let* Ω *be an arbitrary open subset of* $\mathbb{R}^N$. *Let* β *be a maximal monotone graph of* $\mathbb{R}^2$, $\beta = \partial j$, *satisfying* $\beta(0) = 0$, *and*

$$D(\beta) = \mathbb{R} \tag{46}$$

If Ω *is unbounded, we also assume*

$$|\beta^\circ(r)| \geq k|r|^{p-1} , \forall r \in [-R,R] \text{ for some } p>1 , R>0 \text{ and } k>0, \tag{47}$$

Then , for every $g \in W^{-1,p'}(\Omega)$ *there exists* $u \in W_0^{1,p}(\Omega)$ *and* $w \in L^1_{loc}(\Omega) \cap W^{-1,p'}(\Omega)$ *such that*

$$w(x) \in \beta(u(x)) \quad \text{a.e.} \quad x \in \Omega ,$$

$$- \Delta_p u + w = g \quad \text{in} \quad W^{-1,p'}(\Omega). \tag{48}$$

Remark 4.4. Condition (46) is neccesary in order to assure that any element T of $\partial J_0(u)$ (J_0 given by (43)) is in fact a function $T \in L^1_{loc}(\Omega)$ with $T \in \partial j(u)$ a.e. on Ω. When (46) does not hold, it is known (see the mentioned references) that T is a measure of Lebesgue descomposition $T = \gamma + S$, where $\gamma \in L^1_{loc}(\Omega)$ and S is a nontrivial singular measure. The characterization of ∂J_h, when J_h is associated with the obstacle problem ($u \geq \psi$ on Ω) can also be found in Bidaut-Veron [3]. Theorem 1.4 also holds for more general quasilinear operators as, for instance, the given by (29), and even for 2m order quasilinear operators (Bidaut-Veron [2], [3]). We remark that when $g \in L^{p'}(\Omega)$ it is proved (see also section 4.2) that $w \in L^{p'}(\Omega)$ and then $\Delta_p u \in L^{p'}(\Omega)$. □

For a general diffusion operator, such as the one given in (29) , there is another point of view, in order to characterize the solution of the

Variational Inequality (41). It is, essentially, a hilbertian method due to Brezis [3] and Altouch-Damlamian [1]: For the sake of simplicity, let us assume Ω bounded and $h=0$. The main idea is to take $g \in L^2(\Omega)$ and to write (41) as

$$\int_\Omega g(u-v)dx + \phi(v) - \phi(u) \geq 0 \qquad \forall v \in D(\phi), \tag{49}$$

where ϕ is the convex, l.s.c. functional, from $L^2(\Omega)$ into $(-\infty,+\infty]$, defined by

$$\phi(u) = \begin{cases} \frac{1}{p}\int_\Omega |\nabla u|^p dx + \int_\Omega j(x,u)dx, \text{if } u \in W_0^{1,p}(\Omega)\cap L^2(\Omega) \text{and } j(x,u) \in L^1(\Omega) \\ +\infty \qquad\qquad , \text{otherwise.} \end{cases}$$

$$D(\phi) = \{u \in L^2(\Omega) : \phi(u) < +\infty\} .$$

Then, assuming that j is a normal convex integrand, it turns out that ϕ is convex, l.s.c. on $L^2(\Omega)$ and so (49) is equivalent to the equation

$$g \in \partial\phi(u).$$

As remarked before, this approach replaces the dual of the energy space (Orlicz-Sobolev spaces, for general quasilinear equations) by the space $L^2(\Omega)$. We note that the above theorems in this subsection give some surjectivity results for the operator $\partial\phi$. Again, the hard problem is to characterize the operator $\partial\phi$ and, more precisely, its effective domain $D(\partial\phi)$. Such questions are related to the regularity of the solutions of (45). The answers for our particular formulation, coincide essentially with Theorem 4.5, but it is important to point out that general quasilinear operators such as the one given in (29) can also be treated without coercivity assumptions of the type of (17).

To end this subsection we shall give some remarks about the variational approach to some other problems considered in the above Chapters.

Remark 4.5. Problem (30),(31) includes in a obvious way the case in which the operator A is given through a general second order elliptic operator L as the considered in Theorem 1.13. Again the coercivity assumption (35) is assured by the (uniform) ellipticity of L (see (38) or (66) of Section 1.1). Existence results for degenerate elliptic linear operators can by obtained easily if the term $f(\cdot,u)$ is coercive (e.g. $R(f(\cdot,u)) = \mathbb{R}$) (see Murthy-Stampacchia [1] and Alvino-Trombetti [1]). □

Remark 4.6. We point out that, in some cases, the existence of solutions for the Hamilton-Jacobi-Bellman problem (see subsection 2.4b) can be obtained through a suitable variational inequality. (See Brezis-Evans [1]).

Remark 4.7. The variational approach can be also applied to derive existence theorems for other boundary conditions like the considered in Section 2.5. (See e.g. Ladyzhenskaya-Uraltseva [1], J.L.Lions [1], Brezis [5] and Amann [2]). □

4.1c. On the regularity of solutions. L^∞_{loc}-estimates

The question of the regularity of the solution of quasilinear equations is one of the more studied problems in the theory of PDE. After the pioneering results by Bernstein in 1904 giving a partial answer to the nineteenth problem of Hilbert, many important results and methods have been developed by many authors; Hopf, Morrey, Agmon, Douglis, Nirenberg, De Giorgi, Bombieri, Maser, Stampacchia, Ladyzenskaya, Uraltseva, Trudinger... etc. Obviously, in this subsection we shall not make an exposition of all those results, for which the reader can see the books by Morrey [1], Ladyzenskaya-Ural'tseva [1], Stampacchia [2], Gilbarg-Trudinger [1], Friedman [3], Giaquinta [1] etc. We shall only review here some of those regularity results, more or less in connection with the study of the formation and properties of the free boundary $\mathcal{F}(u)$ made in previous chapters.

We shall start with the L^s-regularity. For the sake of simplicity in the exposition we shall only consider weak solutions $u \in W^{1,p}(\Omega)$ with $f(u) \in L^{p'}(\Omega)$ of the equation

$$- \Delta_p u + f(u) = g. \qquad (50)$$

Here the equation is satisfied in the sense that $\forall\zeta \in W_0^{1,p}(\Omega)$ we have

$$\int_\Omega \{|\nabla u|^{p-2} \nabla u\cdot\nabla\zeta + f(u)\zeta\}dx = \int_\Omega \{g_0\zeta - \sum_{i=1}^N g_i \frac{\partial\zeta}{\partial x_i}\}dx ,$$

assumed $g \in W^{-1,p'}(\Omega)$, g given by (13). Our propose is to show that under suitable additional conditions on g, any weak solution u of (50) belongs locally or globaly to the space L^s for some $p \leq s \leq \infty$. The results are of different nature according the assumptions on the term f(u). We first consider the general case of

$$f(r)r \geq 0 \qquad \forall r \in \mathbb{R} , \tag{51}$$

and we center our attention on L^∞-estimates, already used in previous Chapters. First of all we recall that, by the Sobolev and Morrey imbeding theorems (see e.g Adams [1]), we have

$$u \in L^{p^*}(\Omega) \text{ if } p < N , u \in L^s(\Omega) \text{ if } p = N \text{ and } u \in L^\infty(\Omega) \text{ if } p > N,$$

being $p^* = pN/(N-p)$ and s any positive real number. So, it suffices to consider the case $p \leq N$. The following result was established by different authors (see, e.g., Serrin [1]): Let x_0 be a fixed point of Ω and let $\|\cdot\|_{s,R}$ be the L^s norm of a function on the open ball $B_R(x_0)$.

Theorem 4.6. *Let* $u \in W^{1,p}$ *be a weak solution of equation* (50) *defined in some ball* $B_{2R}(x_0) \subset \Omega$. *Assume* $p \leq N$, (51) *and let*

$$g = g_0 + \sum_{i=1}^N \frac{\partial g_i}{\partial x_i} \text{ with } g_0 \in L^{\frac{N}{p-\varepsilon}} \text{ and } g_i \in L^{\frac{N}{p-1}} \text{, for some } \varepsilon > 0. \tag{52}$$

Then

$$\|u\|_{\infty,R} \leq C\, R^{-N/p} (\|u\|_{p,2R} + k\, R^{N/p}) , \tag{53}$$

where C *and* k *are constants,* $C = C(p,N,\varepsilon)$ *and*

$$k = C'(p) (\Sigma \|g_i\| + R^\varepsilon \|g_0\|)^{1/(p-1)} , C'(p) > 0, \tag{54}$$

the norms of the data functions taken in the respective Lebesgue spaces. Moreover, if u *is a weak solution of* (50) *in a domain* $D \subset \Omega$ *and* $|u| \leq M$ *on the boundary of* D, *then* $u \in L^{\infty}(D)$ *and*

$$\|u\|_{L^{\infty}(D)} \leq M + C(\|g_0\| + \sum_i \|g_i\|) \tag{55}$$

where C *is a positive constant depending on* $|D|, p, N$ *and* ε. □

Remark 4.6. If the hypothesis (52) on g is not satisfied but g_0 and g_i belong to some suitable spaces, others L^s-estimates on u may be obtained. So, if $u \in W^{1,p}(\Omega)$ is a weak solution of (50) on Ω with $u \leq M$ on $\partial\Omega$, and $g_i \in L^r(\Omega)$, $g_0 \in L^{r'}(\Omega)$ with $p' \leq r < \frac{N}{p-1}$, then $u \in L^s(\Omega)$ with $s = [r(p-1)]^*$ (See e.g., Serrin[1] Stampacchia [2] and Boccardo-Giachetti [2]).

Remark 4.7. Optimal global L^s-estimates may be obtained by means of rearrangement techniques. See subsection 1.3a and the references indicated there. □

The above results are derived esentially from the information on the diffusion term $\operatorname{div}(|\nabla u|^{p-2}\nabla u)$ but, in fact, in the mentioned references they are stated for more general diffusion terms of the form $\operatorname{div} A(x,u,\nabla u)$. More precise L^s-estimates can be derived when some growing conditions are known on the absorption term $f(u)$. First of all, we recall the general estimate

$$\|f(u)\|_{L^s(\Omega)} \leq \|g\|_{L^s(\Omega)}, \quad 1 \leq s \leq \infty, \tag{56}$$

true for nondecreasing real functions (and even when $f = \beta$ is a maximal monotone graph of $\mathbb{R}^2$) (See Brezis-Strauss [1], Benilan [1] , Boccardo-Giachetti [2] etc).

To illustrate some other sharper L^s-estimates we shall restrict ourselves to the semilinear case $p = 2$. We first remark that, in that case, solutions of the corresponding equation (50) can be introduced by a duality argument and without the condition $u \in H^1(\Omega)$:

Definition 4.5. *A function* $u \in L^1_{loc}(\Omega)$ *is a very weak solution of the equation*

$$- \Delta u + f(u) = g \qquad (57)$$

if $f(u) \in L^1_{loc}(\Omega)$ *and*

$$- \int_\Omega u \, \Delta \zeta + \int_\Omega f(u)\zeta = \int_\Omega g\zeta \qquad (58)$$

for every $\zeta \in D(\Omega)$. Note that the above definition makes sense when $g \in L^1_{loc}(\Omega)$ and even when $g \in M(\Omega)$, the space of bounded Radon measures on Ω. In both cases it is known (Stampacchia [2]) that if u is a very weak solution, $u \in W^{1,s}_{loc}(\Omega)$ with $1 \leqslant s < N/(N-1)$. On the other hand, it is clear that, if u is a very weak solution with $u \in H^1(\Omega)$ and $f(u) \in L^2_{loc}(\Omega)$, u satisfies equation (57) in the standard weak sense.

An example of the sharp L^s-estimates mentioned above is the following:

Theorem 4.7. *Let* $u \in W^{1,1}_0(\Omega)$ *with* $f(u) \in L^1(\Omega)$ *be a very weak solution of equation* (57) *for* $g \in L^s(\Omega)$, *for some* $1 \leqslant s < \infty$. *Assume that* f *is locally Lipschitz continuous on* $\mathbb{R} - \{0\}$, *and that*

$$C \geqslant f'(r)\, |f(r)|^{m-1} \qquad \text{a.e.} \quad r \in \mathbb{R} - \{0\} \qquad (59)$$

for some $C > 0$ *and* $m > 0$. *Then if* $N \geqslant 3$ *we have the estimate*

$$\int_\Omega |f(u)|^s dx + C'\left(\int_\Omega |f(u)|^{(s+m-1)N/(N-2)} dx\right)^{(N-2)/N} \int_\Omega |g|^s \, dx \qquad (60)$$

with C' *constant independent of* s. *Moreover, if* $N(m-1) + 2s > 0$ $f(u) \in L^\infty(\Omega)$.□

Remark 4.8. The above and other similar results were proved by several authors as a preliminar step to show "regularizing effects" for the porous media equation (Benilan [4], Veron [1], Pazy [1], Benilan-Berger [1] etc). The best constant C' may be computed explicitly by rearrangement techniques (Vazquez [3]). Note that here the regularity is obtained "a priori", and the existence of such solutions is given in the next subsection. Finally, regularizing properties as Theorem 4.7 are also available for the pseudo-Laplacian operator $\Delta_p u$ (See Veron[1] and Herrero-Vazquez [1]).□

Remark 4.9. L^s- estimates for the variational inequality (41),(42) can be found, for instance, in Brezis [5] and Boccardo-Giachetti [2].□

Other different regularity properties of weak solutions of (50) are concluded from the information $g \in L^r(\Omega)$ for some $p' \leq r \leq \infty$, or, more exactly, from the fact that $\Delta_p u \in L^r(\Omega)$. The linear case $p = 2$ is, in this point, quite different from the quasilinear one, $p \neq 2$. Indeed, by the results of Agmon-Douglis-Nirenberg [1] it is well-known that $\Delta u \in L^r$ implies that $u \in W^{2,r}_{loc}(\Omega)$ if $1 < r < \infty$, being $u \in W^{2,t}_{loc}(\Omega)$ for every $1 < t < \infty$ if $\Delta u \in L^\infty(\Omega)$. Nevertheless, if, for instance, $u \in W^{1,p}(\Omega)$ and $\Delta_p u \in L^{p'}(\Omega)$ then, $u \in W^{m,p}_{loc}(\Omega)$ with $m = 1 + 1/(p-1) - \varepsilon$ if $p > 2$ (Simon [1]) and $u \in W^{2,p}_{loc}(\Omega)$ if $1 < p \leq 2$ (De Thelin [1]) where the regularity is optimal in both cases, (a sharper regularity is given in terms of Besov spaces). Finally, $\Delta_p u \in L^\infty(\Omega)$ implies that $u \in W^{2,2}_{loc}(\Omega) \cap W^{1,\infty}_{loc}(\Omega)$ if $p \geq 2$ and $u \in W^{2,p}_{loc}(\Omega) \cap W^{1,p+2}_{loc}(\Omega)$ if $p \leq 2$ (Tolksdorf [1]).□

Remark 4.10. The above $W^{2,r}$ - regularity can also used to obtain several different conclusions. So, for instance, from the information $\Delta u \in L^r(\Omega)$ for some $1 < r < \infty$ and the Morrey's theorem we conclude that $u \in C^{1,s}_{loc}(\overline{\Omega})$ when $r > N$ and $s = 1 - N/r$. Moreover, if u is a weak solution of the semilinear equation (57) and we assume $f \in C^{0,\alpha}(\mathbb{R})$ and $g \in C^{0,\beta}(\Omega)$ for some $0 < \alpha, \beta < 1$ we can apply the Schaüder theory to the linear expression $-\Delta u = H(x)$, $H(x) = g(x) - f(u(x))$ and then we obtain that $u \in C^{2,\tilde{\beta}}(\overline{\Omega})$ for some $0 < \tilde{\beta} < 1$. If in fact $f(u)$ is given by $f(u) = |u|^{\alpha-1} u$ with $0 < \alpha < 1$ and $g \equiv 0$, from the only information $\Delta u \in L^1$ we can obtain L^∞- estimates by using an bootstrap argument (See Gallouet-Morel[3]).□

Remark 4.11 The regularity $C^{0,\alpha}$ and $C^{1,\alpha}$ for elliptic quasilinear equations was first obtained by De Giorgi [1] and, after, considered by many others Moser, Serrin , Stampacchia , etc. (See e.g. the books mentioned in the beginning of this subsection where other regularity results are quoted. For quasilinear degenerate equations like (50) the $C^{1,\alpha}$-regularity is due to Uraltseva, Uhlenbeck , Evans , Lewis and more recently , Di-Benedetto [1] and Tolksdorf [1](see also Stredulinsky [1]). We also mention the results of Brezis, Brezis-Stampacchia, Brezis-Kinderlehrer, Caffarelli and many others for variational inequalities (see the books mentioned on this problem) and the ones due to Phillips and Giaquinta-Giusti for singular equations mentioned in Section 2.3. Finally we point out that L^s-estimates and the $C^{0,\alpha}$, and $C^{1,\alpha}$ regularity may be shown for local minimum points of functionals J by working directly with the functional J instead of working with its Euler-Lagrange equations. (see Giaquinta [1] and the references there in).□

We shall end this subsection by collecting an L^∞-estimate on the set Int N(g) , used in Chapter 1.

Theorem 4.8. *Let* $g \in W^{-1,p'}(\Omega)$, $h \in W^{1,p}(\Omega)$ *with* $j(h) \in L^1(\Omega)$, $j(r) = \int_0^r f(t)dt$. *Let* $u \in W^{1,p}(\Omega)$ *be a variational and weak solution of*

$$-\Delta_p u + f(u) = g \quad in \quad \Omega \tag{61}$$

$$u = h \quad on \quad \partial\Omega , \tag{62}$$

where f *is a continuous real function satisfying* (51). *Let* D *be a compact set of* $\mathbb{R}^N$ *such that* $D \subset \text{int } N(g)$ *and let* $d = d(D,\partial N(g))$. *Then* $u \in L^\infty(D)$ *and*

$$||u||_{L^\infty(D)} \leq C(||g||^{p'/p}_{W^{-1,p'}(\Omega)} + ||h||_{W^{1,p}(\Omega)} + ||j(h)||^{1/p}_{L^1(\Omega)}) \tag{63}$$

for some constant C *only dependent on* p,N,Ω *and* d.

Proof.- First we shall deduce the inequality

$$||u||_{W^{1,p}(\Omega)} \leq C(||g||^{p'/p}_{W^{-1,p'}(\Omega)} + ||h||_{W^{1,p}(\Omega)} + ||j(h)||^{1/p}_{L^1(\Omega)}) \tag{64}$$

for some constant C only dependent on p. Due to the special form of the equation (61), u minimizes a functional J (see (12)) on the set $K = \{v \in W^{1,p}(\Omega)\ ,\ v = h \text{ on } \partial\Omega\}$, and so

$$\int_\Omega |\nabla u|^p dx + \int_\Omega j(u)dx - \langle g,u\rangle \leq \int_\Omega |\nabla h|^p dx + \int_\Omega j(h)dx - \langle g,h\rangle. \qquad (65)$$

But by hypothesis (51) $j(r) \geq 0$ for every $r \in \mathbb{R}$. Then, from (65), the characterization of $W^{-1,p'}(\Omega)$ (see (13)) and Young's inequality, we deduce (64) . The conclusion (63) is now obvious from (64) if $p > N$. From $p \leq N$ we apply Theorem 4.6 for $x_0 \in D$ and $2R = d(x_0, \partial N(g))$. Then

$$||u||_{L^\infty(B_R(x_0))} \leq C\, d^{-N/p} \left(||u||_{L^p(B_{2R(x_0)})}\right) \qquad (66)$$

(note that $2R \geq d$ and that in (53) now is $k = 0$ because $g \equiv 0$ on $B_{2R}(x_0)$) . Finally estimate (63) follows from (64) and (66).$\square$

Remark 4.12. The dependence of C on d given in (66) is not sharp. On the other hand, other L^∞-estimates on $N(g)$ may be obtained under some growing conditions on $f(u)$. For the case of $g \in L^1(\Omega)$ see later Theorem 4.18. Finally, we remark that Theorem 4.8. remains true when u is merely a variational solution (see Remark 4.11).$\square$

4.1d. Uniqueness and comparison results. Existence via comparison.

The uniqueness of solution of (1),(2) can be derived by different methods according to the nature of the terms A and f involved in the equation.

When the solution of (1),(2) is obtained by the minimization of some convex l.s.c. functional J on a convex set K of a Banach space V, it is well-knowm that, if J is strictly convex , then the minimum is unique. Indeed, the set of minimum points is a closed convex subset of K and so, if u_1, u_2 are two different elements of K minimizing J , then $(u_1+u_2)/2$ is also a minimum point and, from the strict convexity of J,

$$J\left(\frac{u_1+u_2}{2}\right) < \frac{1}{2}(J(u_1) + J(u_2)) = \text{minimum of } J \text{ on } K.$$

Hence, if for instance in Theorem 4.2 we assume $p > 1$ and $j(x,u)$ convex l.s.c., then the functional J given by (12) is strictly convex and so there exists a unique $u \in K$ minimizing J on K.

By the characterizations mentioned in Subsection 4.1b, the above uniqueness criterion applies to the model equation (10) and, more generally, to the multivalued equation (48). The following result is stronger because it gives the uniqueness as a consequence of the comparison of solutions. In order to include the "obstacle problem" we shall state this result for the variational inequality (28) (note that, in fact, it reduces to the equation (10) when $j \in C^1(\mathbb{R})$).

Theorem 4.9. *Let Ω bounded, $p > 1$ and $j(x,u) : \Omega \times \mathbb{R} \to [0,+\infty]$, $j \not\equiv +\infty$, be measurable on x and convex l.s.c on u. Let $g, \tilde{g} \in W^{-1,p'}(\Omega)$ such that $g \leq \tilde{g}$ in $W^{-1,p'}(\Omega)$ (i.e. $\langle \tilde{g} - g, \zeta \rangle \geq 0 \ \forall \zeta \in W_0^{1,p}(\Omega)$, $\zeta \geq 0$). Let also $h, \tilde{h} \in W^{1,p}(\Omega)$ such that $h \leq \tilde{h}$ on $\partial\Omega$ in the sense of traces. Then, if u and $\tilde{u}$ are functions satysfying the corresponding variational inequality (28) then $u \leq \tilde{u}$ on Ω.*

We first prove the following useful inequalities:

Lemma 4.10 *Let $\xi, \tilde{\xi} \in \mathbb{R}^N$. Then there exists $C > 0$ such that if $p \geq 2$*

$$(|\tilde{\xi}|^{p-2}\tilde{\xi} - |\xi|^{p-2}\xi)\cdot(\tilde{\xi} - \xi) \geq C\,|\tilde{\xi} - \xi|^p \tag{67}$$

and if $1 < p \leq 2$

$$(|\tilde{\xi}|^{p-2}\tilde{\xi} - |\xi|^{p-2}\xi)\cdot(\tilde{\xi} - \xi) \geq C\,\frac{|\tilde{\xi} - \xi|^2}{(|\tilde{\xi}| + |\xi|)^{2-p}} \quad \text{if } |\tilde{\xi}| + |\xi| \neq 0 \tag{68}$$

Proof. By homogeneity and symmetry, it is enough to consider $|\tilde{\xi}| = 1$ and $|\xi| \leq 1$. Moreover, by choosing an adequate coordinate system we can assume $\tilde{\xi} = (1,0,0\ldots0)$ and $\xi = (\xi_1, \xi_2, 0, \ldots, 0)$. If, for instance, $1 < p \leq 2$, (68) is equivalent to

$$\left[\left(1 - \frac{\xi_1}{(\xi_1^2 + \xi_2^2)^{\frac{2-p}{2}}}\right)(1-\xi_1) + \frac{\xi_2^2}{(\xi_1^2 + \xi_2^2)^{\frac{2-p}{2}}}\right] \frac{(1+(\xi_1^2+\xi_2^2)^{\frac{1}{2}})^{2-p}}{(1-\xi_1)^2 + \xi_2^2} \geqslant C \qquad (69)$$

for every $\xi = (\xi_1,\xi_2)$, $\xi_1^2 + \xi_2^2 \leqslant 1$, $\xi \neq (1,0)$. But (67) holds with $C = p-1$, as a consequence of the following inequalities

$$1 - \frac{\xi_1}{(\xi_1^2 + \xi_2^2)^{\frac{2-p}{2}}} \geqslant 1 - \frac{\xi_1}{\xi_1^{2-p}} \geqslant (p-1)(1-\xi_1) \quad \text{if} \quad 0 \leqslant \xi_1 \leqslant 1,$$

$$1 - \frac{\xi_1}{(\xi_1^2 + \xi_2^2)^{\frac{2-p}{2}}} \geqslant 1 - \xi_1 \geqslant (p-1)(1-\xi_1) \quad \text{if} \quad \xi_1 \leqslant 0,$$

$$\frac{1}{(\xi_1^2 + \xi_2^2)^{\frac{2-p}{2}}} \geqslant 1 \quad \text{and} \quad 1 + (\xi_1^2 + \xi_2^2)^{\frac{2-p}{p}} \geqslant 1.$$

The case $p \geqslant 2$ can be treated in an analogous way and we omit the proof. □

<u>Remark 4.13</u>. This proof is due to Simon [1]. Other different proofs of this key lemma can be seen in Glowinski-Maroco [1], Hartman-Stampacchia [1] and Morrey [1] p.208. □

<u>Proof of Theorem 4.9.</u> Consider $v_1 = \min\{u,\tilde{u}\} = u - (u-\tilde{u})^+$ and $v_2 = \max\{u,\tilde{u}\} = \tilde{u} + (u-\tilde{u})^+$. By the truncation result of Stampacchia [2] and the assumption $h \leqslant \tilde{h}$ on $\partial\Omega$ we have that $v_1 - h$ and $v_2 - \tilde{h}$ belong to $W_0^{1,p}(\Omega)$. On the other hand, it is clear that

$$\int_\Omega j(x,v_1(x))dx + \int_\Omega j(x,v_2(x))dx \leqslant \int_\Omega j(x,u(x))dx + \int_\Omega j(x,\tilde{u}(x))dx \quad (70)$$

and so $j(x,v_1)$, $j(x,v_2) \in L^2(\Omega)$. Now, taking $v = v_1$ and $v = v_2$ in the corresponding variational inequalities (28) and adding the two expressions we obtain

$$\int_\Omega (|\nabla u|^{p-2}\,\nabla u - |\nabla\tilde{u}|^{p-2}\,\nabla\tilde{u})\cdot\nabla(u-\tilde{u})^+\,dx \leq 0.$$

Then by Lemma 4.1 and the fact that $\nabla(u-\tilde{u})^+(x) = 0$ and $\nabla(u-\tilde{u})^+(x) = \nabla(u-\tilde{u})(x)$ on the sets $\{x \in \Omega : u(x) \leq \tilde{u}(x)\}$ and $\{x \in \Omega ; u(x) \geq \tilde{u}(x)\}$, respectively , we conclude that

$$C \int_\Omega |\nabla(u-\tilde{u})^+|^p\,dx \leq 0 \qquad \text{if} \quad p \geq 2$$

and

$$C \int_\Omega \frac{|\nabla\,(u-\tilde{u})^+|^2}{(|\nabla u| + |\nabla\tilde{u}|)^{2-p}}\,dx \leq 0 \qquad \text{if} \quad 1 < p \leq 2.$$

Then $(u-\tilde{u})^+$ is a constant function on Ω . But $h \leq \tilde{h}$ on $\partial\Omega$ and so $u \leq \tilde{u}$ a.e. $x \in \Omega$.□

Remark 4.14. With suitable changes in the proof, the above conclusion holds in many other situations. For instance, other choices of the convex K in the variational inequality (28) are possible (Brezis [5]); the diffusion operator $-\Delta_p u$ may be substituted by a general coercive quasilinear operator (Hartman-Stampacchia [1]) or a general second order operator Lu as the given in (64) of Section 1.1 (Stampacchia [2]); Ω may be unbounded and $u \in V^{1,q+1,p}(\Omega)$ (Remark 4.1); and so on .□

It is obvious that if $g = \tilde{g}$ and $h = \tilde{h}$ then we get that $u = \tilde{u}$, which gives the uniqueness of solutions for the variational inequality (28) and so, for its corresponding single or multivalued Euler-Lagrange equations (10) or (48) . We also note that taking $g \leq 0$ (resp. $g \geq 0$) and $h \leq 0$ (resp. $h \geq 0$) then $u \leq 0$ (resp. $u \geq o$), property known as the weak maximum (minimum) principle. We recall that an optimal version of the strong maximum principle was given in Subsection 1.2a. (In fact, a strong maximum principle for the difference function $\tilde{u} - u$ is available if, for instance, f is Lipschitz continuous : See Vazquez [5], and Brezis [5] for variational inequalities).

The comparison principle may be obtained for the general class of problems (1),(2) without the coercivity assumption on the diffusion term div $A(x,u,\nabla u)$.

Theorem 4.11 *Let* Ω *be a bounded open set. For* $j = 1,2$, *let* $h^j \in W^{1,p'}(\Omega)$ *and* $g^j \in W^{-1,p'}(\Omega)$ *with*

$$g^j = g_0^j + \sum_{i=1}^{N} \frac{\partial}{\partial x_i} g_i^j \quad , \qquad g_0^j,\ g_i^j \in L^{p'}(\Omega).$$

Let $A(x,r,\xi)$ *and* $f(x,r)$ *be Caratheodory functions such that*

$$\left.\begin{array}{l} A(x,r,\xi) \text{ is monotone with respect to } \xi \text{ and Lipschitz} \\ \text{continuous with respect to } r , \end{array}\right\} \quad (71)$$

$$f(x,r) \text{ is nondecreasing in } r,\ \ f(x,0) = 0, \text{for a.e. } x \in \Omega. \quad \} \quad (72)$$

Let $u^j \in W^{1,p}(\Omega)$ *with* $A(x,u^j,\nabla u^j)$, $f(x,u^j) \in L^{p'}(\Omega)$, *be weak solutions of* (1),(2) *corresponding to* $g = g^j$ *and* $h = h^j$. *Then, if*

$$h^1 \leqslant h^2 \ \text{ on } \ \partial\Omega \text{ , and , } \frac{\partial g_i^1}{\partial x_i} \leqslant \frac{\partial g_i^2}{\partial x_i} \ \text{ in } \ D'(\Omega) \text{ , } \forall 1 \leqslant i \leqslant N, \qquad (73)$$

the following estimate holds

$$\| [f(x,u^1) - f(x,u^2)]^+ \|_{L^1(\Omega)} \leqslant \| [g_0^1 - g_0^2]^+ \|_{L^1(\Omega)} . \qquad (74)$$

If, in particular, $f(x,u)$ *is strictly increasing in* u *for a.e.* $x \in \Omega$, *and* $g_0^1 \leqslant g_0^2$, *we conclude that* $u^1 \leqslant u^2$ *on* Ω .

Proof. Let $T \in W^{1,\infty}(\mathbb{R})$ be a nondecrasing function such that $T(r) = 0$ if $r \leqslant 0$ (i.e. $T' \geqslant 0$ a.e.). By well-known results (Stampacchia [2]), $\zeta = T(u^1 - u^2)$ is a test function , $\zeta \in W_0^{1,p}(\Omega) \cap L^\infty(\Omega)$ and hence

$$\int_\Omega (A(x,u^1,\nabla u^1) - A(x,u^2,\nabla u^2)) \cdot \nabla(u^1-u^2) T'(u^1-u^2)dx +$$
$$+ \int_\Omega (f(x,u^1) - f(x,u^2)) \ T(u^1-u^2)dx \leqslant \int_\Omega (g_0^1-g_0^2) \ T(u^1-u^2)dx.$$

But,from the assumption (71),

$$\int_\Omega (A(x,u^1,\nabla u^1) - A(x,u^2,\nabla u^2))\cdot \nabla(u^1-u^2)T'(u^1-u^2)\,dx =$$

$$\int_\Omega (A(x,u^1,\nabla u^1)-A(x,u^1,\nabla u^2))\cdot\nabla(u^1-u^2)dx + \int_\Omega (A(x,u^1,\nabla u^2)-A(x,u^2,\nabla u^2)\cdot$$

$$\cdot\ \nabla(u^1-u^2)T'(u^1-u^2)dx \leqslant C\int_\Omega |\nabla(u^1-u^2)|\,|u^1-u^2|\ T'(u^1-u^2)dx.$$

Now, for $n \in \mathbb{N}$, choose $T(r) = T_0(nr)$ with $n \in \mathbb{N}$ and $T_0 \in W^{1,\infty}(\mathbb{R})$, $T_0' \geqslant 0$ and $T_0 = 0$ on $r \leqslant 0$ and $T_0 = 1$ on $r > 2$. Then $rT'(r) = nrT_0'(n,r)$

$$\int_\Omega (f(x,u^1) - f(x,u^2))T_n(u^1-u^2)dx \leqslant \int_\Omega (g_0^1-g_0^2)\ T_n(u^1-u^2) +$$

$$+\ 2C\int_{\{|u^1-u^2|<\frac{2}{n}\}} |\nabla(u^1-u^2)|\ T_0'(n(u^1-u^2))dx.$$

Letting $n \to \infty$, we have that $T_n(r) \to \mathrm{sign}_0^+(r)$ (defined by 0 if $r \leqslant 0$ and 1 if $r > 0$) and, by (72) and Fatou Lemma,

$$\int_{\{u^1\geqslant u^2\}} (f(x,u^1)-f(x,u^2))dx \leqslant \int_\Omega (g_0^1-g_0^2)^+dx\ ,$$

which proves the estimate (74). $\square$

The Lipschitz dependence in assumption (71) may be replaced by a weaker hypothesis in some cases. We shall illustrate it for the equation

$$-\Delta_p u + \mathrm{div}\,B(x,u) + f(x,u) = g \qquad \text{in} \quad \Omega \tag{75}$$

<u>Theorem 4.12</u>. *Let* $1 < p < \infty$ *and assume* f, g^j *and* h^j *as in Theorem 4.11. Let* $B : \Omega \times \mathbb{R} \to \mathbb{R}^N$ *be a Caratheodory function which is Holder continuous of exponent* $\alpha \geqslant 1/p'$, *with respect to* u, *i.e.* , *such that*

$$|B(x,r) - B(x,s)| \leqslant C|r-s|^\alpha \quad \forall r,s \in \mathbb{R} \quad \text{a.e. } x\in\Omega,\ \frac{1}{p'}\leqslant\alpha\leqslant 1 \tag{75}$$

Let $u^j \in W^{1,p}(\Omega)$ *with* $B(x,u^j) \in L^{p'}(\Omega)$ *and* $f(x,u^j) \in L^1(\Omega)$, *satisfying* (75) *and* $u^j = h^j$ *on* $\partial\Omega$. *Then, estimate* (73) *holds, assumed* (72) .

Proof. As before, we have that

$$\int_\Omega (|\nabla u^1|^{p-2}\nabla u^1 - |\nabla u^2|^{p-2}\nabla u^2)\cdot(u^1-u^2)T'(u^1-u^2)dx +$$

$$+ \int_\Omega (f(x,u^1)-f(x,u^2))T(u^1-u^2)dx$$

$$\leq \int_\Omega (g_o^1-g_o^2)T(u^1-u^2)dx + \int_\Omega (B(x,u^1)-B(x,u^2))\cdot\nabla(u^1-u^2)T'(u^1-u^2)dx,$$

for every $T \in W^{1,\infty}(\mathbb{R})$, $T' \geq 0$, $T(r) = 0$ if $r \leq 0$. Now, for $n \in \mathbb{N}$, choose $T(r) = T_n(r)$ given by $T_n(r) = 0$ if $r \leq 0$, $T_n(r) = 1$ if $r \geq \frac{2}{n}$, $T_n'(r) = n^2r$ if $r \in [0,\frac{1}{n}]$ and $T_n'(r) = -n^2r + 2n$ if $r \in [\frac{1}{n}, \frac{2}{n}]$. Using (76) and Young's inequality ($ab \leq (\varepsilon/p)a^p + (C\varepsilon/p')b^{p'}$, $C_\varepsilon = \varepsilon^{-p'/p}$) we obtain that

$$I_n \equiv \int_\Omega (B(x,u^1) - B(x,u^2))\cdot\nabla(u^1-u^2)T_n'(u^1-u^2)dx$$

$$\leq C \int_\Omega |u^1-u^2|^{\alpha p'}|\nabla(u^1-u^2)|^p T_n'(u^1-u^2)$$

$$\leq \frac{\varepsilon}{p}\int_\Omega |\nabla(u^1-u^2)|^p T_n'(u^1-u^2)dx + \frac{CC_\varepsilon}{p'}\int_\Omega |u^1-u^2|^{\alpha p'}T_n'(u^1-u^2)dx$$

$$\leq \frac{\varepsilon}{p}\int_\Omega |\nabla(u^1-u^2)|^{p'}T_n'(u^1-u^2)dx + \frac{CC_\varepsilon}{p'}\left(\frac{2}{n}\right)^{\alpha p'} n \text{ meas }\{x\in\Omega: 0<u^1-u^2<\tfrac{2}{n}\}.$$

Using Lemma 4.10, we get

$$(\tilde{C} - \frac{\varepsilon}{p})\int_\Omega |\nabla(u^1-u^2)|^p T_n'(u^1-u^2)dx + \int_\Omega (f(x,u^1)-f(x,u^2))T_n(u^1-u^2)dx$$

$$\leq \int_\Omega (g_o^1-g_o^2)T_n(u^1-u^2)dx + \frac{CC_\varepsilon}{p'}\left(\frac{2}{n}\right)^{\alpha p'} n \text{ meas }\{x \in \Omega: 0 < u^1-u^2 < \tfrac{2}{n}\}.$$

Letting $n \to +\infty$ and using that $\alpha \geq 1/p'$ we conlcude as in Theorem 4.11. □

Remark 4.15. It is not difficult to show the same conclusion for every $1 < p < \infty$ when $N = 1$ (Benilan [5]). We also mention the uniqueness result for equation (71) without the strict monotonicity assumption on f but for $p = 2$; this is due to Carrillo-Chipot [2] (they also assume $\alpha \geq 1/2$ unless if $N = 1$). The proof of Theorem 4.11 also applies, for instance, to a linear elliptic second order operator L not necessarily uniformly elliptic (i.e. eventually degenerate). □

Remark 4.16. The comparison of solutions also holds for other boundary conditions. Indeed, slight changes in the proof of Theorems 4.9 and 4.11 allow us to obtain the same conclusions if, for instance $\partial\Omega = \partial_1\Omega \cup \partial_2\Omega$ and we know that

$$u^1 \leqq u^2 \text{ on } \partial_1\Omega \quad \text{and} \quad \frac{\partial u^1}{\partial n} \leqq \frac{\partial u^2}{\partial n} \text{ on } \partial_2\Omega$$

(see, e.g., Ladyzhenskaya-Uraltseva [1], Brezis [5] and Diaz [2]). □

Remark 4.17. When $h \equiv \tilde{h} \equiv 0$, Theorems 4.9 and 4.11 follow from some abstract results stated for certain classes of operators A. In the case of Theorem 4.9, the conclusion is derived from the comparison for T-monotone operators from a Banach lattice V into V' (or $P(V')$), i.e. satisfying

$$\langle Au-Av,(u-v)^+\rangle_{V'V} \geqq 0 \qquad \forall u,v \in V$$

(Haugazeau [1], Brezis-Stampacchia [1], Brezis [5], Tartar [1] etc.). (In fact the proof of Theorem 4.9 shows that the corresponding operator is T-monotone.) Theorem 4.11 is associated with the so-called T-accretive operators from a Banach lattice X into X (or $P(X)$) which will be considered in the next section. □

Up to this moment, the absorption term f(x,u) in the equations has been assumed to be a nonincreasing function of u. Some uniqueness results are still true for more general semilinear problems

$$-\Delta u + f(x,u) = 0 \text{ in } \Omega$$

$$u = 0 \qquad\qquad \text{on } \partial\Omega ,$$

assumed f(x,u)/u increasing and $f(x,u) \geqq -C(u+1)$. See Keller [1], Cohen-Laetsch [1], Laetsch [1], Friedman-Phillips [1] and Brezis-Oswald [1]. For a quasilinear version of those results, see Diaz-Saa [1] .

Another useful comparison result is related to the case of solutions corresponding to equations with different absorption terms.

Theorem 4.13. *Let g, $\tilde{g}$ and h, $\tilde{h}$ be regular as in Theorem 4.9. Let $u \in W^{2,p}(\Omega)$ be a weak solution of (6.1), (6.2), where f is only assumed to be continuous and such that $f(s)s \geqq 0$ $\forall s \in \mathbb{R}$. Let $\tilde{u} \in W^{1,p}(\Omega)$ be the weak* *solution of (6.1), (6.2) substituting* g *by* $\tilde{g}$, h *by* $\tilde{h}$ *and* f *by* $\tilde{f}$, *where* $\tilde{f}$ *is a nondecreasing continuous function such that*

$$|\tilde{f}(s)| \leq |f(s)| \qquad \forall s \in \mathbb{R} \tag{77}$$

Finally, assume that u *and* $\tilde{u}$ *have constant sign in* Ω. *Then* $0 \leq g \leq \tilde{g}$ *and* $0 \leq h \leq \tilde{h}$ *implies* $0 \leq u \leq \tilde{u}$ (*resp.* $0 \geq g \geq \tilde{g}$, $0 \geq h \geq \tilde{h}$ *implies* $0 \geq u \geq \tilde{u}$).

Proof. Assume $0 \leq g \leq \tilde{g}$ and $0 \leq h \leq \tilde{h}$, so, by Theorem 4.9, $\tilde{u} \geq 0$. Then $u \geq 0$ and, by the assumptions on f and $\tilde{f}$, $0 \leq \tilde{f}(u) \leq f(u)$. Moreover,

$$-\Delta_p u + \tilde{f}(u) = g - f(u) + \tilde{f}(u) \leq \tilde{g}$$

in the sense of distributions on Ω. Finally, applying again Theorem 4.9 for the equation associated to $\tilde{f}$, we obtain the conclusion. The proof of the other inequality is analogous. □

Remark 4.18. If $\tilde{f}$ is such that the strong maximum principle holds for the associated equation (61) (for $p = 2$, $\tilde{f}$ Lipschitz continuous is enough), and if the inequality in (77) is strict, we conclude that $u < \tilde{u}$ (resp. $u > \tilde{u}$) on the set $\{x \in \Omega : u > 0\}$ (resp. on the set $\{x \in \Omega : u < 0\}$ (see the proof of Theorem 2.3)). We remark that the conclusion of Theorem 4.12 can also be obtained through Theorem 4.11. We shall use this later (see Theorem 4.18) to obtain L^∞-estimates in the L^1-framework. □

Remark 4.19. All the comparison results given above are stated under some regularity on the domain (Ω is assumed to be of class C^1) and the data ($g \in W^{-1,p'}(\Omega)$, and $h \in W^{1,p}(\Omega)$ or, equivalently, $h \in W^{1-1/p,p}(\partial\Omega)$). Nevertheless, a careful revision of the proofs shows that they remain true if $\partial\Omega$ is locally Lipschitz and the boundary data h may be approximated by h_n with $h_n \in W^{1-1/p,p}(\partial\Omega)$. □

The existence of solutions of quasilinear equations such as, for instance, that one given in (1), can be obtained by means of comparison arguments. These are the so-called super and subsolutions methods. First we recall the notion of such functions:

Definition 4.5. *A function* $\phi \in W^{1,p}(\Omega)$ *is a supersolution of (1), (2) if* $A(x,\phi,\nabla\phi)$, $f(x,\phi) \in L^{p'}(\Omega)$, $\phi \geq h$ *in the sense of traces on* $\partial\Omega$ *and*

$$\int_\Omega A(x,\phi,\nabla\phi)\cdot\nabla\zeta \, dx + \int_\Omega f(x,\phi)\zeta \, dx \geq \int g\zeta \, dx$$

for every $\zeta \in W_0^{1,p}(\Omega)$, $\zeta \geq 0$ a.e. *on* Ω .

Similarly, a subsolution of (1), (2) is defined by the reverse inequality. For the next result we need the operator A associated with the expression $-\mathrm{div}\, A(x,u,\nabla u)$ to be a coercive pseudo-monotone and T-monotone operator from $W_0^{1,p}(\Omega)$ into its dual. This is true if, for instance,

$$A(x,u,\xi) = A(x,\xi), \text{ A satisfies (1.7) and } A(x,\nabla u) \in L^{p'}(\Omega) \ \forall u \in W^{1,p}(\Omega). \tag{78}$$

However, no sign condition on $f(x,u)$ will be assumed.

<u>Theorem 4.14</u>. *Let* Ω *bounded,* $g \in W^{-1,p'}(\Omega)$ *and* $h \in W^{1,p}(\Omega)$. *Assume (78) and let* $f(x,u)$ *be a Caratheodary function such that*

$$f(\cdot,u) \in C^{0,\alpha}(\mathbb{R}) \text{ for some } \alpha \in (0,1) \text{ and } |f(x,\cdot)| \leq \kappa(x), \kappa \in L^{p'}(\Omega). \tag{79}$$

Suppose that ϕ *and* ψ *are bounded super and subsolutions of (1), (2) with* $\phi \geq \psi$ *a.e. in* Ω. *Then problem (1), (2) has at least one solution* $u \in W^{1,p}(\Omega)$ *such that* $\psi \leq u \leq \phi$ *a.e. in* Ω. *Moreover, there exist maximal and minimal solutions* $\bar{u}$, $\underline{u}$, *in the sense that if* u *is any solution of (1), (2) with* $\psi \leq u \leq \phi$, *then*

$$\psi \leq \underline{u} \leq u \leq \bar{u} \leq \phi \quad \text{a.e. in } \Omega. \quad \square$$

The main idea in the proof of the theorem is the following: Let $M > 0$ such that the function

$$\tilde{f}(x,u) = f(x,u) - M|u|^{\alpha-1}u, \quad 0 < \alpha \leq 1,$$

is nonincreasing on the set $[\inf\psi, \sup\phi]$. Define $K = \{v \in L^p(\Omega) : \psi \leq v \leq \phi$ a.e. on $\Omega\}$ and consider the operator $T: K \to W^{1,p}(\Omega)$ given by $T(v) = w$, where w is the solution of the problem

$$-\mathrm{div}\, A(x,\nabla w) + M|w|^{\alpha-1}w = -\tilde{f}(x,v) \text{ in } \Omega$$

$$w = h \quad \text{on } \partial\Omega.$$

Then, it is shown that $T(K) \subset K$ and that T is an order preserving operator. Taking $\bar{u}_0 = \phi$, $\bar{u}_i = T(\bar{u}_{i-1})$, $\underline{u}_0 = \psi$ and $\underline{u}_i = T(\underline{u}_{i-1})$, the sequences $\{\bar{u}_i\}$ and $\{\underline{u}_i\}$ converge monotonically from above or below, respectively, to the solutions $\bar{u}$ and $\underline{u}$ and

$$\psi \leqq \underline{u}_1 \leqq \dots \leqq \underline{u}_i \leqq \dots \leqq \bar{u}_i \leqq \dots \leqq \bar{u}_2 \leqq \bar{u}_1 \leqq \phi \quad \text{a.e. in } \Omega.$$

Finally, it is shown that T admits at least one fixed point u, with $\underline{u} \leqq u \leqq \bar{u}$. □

Remark 4.20. The idea of introducing the iterations $\underline{u}_i = T^i(\underline{u}_0)$ and $\bar{u}^i = T^i(\bar{u}_0)$ is a well-known numerical procedure that was already applied in the pioneering book of Courant-Hilbert [1, pp. 369-372]. This method was improved later in a series of works by Keller, Cohen, Laetsch and Amann (see references in Amann [1]). Many extensions are today available in the literature: see, e.g., Stuart [1] for nonlinear equations with discontinuous terms and Berestycki-Lions [1] for a local approach. In fact, the method of super and subsolutions also applies to more general equations of the form

$$-\text{div } A(x,u,\nabla u) + B(x,u,\nabla u) = g$$

when $|B(x,u,\xi)| \leqq c(|u|)(1 + |\xi|^p)$; however, in this case, the existence of maximal and minimal solutions cannot, in general be assured. See Puel [1], Duel-Hess [1] and Boccardo-Murat-Puel [1], [2]. (The multiplicity of solutions is studied in Amann-Crandall [1].) The case of unbounded domains is considered in Hess [1], Cac [1], [2], Donato-Migliaccio-Schianchi [1] and, more recently, Donato-Giachetti [1]. □

4.2. SOLUTIONS OUTSIDE THE ENERGY SPACE

In some physical situations the assumption $g \in W^{-1,p'}(\Omega)$ is not natural in the context, and one can consider the question of solving problem (1), (2) of Section 4.1 when g is not in such a space but merely an integrable function, $g \in L^1(\Omega)$.

It seems that Stampacchia [1] was one of the first to solve the homogeneous Dirichlet problem in $L^1(\Omega)$ for linear second order operators L,

$$Lu = \sum_{ij=1}^{N} \frac{\partial}{\partial x_i}\left(a_{ij}\frac{\partial u}{\partial x_j}\right) - \sum \frac{\partial}{\partial x_i}(b_i u) - au \tag{1}$$

assumed, for instance, that

$$a_{ij},\ b_i \in C^1(\bar{\Omega}),\ a \in L^\infty(\Omega),$$

$$\Sigma a_{ij}\xi_i\xi_j \geqq \lambda\,|\xi|^2 \text{ with } \lambda > 0,$$

$$a \geqq 0,\ a + \sum_i \frac{\partial b_i}{\partial x_i} \geqq 0 \quad \text{a.e.}\ .$$

To solve the problem

$$-Lu = g \text{ in } \Omega \tag{2}$$

$$u = 0 \text{ on } \partial\Omega \tag{3}$$

when $g \in L^1(\Omega)$ and Ω is bounded, he uses a simple duality argument: let L^* be the formal adjoint operator of L given by

$$L^*\phi = \sum_{i,j} a_{ij} \frac{\partial^2\phi}{\partial x_i \partial x_j} + \frac{\partial a_{ij}}{\partial x_i}\frac{\partial\phi}{\partial x_j} + \sum_i b_i \frac{\partial\phi}{\partial x_i} - a\,\phi\,.$$

As in Definition 4.5, a very weak solution of (2), (3) is defined as any $u \in W_o^{1,1}(\Omega)$ such that

$$\int_\Omega u(-L^*\phi)dx = \int_\Omega g\phi\ dx$$

for any $\phi \in H_o^1(\Omega) \cap L^\infty(\Omega)$ with $L^*\phi \in \mathcal{D}(\Omega)$, i.e.,

$$\int_\Omega u\psi\ dx = \int_\Omega gG^*\psi\ dx$$

for any $\psi \in \mathcal{D}(\Omega)$, where G^* is the Green operator associated with $-L^*$ with homogeneous boundary conditions. By L^∞-regularity (Theorem 4.6), G^* is a linear and continuous operator from $W^{-1,p}(\Omega)$ into $L^\infty(\Omega) \cap H^1(\Omega)$, assumed $p > N$. Then, calling G the dual operator of G^*, G is a linear and continuous operator from $L^1(\Omega)$ into $W_o^{1,q}(\Omega)$ with $1 \leqq q < N/(N-1)$ and so $u = Gg$ is the unique very weak solution of (81), (82).

The study of nonlinear equations in the L^1-setting is much more complicated. In Subsection 4.2a we start considering semilinear equations for which the key result is the Brezis-Strauss theorem for bounded domains. The case of $\Omega = \mathbb{R}^N$ is also discussed. Later, in Subsection 4.2b, we review some abstract results allowing the consideration of quasilinear equations in $L^1(\Omega)$. Those results are stated in terms of the so-called accretive operators and are of interest in the treatment of nonlinear parabolic equations.

4.2a. Semilinear equations in $L^1(\Omega)$ and other spaces

For the sake of simplicity in the exposition we shall be concerned with the semilinear problem

$$-\Delta u + f(x,u) = g \text{ in } \Omega \tag{4}$$

$$u = 0 \text{ on } \partial\Omega. \tag{5}$$

We start by assuming that Ω is a bounded open set. A natural way of trying to solve (4), (5) is by approaching $g \in L^1(\Omega)$ by $g_n \in L^2(\Omega)$. So, assume for the moment $g \in L^2(\Omega)$. In this case, if $f(x,u)$ satisfies assumption (11) or (33), (34) of Section 4.2, by the variational approach we know that there exists $u \in H^2(\Omega) \cap H_0^1(\Omega)$, solution of (4), (5) (and unique if $f(\cdot,u)$ is non-increasing). To pass to the limit, the following estimate is crucial:

$$\int_{\{x\in\Omega: |u|\geq t\}} |f(x,u)|\,dx \leq \int_{\{x\in\Omega: |u|\geq t\}} |g|\,dx \tag{6}$$

This was first proved in Brezis-Strauss [1] when $f(\cdot,u)$ is a maximal monotone graph of $\mathbb{R}^2$. Note that, if we make $t = 0$ in (6), this coincides with (56) of Section 4.1, taking $s = 1$. Several generalizations of the Brezis-Strauss result have been published, and they will be given later. In any case, we note that if $g \in L^2(\Omega)$, (6) is proved as in Theorem 4.11 or 4.12 with obvious modifications (in fact (6), with $t = 0$, can be derived from estimate (74) of Section 4.1). To pass to the limit when $g_n \to g$ in $L^1(\Omega)$ there are two possibilities: when $f(x,u)$ is nondecreasing in u, the main idea is to use the fact that $-\Delta$ is coercive in L^1, the maximum principle and estimate (6) or, more specifically, (74) (see Brezis-Strauss [1]). When $f(x,u)$ is not monotone but satisfies (33), (34) of Section 4.1, the idea is that, from (6), we derive

$$\|\Delta u_n\|_{L^1} \leq 2\|g_n\|_{L^1}.$$

Using the continuous embedding $L^1(\Omega) \to W^{-1,p}(\Omega)$, for $1 \leq p < N/(N-1)$, this proves that $\|u_n\|_{W^{1,p}(\Omega)}$ is bounded and so $u_n \to u$ in $L^1(\Omega)$, $W_0^{1,p}(\Omega)$-weak and almost everywhere. Finally, by truncation of $f(\cdot,r)$ by $f_n(\cdot,r)$, estimate (6) shows the equi-integrability of the sequence $f_n(\cdot,u_n)$ and then, by Vitali's theorem and Fatou's lemma, it is shown that $f_n(\cdot,u_n(x)) \to f(\cdot,u)$ in $L^1_{loc}(\Omega)$

and that u is a solution of (4), (5) (see Gallouet-Morel [1], [2]). The following result summarizes some of the conclusions of the articles mentioned.

<u>Theorem 4.15.</u> *Let* Ω *be a bounded open regular set of* $\mathbb{R}^N$ *and let* $f(x,s):\Omega\times\mathbb{R}\to\mathbb{R}$ *be a continuous function in* s, *measurable in* x *and such that*

$$f(x,s)s \geqq 0 \text{ for all } s \in \mathbb{R}, \tag{7}$$

$$\sup\{|f(\cdot,s)|,\ |s| \leq t\} \in L^1_{loc}(\Omega) \text{ for all } t \in \mathbb{R}^+ \tag{8}$$

Then, for all $g \in L^1(\Omega)$, *problem* (4), (5) *has at least one solution* u, $u \in W_0^{1,1}(\Omega)$, $f(\cdot,u) \in L^1(\Omega)$ *and satisfying* (4) *in* $\mathcal{D}'(\Omega)$. *In the case of* $f(x,s) = \beta(x,s)$, *maximal monotone graph of* $\mathbb{R}^2$, for a.e. $x \in \Omega$, *the existence of a unique* $u \in W_0^{1,1}(\Omega)$ *satisfying* (4) *in the sense that* $\Delta u \in L^1(\Omega)$ *and* $\Delta u(x) + g(x) \in \beta(x,u(x))$ *a.e.* $x \in \Omega$ *is assured for* $g \in L^1(\Omega)$ *given, without any additional condition if* $\beta(x,r) = \beta(r)$ *and if* $D(\beta(x,\cdot)) = \mathbb{R}$ *otherwise. Finally, if* $f(x,r) = \beta(x,r)$, *we have the estimate*

$$\|[g + \Delta u - \tilde{g} - \Delta\tilde{u}]^+\|_{L^p(\Omega)} \leq \|[g-\tilde{g}]^+\|_{L^p(\Omega)} \tag{9}$$

for every $1 \leq p \leq \infty$, *if* $u,\tilde{u}$ *are the solutions associated with* $g,\tilde{g}$.

<u>Remark 4.21.</u> If f(x,s) satisfies (7), (8), as a direct consequence of Theorem 4.15 we can solve equation (4) with the boundary condition

$$u = h \text{ on } \partial\Omega,$$

assumed that

$$h \in L^1_{loc}(\Omega),\ f(x,h(x)) \in L^1(\Omega) \text{ and } f(x,t+h(x)) \in L^1_{loc}(\Omega)\ \forall t \in \mathbb{R}. \tag{10}$$

Indeed, it suffices to apply Theorem 4.15 with $\tilde{f}(x,s) = f(x,s+h(x)) - f(x,h(x))$ instead of f(x,s), and $\tilde{g}(x) = g(x) - f(x,h(x))$ instead of g(x). Note that now the conclusion is that $u-h \in W_0^{1,1}(\Omega)$, $\Delta(u-h) \in L^1$, $f(x,u) \in L^1(\Omega)$ and $-\Delta u + f(x,u) = g - \Delta h$. We also recall the results of Bidaut-Veron [3] and Diaz [5] for the obstacle problem ($\beta(x,r) = \beta(r,\psi(x))$ with $\psi \in L^2(\Omega)$ and $D(\beta) = [0,\infty)$). □

<u>Remark 4.22.</u> By applying the above remark to the case $f(x,u) = |u|^{q-1}u$, with $q > 1$, Gallouet-Morel [3] give a necessary and sufficient condition

on $\mu \in \mathcal{D}'(\Omega)$ in order that the problem

$$\left.\begin{aligned} -\Delta u + |u|^{q-1}u &= \mu \text{ in } \Omega \\ u &\in W_o^{1,1}(\Omega) \cap L^q(\Omega) \end{aligned}\right\} \qquad (11)$$

has a solution. In particular, if μ is a measure, (11) has a solution if and only if $\mu \in L^1(\Omega) + W^{-2,q}(\Omega)$. Another characterization is given in Baras-Pierre: (11) has a solution if and only if $|\mu|(A) = 0$ for every subset A of Ω whose $W^{2,q'}$-capacity is zero. The existence of solutions of (11) is related to the problem of the study of removable singularities, according to the values of N, q and the nature of the singularity, here represented by μ (see Baras-Pierre [1] and the references therein to the works by Brezis-Veron, Veron, and others). □

The study of semilinear equations on unbounded domains is more complicated. One of the difficulties comes from the fact that now the operator $-\Delta$ is not coercive in $L^1(\Omega)$ if Ω is unbounded. The first progress in this direction is due to Benilan-Brezis-Crandall [1] and concerns the equation

$$-\Delta u + f(x,u) = g \text{ in } \mathbb{R}^N. \qquad (12)$$

Again, the idea is to approximate (12) by easier problems

$$-\Delta u_\varepsilon + \varepsilon u_\varepsilon + f_\varepsilon(x,u_\varepsilon) = g_\varepsilon \text{ in } \mathbb{R}^N$$

where f_ε is a truncation of f and, for instance, $g_\varepsilon \in L^2(\mathbb{R}^N)$. The estimate (6) is proved as in the case of bounded domains but now replacing $f(\cdot,r)$ by $\varepsilon r + f_\varepsilon(\cdot,r)$. If $f(x,u) = \beta(u)$, where β is a maximal monotone graph of $\mathbb{R}^2$ with $0 \in \beta(0)$, the passing to the limit can be obtained throughout (6) and the estimate

$$\|w_\varepsilon(x+h) - w_\varepsilon(x)\|_{L^1} \leq \|g(x+h)-g(x)\|_{L^1}, \quad \forall h \in \mathbb{R}^N,$$

when $w_\varepsilon = g_\varepsilon + \Delta u_\varepsilon$. This comes from estimates of the type (74) of Section 4.1 and the fact that the equation is invariant by translations. Thus, by the compactness theorem of Kolmogorov, the set $\{w_\varepsilon : \varepsilon > 0\}$ is precompact in $L^1_{loc}(\mathbb{R}^N)$ and $w_\varepsilon \to w$ is $L^1_{loc}(\mathbb{R}^N)$ when $\varepsilon \to 0$. In fact, this same conclusion can be derived for f(x,u) satisfying only (7) and (8), by using estimate (6) and Vitali's theorem (see Gallouet-Morel [1]). Nevertheless, in this case the estimate

$$\|\Delta u_\varepsilon\|_{L^1(\mathbb{R}^N)} \leq 2\,\|g\|_{L^1(\mathbb{R}^N)}$$

is not enough to find an estimate (independent of ε) of the norm in $W^{1,1}(\mathbb{R}^N)$ of u_ε, and the properties of $-\Delta^{-1}$ in $L^1(\mathbb{R}^N)$ need to be carefully taken into account. We recall that, if $z \in L^1(\mathbb{R}^N)$ is such that

$$-\Delta z = F,\ F \in L^1(\mathbb{R}^N),$$

then $z = E_N * F$, where E_N is the fundamental solution for $-\Delta$.

$$E_N(x) = \begin{cases} \dfrac{1}{(N-1)\omega_N |x|^{N-2}} & \text{if } N \geq 3 \\ \\ \dfrac{1}{2}\log\dfrac{1}{|x|} & \text{if } N = 2, \end{cases}$$

where ω_N is the volume of the unit ball in $\mathbb{R}^N$. It turns out (see the appendix of Benilan-Brezis-Crandall [1]) that $E_N \notin W^{1,1}(\mathbb{R}^N)$ and so we need to go outside the space $L^1(\Omega)$ and use the Marcinkiewicz (or weak-L^p) space.

Definition 4.6. *Let* u *be a measurable function on* $\mathbb{R}^N$, $1 < s < \infty$ *and* $1/s + 1/s' = 1$. *Then we define*

$$\|u\|_{M^s} = \min\{C \in [0,\infty]: \int_K |u(x)|dx \leq C(\text{meas } K)^{1/s'} \text{ for all measurable } K \subset \mathbb{R}^N\}.$$

Finally, we denote by $M^s(\mathbb{R}^N)$ *the set of measurable functions* u *on* $\mathbb{R}^N$ *satisfying* $\|u\|_{M^s} < \infty$.

It is easy to verify that $M^s(\mathbb{R}^N)$ is a Banach space and that $M^s(\mathbb{R}^N) \subset L^{s'}_{loc}(\mathbb{R}^N)$ with continuous injection $1 \leq s' < s$. On the other hand, it is proved that $E_N \in W^{1,1}_{loc}(\mathbb{R}^N)$ and $E_N \in M^{N/(N-2)}(\mathbb{R}^N)$ for $N \geq 3$ and $|\text{grad } E_N| \in M^{N/(N-1)}(\mathbb{R}^N)$ if $N \geq 2$. Thus, if $z \in L^1(\mathbb{R}^N)$ and $\Delta z \in L^1(\mathbb{R}^N)$ we have that $z \in M^{N/(N-2)}(\mathbb{R}^N)$ if $N \geq 3$.

The following theorem summarizes the results of the articles mentioned on equation (12):

Theorem 4.16. *Let* $g \in L^1(\mathbb{R}^N)$. *Suppose* $N \geq 3$ *and* $f(x,s)$ *satisfying (7), (8), then there exists at least one function* $u \in M^{N/(N-2)}(\mathbb{R}^N)$ *with* $\Delta u \in L^1(\mathbb{R}^N)$ *and* $f(\cdot,u) \in L^1(\mathbb{R}^N)$ *solution of (12). Moreover, if* $f(x,u) = \beta(u)$, *with* β

maximal monotone graph of $\mathbb{R}^2$ *with* $0 \in \beta(0)$, *there exists a unique solution* $u \in M^{N/(N-2)}(\mathbb{R}^N)$ *with* $\Delta u \in L^1(\mathbb{R}^N)$ *and* $g + \Delta u \in L^1(\mathbb{R}^N)$. *If* $N = 2$ (*resp.* $N = 1$) *the existence of, at least one, solution* $u \in W^{1,1}_{loc}(\mathbb{R}^2)$ *with* $|\nabla u| \in M^2(\mathbb{R}^2)$ *and* $\Delta u \in L^1(\mathbb{R}^2)$ (*resp.* $u \in W^{1,\infty}(\mathbb{R})$ *with* $\frac{d^2u}{dx^2} \in L^1(\mathbb{R})$) *is assured if, in addition, for some* $c_1, c_2 \in [0,\infty)$ *one has*

$$\text{meas } \{x, f(x,c_1) \leq c_2 \text{ or } f(x_1-c_1) \geq -c_2\} < \infty$$

or $f(x,u) = \beta(u)$ *and* β *satisfies* $0 \in \text{int } \beta(0)$. *In this last case, two solutions, in their respective class, differ by a constant. Finally, if* $f(x,u) = \beta(x,r)$ *is a maximal monotone graph for a.e.* $x \in \Omega$

$$\|[g-\Delta u-\tilde{g}+\Delta\tilde{u}]^+\|_{L^1(\mathbb{R}^N)} \leq \|[u-\tilde{u}]^+\|_{L^1(\mathbb{R}^N)}$$

and, if $g \leq \tilde{g}$ *a.e.*, $g \neq \tilde{g}$, *then* $u \leq \tilde{u}$ *a.e. on* $\mathbb{R}^N$.

Remark 4.23. There are several generalizations and variants of the above theorem in the literature concerning other conditions for the dependence on x of f(x,u) (Schatzman [1], Di Blasio [1]), or maximal monotone graphs β without the condition $0 \in \beta(0)$ (works by Crandall-Evans, Fisher, Kurtz and Vazquez, see references in Vazquez [2]). On the other hand, as in the case of bounded domains, equation (12) can be solved when g is merely a distribution (Gallouet-Morel [1]) or a bounded measure (Benilan-Brezis (see Brezis [9]), and Vazquez [4]). Finally, we mention that abstract semilinear equations in $L^1(\Omega)$, when Ω is a general unbounded set, are treated in Benilan [6]. □

Recently, the study of semilinear equations in $\mathbb{R}^N$ for data $g \in L^1_{loc}(\mathbb{R}^N)$ has been undertaken by Brezis [10] and Gallouet-Morel [3]. When considering the equation

$$-\Delta u + |u|^{q-1}u = g \quad \text{in } \mathcal{D}'(\mathbb{R}^N) \tag{14}$$

the results depend on the value of $q > 0$. So, if $q > 1$, no limitation on the growth at infinity of g is required for the existence of a solution: for every $g \in L^1_{loc}(\mathbb{R}^N)$ there exists a unique $u \in L^q_{loc}(\mathbb{R}^N)$ satisfying (14); moreover, if $g \geq 0$ then $u \geq 0$ (Brezis [10]). The case of $0 < q < 1$ is different. Let $g \in L^1_{loc}(\mathbb{R}^N)$, $g \geq 0$; then there exist $C_1 > 0$ and $C_2 > 0$ depending only on N and q such that:

$$\limsup_{R\to\infty} R^{-N-2q/(1-q)} \int_{B_R(0)} g\,dx < C_1 \Rightarrow \text{(93) has a positive solution}$$

$$u \in L^1_{loc}(\mathbb{R}^n) \qquad (15)$$

$$\text{(93) has a positive solution } u \in L^1_{loc}(\mathbb{R}^N) \Rightarrow \limsup_{R\to\infty} R^{-N-2q/(1-q)}$$

$$\int_{B_R(0)} g\,dx < C_2 \qquad (16)$$

(see Gallouet-Morel [3]). The proof of this last result is carried out by the method of super and subsolutions. We point out that the existence of a nontrivial supersolution is obtained throughout sharp estimates on the location of the support of solutions of (93) when g is assumed to have compact support. The existence in the case $q > 1$ is based on a different localization property which can be compared with other localization effects such as the existence of the free boundary $\mathcal{F}(u)$. The following result shows such a property (Baras-Pierre [1], Brezis [10]):

Proposition 4.17. *Let* $R < R'$ *and assume* $u \in L^q_{loc}(B_{R'}(0))$ *satisfying equation (4) in* $\mathcal{D}'(B_{R'}(0))$, *with* $g \in L^1(B_{R'}(0))$. *Then*

$$\int_{B_R(0)} |u|^q dx \leq G(1 + \int_{B_{R'}(0)} |g|dx), \qquad (17)$$

where depends only on q, R *and* R'.

Proof. From Kato's inequality (see Kato [1]),

$$\Delta|u| \geq (\Delta u)\,\text{sign}\,u.$$

Then, multiplying by sign u in (93), we have

$$-\Delta|u| + |u|^q \leq |g| \text{ in } \mathcal{D}'(B_{R'}(0)). \qquad (18)$$

Let $\zeta \in \mathcal{D}(B_{R'}(0))$ be such that $0 \leq \zeta \leq 1$ and $\zeta = 1$ on $B_R(0)$. Multiplying (97) through by ζ^α, where is an integer, and integrating, we find

$$\int |u|^q \zeta^\alpha dx \leq \int |g|dx + C\int |u|\zeta^{\alpha-2}dx \leq \int |g|dx + C\int |u|\zeta^{\alpha/q}dx, \qquad (19)$$

provided $\alpha-2 \geq \alpha/q$, i.e., $\alpha \geq 2q/(q-1)$ and we fix any such α. The conclusion now follows easily from (19). □

Remark 4.24 Proposition 4.17 and, in fact, the existence of solutions for $g \in L^1_{loc}(\mathbb{R}^N)$ can be extended to more general semilinear equations such as (12). If, for instance, $f(x,u) = f(u)$, f must be assumed convex on $\mathbb{R}^+$, $f(s)s \geqq 0$, and

$$\int^{\infty} \frac{ds}{F(t)^{1/2}} < \infty \quad , \quad F(t) = \int_0^t f(s)ds$$

(Gallouet-Morel [3]). Note that in the case of Proposition 4.17 the strong maximum principle holds and so if, for instance, $g = 0$ in $B_{R'}(0)$ and $g > 0$ in $\mathbb{R}^N - B_{R'}(0)$, then $u > 0$ in $\mathbb{R}^N$. Nevertheless, Proposition 4.17 shows that, in fact, the values of g in $\mathbb{R}^N - B_{R'}(0)$ affect u only mildly in $B_R(0)$: u may be estimated on $B_R(0)$ independently of the values of g on $\mathbb{R}^N - B_{R'}(0)$, even if $g(x) \to +\infty$ when $|x| \to \infty$. □

We shall end this subsection by giving an L^∞-estimate on the set Int N(g) similar to Theorem 4.8.

Theorem 4.18. *Let $g \in L^1(\Omega)$ and $h \in W^{1,1}(\Omega)$. Assume f to be a continuous real function satisfying $f(t)t \geqq 0$ for $t \geqq 0$. Let $u \in W^{1,1}(\Omega)$ be a L^1-solution (see Remark 4.21) of*

$$-\Delta u + f(u) = g \text{ in } \Omega \tag{20}$$

$$u = h \qquad \text{on } \partial\Omega. \tag{21}$$

Let D be a compact set of $\mathbb{R}^N$ such that $D \subset \text{int } N(g)$, and let $d = d(D, N(g))$. Then $u \in L^\infty(D)$ and

$$||u||_{L^\infty(D)} \leqq C(\,||g||_{L^1(\Omega)} + ||h||_{L^1(\partial\Omega)}) \tag{22}$$

for some constant C only dependent on N, Ω and d.

Proof. Let g_+ and h_+ (respectively g_-, h_-) be the positive (resp. negative) parts of functions g and h. On the other hand, let v_+ (resp. v_-) be the unique solution of the problem

$$-\Delta v = \tilde{g} \text{ in } \Omega \tag{23}$$

$$v = \tilde{h} \text{ on } \partial\Omega \tag{24}$$

where $\tilde{g} = g_+$, $\tilde{h} = h_+$ (resp. $\tilde{g} = -g_-$, $\tilde{h} = -h_-$). Then, by Theorem 4.13 and Theorem 4.15 (or Theorem 4.11 after passing to the limit in L^1), if u is any solution of (20), (21) then $v_- \leq u \leq v_+$ a.e. in Ω. Thus, it is enough to prove the theorem for v solution of (23), (24) when $\tilde{g} \in L^1(\Omega)$ and $\tilde{h} \in W^{1,1}(\Omega)$. We shall first obtain an estimate of $\|v\|_{L^1(\Omega)}$ by following an idea of Gallouet-Morel [3]. Let $\tilde{g}_n \in C^\infty(\bar{\Omega})$ and $\tilde{h}_n \in C^\infty(\partial\Omega)$ such that $\tilde{g}_n \to \tilde{g}$ in $L^1(\Omega)$ and $\tilde{h}_n \to \tilde{h}$ in $L^1(\partial\Omega)$. Let $v_n \in C^\infty(\bar{\Omega})$ be the unique solution of (23), (24) corresponding to the data $\tilde{g}_n$ and $\tilde{h}_n$. Let $\phi \in W^{2,p}(\Omega)$, $\phi = 0$ on $\partial\Omega$, $1 < p < \infty$. One has

$$\int_\Omega \tilde{g}_n \, dx = - \int_\Omega \Delta \, v_n \phi dx = - \int_\Omega v_n \Delta\phi dx + \int_{\partial\Omega} \tilde{h}_n \frac{\partial\phi}{\partial n} \, d\sigma.$$

Then

$$\int_\Omega v_n \Delta\phi dx = - \int_\Omega \tilde{g}_n \phi dx + \int_{\partial\Omega} \tilde{h}_n \frac{\partial\phi}{\partial n} \, d\sigma. \tag{25}$$

For $\psi \in L^\infty(\Omega)$ fixed, let $\phi \in W^{2,p}(\Omega)$, $1 \leq p < \infty$ such that $-\Delta\phi = \psi$ in Ω, $\phi = 0$ on $\partial\Omega$. By the $W^{2,p}$-regularity (see Subsection 4.1c) we have that

$$\|\phi\|_{W^{2,p}} \leq C\|\psi\|_{L^p} \leq C\|\psi\|_{L^\infty},$$

where C denotes some constants dependent only on Ω and p. Taking some $p > N$, by the Sobolev theorem we conclude

$$\|\phi\|_{L^\infty} \leq C\|\psi\|_{L^\infty} \quad \text{and} \quad \left\|\frac{\partial\phi}{\partial n}\right\|_{L^\infty} \leq C\|\psi\|_{L^\infty}.$$

Then, for all $\psi \in L^\infty(\Omega)$,

$$\left|\int_\Omega v_n \psi dx\right| \leq C(\|\tilde{g}_n\|_{L^1} + \|\tilde{h}_n\|_{L^1}) \; \|\psi\|_{L^\infty}.$$

Thus

$$\|v_n\|_{L^1} \leq C(\|\tilde{g}_n\|_1 + \|\tilde{h}_n\|_{L^1}). \tag{25'}$$

From this inequality (with $v_n - v_m$ instead of v_n) we conclude that $\{v_n\}$ is a

Cauchy sequence in $L^1(\Omega)$. Then there exists a w such that $v_n \to w$ in $L^1(\Omega)$. To prove that w = v a.e., we note that, by passing to the limit in (25),

$$\int_\Omega w\Delta\phi\, dx = -\int_\Omega \tilde{g}\phi\, dx + \int_{\partial\Omega} \frac{\partial\phi}{\partial n}\tilde{h}\, d\sigma.$$

On the other hand, it is easy to show the existence of $\tilde{v}_n \in \mathcal{D}(\bar{\Omega})$ such that $\tilde{v}_n \to v$ in $W^{1,1}(\Omega)$ and $\Delta\tilde{v}_n \to \Delta v$ in $L^1(\Omega)$. Then,

$$-\int_\Omega \Delta\phi\ \tilde{v}_n dx = -\int_{\partial\Omega} \frac{\partial\phi}{\partial n}\tilde{v}_n\ dx - \int_\Omega \Delta\tilde{v}_n\phi\, dx,$$

and as $-\Delta\tilde{v}_n \to \tilde{g}$ in $L^1(\Omega)$ and $\tilde{v}_n \to \tilde{h}$ in $L^1(\partial\Omega)$ (since $\tilde{v}_n \to v$ in $W^{1,1}(\Omega)$), we conclude that

$$\int_\Omega (v-w)\Delta\phi dx = 0$$

for all $\phi \in C^2(\bar{\Omega})$ with $\phi = 0$ on $\partial\Omega$. Again, this implies that $\int_\Omega (v-w)\psi dx = 0$ for all $\psi \in \mathcal{D}(\Omega)$ and, therefore, w = v a.e. In consequence,

$$\|v\|_{L^1(\Omega)} \leq C(\|\tilde{g}\|_{L^1(\Omega)} + \|\tilde{h}\|_{L^1(\partial\Omega)}).$$

Finally, let D be compact and $D \subset \mathrm{Int}\, N(g)$. Then, $v \in W^{2,p}(D)$ for every $1 \leq p < \infty$ (since $\Delta v = 0$ in D) and, by the results of Agmon-Douglis-Nirenberg [1],

$$\|v\|_{W^{2,p}(D)} \leq C(\|\Delta v\|_{L^\infty(N(\tilde{g}))} + \|v\|_{L^1(N(\tilde{g}))}) \leq C\,\|v\|_{L^1(\Omega)}.$$

Taking $p > N$, we conclude the proof by applying the Sobolev Injection Theorem. □

<u>Remark 4.25</u>. As in Remark 4.12, we point out that the estimate (22) is not sharp. For instance, better estimates can be obtained under growing conditions on f(u) (see Gallouet-Morel [3]). Finally, we note that the duality arguments used in the proof of Theorem 4.18 are also available for a general second order linear operator such as that given in (1) (see Kato [1], Baras-Pierre [1]). □

4.2b. Abstract results. Accretive operators. Application to quasilinear equations

As already indicated, the existence of solutions in $L^1(\Omega)$ of the monotone semilinear equation (4) (Theorem 4.15) was given in Brezis-Strauss [1] for an abstract formulation: let Ω be any measure space and let A be an unbounded linear operator from $L^1(\Omega)$ into $L^1(\Omega)$ which satisfies the following conditions:

A is closed, D(A) is dense in $L^1(\Omega)$ and, for any $\lambda > 0$, $I + \lambda A$ maps D(A) one-to-one onto $L^1(\Omega)$, and $(I+\lambda A)^{-1}$ is a contraction in $L^1(\Omega)$. (26)

For any $\lambda > 0$ and $g \in L^1(\Omega)$, $\sup_\Omega (I+\lambda A)^{-1}g \leq \max \{0, \sup_\Omega g\}$ (if $\sup g = +\infty$, this assumption is empty). (27)

There exists $\alpha > 0$ such that $\alpha\|u\|_{L^1} \leq \|Au\|_{L^1}$ for all $u \in D(A)$. (28)

Theorem 4.19. *Let β be a maximal monotone graph of $\mathbb{R}^2$ with $0 \in \beta(0)$. Then, for every $g \in L^1(\Omega)$ there exists a unique $u \in D(A)$ such that*

$$Au(x) + \beta(u(x)) \ni g(x) \quad \text{a.e.} \tag{29}$$

Moreover, if $g, \hat{g} \in L^1(\Omega)$ and u, $\hat{u}$ are the corresponding solutions of (29) then

$$\|[(\hat{g}-A\hat{u}) - (g-Au)]^+\|_{L^1} \leq \|[\hat{g}-g]^+\|_{L^1}. \tag{30}$$

If in addition $\hat{g} \leq g$ a.e. then $\hat{g}-A\hat{u} \leq g-Au$ a.e. and $\hat{u} \leq u$ a.e. Finally, if $g \in L^p(\Omega) \cap L^1(\Omega)$, $1 \leq p \leq \infty$ then $\|g-Au\|_{L^p} \leq \|g\|_{L^p}$. □

For the application to second order semilinear elliptic equations, we consider that Ω is an open bounded regular set of $\mathbb{R}^N$ and L is the differential operator given in (1) with coefficients a_{ij}, b_i and a satisfying the conditions there indicated. We define the realization of L on $L^1(\Omega)$ by

$$D(A) = \{u \in W_0^{1,1}(\Omega) : Lu \in L^1(\Omega) \text{ in the "very weak sense"}\}$$

$$Au = Lu \text{ for } u \in D(A).$$

We also define the natural realization of L on $L^p(\Omega)$ for $1 < p < \infty$ by $D(A_p) = W^{2,p}(\Omega) \cap W_0^{1,p}(\Omega)$, $A_p u = Lu$, for $u \in D(A_p)$. In Brezis-Strauss [1] is proved:

Theorem 4.20. *The operator A satisfies conditions (26) and (27). Moreover,* $D(A) \subset W_0^{1,q}(\Omega)$ *for* $1 \leq q < N/(N-1)$ *and for some* $\alpha = \alpha(q) > 0$

$$\alpha\|u\|_{W^{1,q}} \leq \|Au\|_{L^1} \text{ for } u \in D(A). \tag{31}$$

Finally, A is the closure in $L^1(\Omega)$ *of the operator* A_2.

Remark 4.26. Theorem 4.20 is proved by using some duality arguments. In fact the same conclusion holds for Neumann or Robin boundary conditions (see Brezis-Strauss [1], Pazy [2], Fattorini [1] and Amann [3]). We also mention here the treatment made in Benilan [6] for abstract semilinear operators. □

The abstract result given in Theorem 4.19 has several possible generalizations in the case in which A is a nonlinear operator. Those generalizations use the notion of accretive operator on a Banach space X. Such a class of operators is defined, in contrast with the class of monotone (or pseudo-monotone) operators, from $D(A) \subset X$ into $\mathcal{P}(X)$. Nevertheless, the two notions of operators are not very far apart: indeed, in the particular case of X = H, a Hilbert space, both classes of operators are the same. There exist several equivalent definitions of accretive operators. One of them has its starting-point in Definition 4.4. For monotone operators on a Hilbert space X = HA is monotone if

$$\langle v-\tilde{v}, u-\tilde{u}\rangle_{HH} \geq 0 \quad \forall u,\tilde{u} \in D(A),\ v \in Au,\ \tilde{v} \in A\tilde{u}. \tag{32}$$

It is easily seen that this is equivalent to saying that $(I+\lambda A)^{-1}$ is a contraction on H, for every $\lambda > 0$, i.e.,

$$\|u-\tilde{u}\|_H \leq \|u-\tilde{u} + \lambda(v-\tilde{v})\|_H \quad \forall u,\tilde{u} \in D(A),\ v \in Au,\ \tilde{v} \in A\tilde{u}. \tag{33}$$

Expression (33) makes sense for a general Banach space X:

Definition 4.6. *An operator* $A:D(A) \subset X \to \mathcal{P}(X)$ *is called accretive if for every* $u,\tilde{u} \in D(A)$, $v \in Au$, $\tilde{v} \in A\tilde{u}$ *and* $\lambda > 0$

$$\|u-\tilde{u}\|_X \leq \|u-\tilde{u} + \lambda(v-\tilde{v})\|_X \tag{34}$$

holds. If X is a Banach lattice, then A is called T-accretive if

$$\|[u-\tilde{u}]^+\|_X \leq \|[u-\tilde{u} + \lambda(v-\tilde{v})]^+\|_X . \tag{35}$$

Remark 4.27. In fact, if X is a normal Banach lattice (i.e. such that $\|x^+\| \leq \|y^+\|$ and $\|x^-\| \leq \|y^-\|$ implies $\|x\| \leq \|y\|$) then any T-accretive operator is accretive. This is the case for $X = L^p(\Omega)$ and $X = C^0(\Omega)$. On the other hand, by defining the bracket [,] on $X \times X$ by

$$[x,y] = \lim_{\lambda \downarrow 0} \frac{\|x+\lambda y\| - \|x\|}{\lambda} = \inf_{\lambda>0} \frac{\|x+\lambda y\| - \|x\|}{\lambda} , \tag{36}$$

it is easy to check that A is accretive if and only if

$$[v-\tilde{v}, u-\tilde{u}] \geq 0 \quad \forall u,\tilde{u} \in D(A), \quad v \in Au, \tilde{v} \in A\tilde{u}. \tag{37}$$

It is not difficult to compute such a bracket when $X = L^p(\Omega)$, $1 \leq p < \infty$ or $X = C^0(\Omega)$ (see e.g. Sato [1]). □

We point out that in assumption (26) the linear operator A is supposed to be accretive in $L^1(\Omega)$ ($(I+\lambda A)^{-1}$ is a contraction)) and that assumption (27) coincides with the condition of T-accretiveness in $L^\infty(\Omega)$ for linear operators. One of the possible generalizations of Theorem 4.19 will be given for the class of nonlinear operators A which are accretive in $L^1(\Omega)$ and T-accretive in $L^\infty(\Omega)$. A general sample of this class of operators can be obtained from the subdifferential of suitable convex l.s.c. functions on $L^2(\Omega)$, as the following result, due to Benilan-Picard [1], shows.

Theorem 4.21. *Let Ω be a bounded open set of $\mathbb{R}^N$ and let $\phi: L^2(\Omega) \to (-\infty,+\infty]$ convex, l.s.c., $\phi \not\equiv +\infty$. Then the following properties are equivalent:*

(i) *let A_p be the operator in $L^p(\Omega) \times L^p(\Omega)$, $1 \leq p < \infty$, defined by $D(A_p) = \{u \in L^p(\Omega) \cap D(\phi) : \partial\phi(u) \cap L^p(\Omega) \neq \emptyset\}$, and*

$$A_p u = \partial\phi(u) \cap L^p(\Omega), \text{ if } u \in D(A_p). \tag{38}$$

Then A_p is accretive (resp. T-accretive) in $L^p(\Omega)$.

(ii) *For every* $u, \tilde{u} \in L^2(\Omega)$ *and for any* $T \in W^{1,\infty}_{loc}(\mathbb{R})$ *with* $0 \leq T' \leq 1$ *being odd (resp.* $T(r) = 0$ *if* $r \leq 0$*)*

$$\phi(u - T(u-\tilde{u})) + \phi(\tilde{u} + T(u-\tilde{u})) \leq \phi(u) + \phi(\tilde{u}). \tag{39}$$

holds.

(iii) *For every* $u, \tilde{u} \in D(\partial\phi)$, $v \in \partial\phi(u)$, $\tilde{v} \in \partial\phi(\tilde{u})$, *and for any* $T \in W^{1,\infty}_{loc}(\mathbb{R})$ *with* $0 \leq T' \leq 1$ *being odd (resp.* $T(r) = 0$ *if* $r \leq 0$*)*

$$\int_\Omega (v-\tilde{v})T(u-\tilde{u})dx \geq 0 \tag{40}$$

holds. □

To generalize Theorem 4.19 some "range condition" will also be needed:

Definition 4.7. *An operator* $A: D(A) \subset X \to \mathcal{P}(X)$ *satisfies the range condition if* $\mathcal{R}(I+\lambda A) \supseteq \overline{D(A)}$ *for every* $\lambda > 0$. *If, in fact,* $\mathcal{R}(I+\lambda A) = X$, *A is called m-accretive in X.*

As we have already mentioned in (44) of Section 4.1, when $X = H$ is a Hilbert space, any maximal monotone operator A satisfies $\mathcal{R}(I+\lambda A) = H$ and so it is m-accretive. The following result (taken from Benilan-Crandall-Pazy [1]) shows that, for general Banach spaces, any m-accretive operator is maximal (in the sense of inclusion of graphs on $X \times X$). The reciprocal turns out to be not true in general (see Crandall-Liggett [1]).

Theorem 4.22. *Let X be a Banach space and let A be an accretive operator in X. Denote by* $\bar{A}$ *the operator whose graph is the closure of the graph of A in* $X \times X$. *Then:*

(i) $\bar{A}$ *is accretive,*

(ii) *A is closed (i.e.* $\bar{A} = A$*) if and only if* $\mathcal{R}(I+\lambda A)$ *is closed in X,*

(iii) *if B is an accretive operator in X such that* $B \supset A$ *and* $\mathcal{R}(I+B) \subset (I+A)$, *then* $A = B$. □

It is a direct consequence of Theorem 4.21 that the operator A_p, given in (38), is m-accretive in $L^p(\Omega)$, assumed that ϕ (resp. $\partial\phi$) satisfies (39) (resp. (40)). A special example of such a functional ϕ, of interest in the applications, is the following (see Benilan [3], Attouch-Damlamian [1]):

Proposition 4.23. *Let* $J: \Omega \times \mathbb{R}^N \to [0,\infty]$ *such that* $J(x,0) = 0$, $J(x,\xi)$ *is convex and l.s.c. on* ξ *and measurable in* x. *Let* $j:\Omega \times \mathbb{R} \to [0,\infty]$ *such that* $j(x,0) = 0$, $j(x,r)$ *is convex and l.s.c. on* r *and measurable in* x. *Consider now* $\phi: L^2(\Omega) \to [0,\infty]$ *given by*

$$\phi(u) = \begin{cases} \int_\Omega J(x,\nabla u)dx + \int_\Omega j(x,u)dx & \text{if } u \in L^2(\Omega) \cap W_0^{1,1}(\Omega) \\ +\infty & \text{otherwise.} \end{cases}$$

Then ϕ *is a convex function satisfying* (39). *Moreover, if*

$$\forall \varepsilon > 0 \quad \exists a \in L^1(\Omega) \text{ such that } |\xi| \leq a(x) + \varepsilon J(x,\xi) \text{ a.e. } x \in \Omega, \forall \xi \in \mathbb{R}^N, \tag{41}$$

then ϕ *is l.s.c.* □

As a consequence of Proposition 4.23 and Theorem 4.21 the operators A_p defined from $\partial\Phi$ are m-accretive operators in $L^p(\Omega)$. Nevertheless, the identification of the subdifferential $\partial\Phi$ is not easy, in general. If, for instance, J and j do not depend on x, Attouch-Damlamian [1] showed that the graph of $\partial\Phi$ is the operator of graph given by

$$\{[u,v] \in (W_0^{1,1}(\Omega) \cap L^2(\Omega)) \times L^2(\Omega) : \exists\, A \in L^1(\Omega)^N \text{ with } A \in \partial J(\nabla u) \text{ a.e.,}$$
$$\operatorname{div} A \in L^2(\Omega), \int_\Omega (u \operatorname{div} A + A.\nabla u) = 0 \text{ and}$$
$$v \in \operatorname{div} A + \partial j(u) \text{ a.e. on } \Omega\}.$$

In the particular case $J(\xi) = \frac{1}{p}|\xi|^p$ and $j \equiv 0$ we have that $\partial\Phi(u) = -\Delta_p u$ for $u \in D(\partial\Phi)$ and so, formally, the operator $\Delta_p u$ is m-accretive in $L^1(\Omega)$ as well as in $L^\infty(\Omega)$. A more exact result for this operator is due to Gariepy-Pierre [1]:

Proposition 4.24. *Let* $A_2 = \partial\phi$ *corresponding to* $J(\xi) = \frac{1}{p}|\xi|^p$ *and* $j \equiv 0$, *then the operator* $A_1 = \overline{A_2}^{L^2}$ *(closure in* $L^1(\Omega)$ *of* A_2*) is an m-accretive operator in* $L^1(\Omega)$ *and T-accretive in* $L^\infty(\Omega)$. *Moreover,*

$$D(A_1) \subset \{u \in W_0^{1,1}(\Omega) : |\nabla u|^{p-2}\nabla u \in L^1(\Omega)^N \text{ and } \Delta_p u \in L^1(\Omega)\}. \tag{42}$$

Finally, if $u \in D(A_1)$ *then* $u \in W_0^{1,s}(\Omega)$ *for* $1 \leq s < N(p-1)/(N-1)$, *and there exists* $\alpha = \alpha(s,p) > 0$ *such that*

$$\alpha\|u\|^{p-1}_{W^{1,s}} \leq \|\Delta_p u\|_{L^1} \tag{43}$$

for every $u \in D(A_1)$.

Remark 4.28. Proposition 4.24 may be interpreted as a partial generalization of Theorem 4.20. We point out that the question of knowing the exact realization of $\Delta_p u$ in L^1 seems to be an open question (note that in (42) there is only an inclusion). We also mention that, in the works mentioned more general examples are considered, including, for instance, the case of non-linear boundary conditions. □

The class of m-accretive operators A on L^1 which are also T-accretive in L^∞ has been studied in Lê [1] and allows generalization of Theorem 4.19 under different "coercivity" assumptions on A or on the graph . See Benilan [2], [3], [4]. One of these generalizations is the following:

Theorem 4.25. *Let* A *be an m-accretive operator in* $L^1(\Omega)$ *being also T-accretive in* $L^\infty(\Omega)$. *Assume that*

$$A^{-1} \textit{ is bounded and continuous} \tag{44}$$

Let β *be a maximal monotone graph of* $\mathbb{R}^2$ *with* $0 \in \beta(0)$. *Then, for every* $g \in L^1(\Omega)$ *there exists a unique* $u \in D(A)$ *such that*

$$Au(x) + \beta(u(x)) \ni g(x) \quad \text{a.e..} \tag{45}$$

Moreover, if $g, \hat{g} \in L^1(\Omega)$ *and* $u, \hat{u}$ *are the corresponding solutions of* (45) *then*

$$\|[(\hat{g}-A\hat{u}) - (g-Au)]^+\|_{L^1} \leq \|[\hat{g}-g]^+\|_{L^1}.$$

If in addition $\hat{g} \leq g$ *a.e. then* $\hat{g}-A\hat{u} \leq g-Au$ *a.e. and* $\hat{u} \leq u$ *a.e. Finally, if* $g \in L^p(\Omega) \cap L^1(\Omega)$, $1 \leq p \leq \infty$, *then* $\|g-Au\|_{L^p} \leq \|g\|_{L^p}$. □

Remark 4.29. Theorem 4.25 may be applied to the operator A of Proposition 4.24 (the proof of (44) is similar to that of (43)). In consequence, we have existence, in L^1, of solutions of

$$\begin{aligned} -\Delta_p u + \beta(u) &\ni g \quad \text{in } \Omega \\ u &= 0 \quad \text{on } \partial\Omega. \end{aligned}$$

The case of nonhomogeneous boundary conditions is also available from the results of Benilan [4]. □

Remark 4.30. The theory of accretive operators is of great interest when solving abstract Cauchy problems. From the Crandall-Liggett theorem (Crandall-Liggett [1]) it is enough to assume that A is an accretive operator satisfying the range condition. In this way, the solution of an evolution problem is reduced to the consideration of the associated stationary problem. For this reason, the results of this section will be used in the consideration of the associated nonlinear parabolic equations (Diaz [7]). The results will also be useful in studying the porous media equation (see the Introduction). We note that the range condition for the realization in $L^1(\Omega)$ of the operator $-\Delta\phi(u)$ is reduced, by a trivial change of variable; to the study of the semilinear problem (4), (5). □

Remark 4.31. We make several comments on the different equations considered in Section 2.4 devoted to nonisotropic equations and their relations with accretive operators. First of all, we note that the operation associated to equation (8) of this section is again an m-T-accretive operator in $L^p(\Omega)$, for every $1 \leqq p \leqq \infty$ (see Proposition 4.23) (see also Cortazar [1]). With respect to the equation with a convection term (equation (8) of Section 2.4), we remark that the accretiveness in $L^1(\Omega)$ of the associated operator is a direct consequence of Theorems 4.11 and 4.12. A more detailed study of such operators is made by Benilan [5], Benilan-Touré [1] and Wolanski [1]. Finally, with respect to the fully nonlinear equation (34) of Section 2.4) we mention that the accretiveness in $L^\infty(\Omega)$ of the associated operator was already pointed out by Evans [1], [2], and Benilan-Catté [1] and Lions-Nisio [1] for the special case of the operator associated to the Hamilton-Jacobi-Bellman equation. We remark that the proof of Proposition 2.43 is similar to that of Theorem 4.13.

4.3. BIBLIOGRAPHICAL NOTES

The proof of Theorem 4.2 is standard; the one given here is taken from Bernis [2].

The local L^∞-estimates on the interior of the set N(g) (Theorems 4.8 and 4.18) are the consequence of well-known results but no explicit reference concerning the statements proved here seems to exist in the literature.

The proof of Theorem 4.11 is adapted from Benilan [5], and that of Theorem 4.12 is due to Diaz-Pierre [1] (see also Gagneux [1]). Note that the T-accretiveness in $L^1(\Omega)$ of the corresponding operators is a direct consequence of the L^1-estimates given there.

Another kind of localization property for bounded domains is given by Lasry-Lions [1]. Finally, fully nonlinear equations in $\mathbb{R}^N$ without growing conditions at the infinite are studied by G. Diaz [5].

Recently J.M. Morel and S. Solimini have proven that Theorems 4.15 and 4.18 remain true assumed $f(x,t)t \geqq -\lambda_1 t$, for a.e. $x \in \Omega$ and any $t > 0$, where λ_1 is the first eigenvalue of the Laplacian with Dirichlet boundary condition in Ω.

References

Adams, R.A.

[1] *Sobolev Spaces*. Academic Press, New York (1975).

Agmon, S., Douglis, A. and Nirenberg, L.

[1] Estimates near the boundary for solutions of elliptic partial differential equations satisfying general boundary conditions, I and II, *Comm. Pure Appl. Math.* *12* (1959), 623-727, and *17* (1964), 35-92.

Ahmed, N. and Sunada, D.K.

[1] Nonlinear flow in porous media, *J. Hydraulics Division Proc. Am. Sc. Civil Eng.* *95* (1969), 1847-1857.

Alt, H.W.

[1] The dam problem. In *Free Boundary Problems: Theory and Applications*. A. Fasano and M. Primicerio, eds. Pitman, London (1983), 52-69.

Alt, H.W. and Caffarelli, L.A.

[1] Existence and regularity for a minimum problem, *J. Reine Angew. Math.* *105* (1981), 105-144.

Alt, H.W. and Phillips, D.

[1] A free boundary problem for semilinear elliptic equations (to appear).

Alvino, A., Lions, P.L. and Trombetti, G.

[1] A remark on comparison results via symmetrization (to appear).

Alvino, A. and Trombetti, G.

[1] Sulle migliori constanti di maggiorazione per una classe di equazioni ellittiche degeneri. *Richerche di Mat.* *27* (1978), 193-212.

[2] Equazioni ellittiche con termini di ordine inferiore e viordinamenti. *Rend. Acc. Naz. Lincei* (8) *66* (1979), 1-7.

[3] Sulle migliori constanti di maggiorazione per una classe di equazioni ellittiche degeneri e non. *Ric. di Mat.* *30* (1981), 15-33.

Amann, H.

[1] On the existence of positive solutions of nonlinear elliptic boundary value problems, *Indiana Univ. Math. J.* *21* (1971), 125-146.

[2] Nonlinear elliptic equations with nonlinear boundary conditions. *Proc. of the Second Scheveningen Conference on Differential Equations*, W. Eckhaus (editor), North Holland, Amsterdam (1976).

[3] Dual semigroups and second order linear elliptic boundary value problems. *Israel J. Math.*, *45* (1983), 225-254.

Amann, H. and Crandall, M.G.

[1] On some existence theorems for semi-linear elliptic equations. *Indiana Univ. Math. J.*, *27* (1978), 779-790.

Ames, W.F.

[1] *Nonlinear Partial Differential Equations in Engineering.* Volumes I and II. Academic Press, New York (1965) and (1972).

Amundson, N.R. and Luss, D.

[1] On a conjecture of Aris: proof and remarks. *A.I.Ch.E.H. 13* (1967), 759.

Antoncer, S.N.

[1] *Dinamika Splošnoi Sredy Vyp. 39* (1979) 23 (Russian).

[2] On the localization of solutions of nonlinear degenerate elliptic and parabolic equations, *Soviet Math. Dokl. 24* (1981), 420-424.

[3] *Zap. Naučn. Seum. Leningrad. Otdel. Mat. Inst. Steklov.* (*LOM1*) *96* (1980) 3.

Apostol, T.M.

[1] *Mathematical Analysis.* Addison-Wesley, Reading, Massachusetts (1957).

Aris, R.

[1] *The Mathematical Theory of Diffusion and Reaction in Permeable Catalysts.* Volumes I and II. Clarendon Press, Oxford (1975).

Aronson, D.A.

[1] Bifurcation phenomena associated with nonlinear diffusion mechanisms. In: *Partial Differential Equations and Dynamical Systems*, W.E. Fitzgibbon III, Ed., Pitman, London (1984).

Aronson, D.A., Crandall, M.G. and Peletier, L.A.

[1] Stabilization of solutions of a degenerate nonlinear diffusion problem. *Nonl. Anal. TMA 6* (1982), 1001-1022.

Astarita, G. and Marruci, G.

[1] *Principles of Non-Newtonian Fluid Mechanics*, McGraw-Hill, New York (1974).

Athanasopoulus, I.

[1] Stability of the coincidence set for the Signorini problem. *Indiana Univ. Math. J. 30* (1981), 235-247.

Attouch, H. and Damlamian, A.

[1] Application des méthodes de convexité et monotonie à l'étude de certains équations quasilineaires. *Proc. Royal Soc. Edinburgh 79A,* (1977), 107-129.

Aubin, T.

[1] *Nonlinear Analysis in Manifolds, Monge-Ampere Equation,* Springer-Verlag Berlin (1982).

Auchmuty, J.F.G. and Beals, R.

[1] Variational solutions of some nonlinear free boundary problems, *Arch. Ration. Mech. Anal. 43* (1971), 255-271.

Badii, M., Diaz, J.I. and Tesei, A.

[1] Existence and attractivity results for a class of degenerate functional-parabolic problems (to appear).

Baiocchi, C., Gastaldi, F. and Tomarelli, F.

[1] Inequations variationnelles non coercives. *C.R. Acad. Sci. 229* (1984), 647-650.

Baiocchi, C. and Capelo, A.

[1] *Variational and Quasivariational Inequalities.* J. Wiley, New York (1984).

Ball, J.M.

[1] Convexity conditions and existence theorems in nonlinear elasticity, *Arch. Rat. Mech. Anal. 63* (1977), 337-403.

Bamberger, A.

[1] Etude d'une equation doublement nonlineaire. *J. Funct. Anal. 24* (1977), 148-155.

Bandle, C.

[1] *Isoperimetric Inequalities and Applications.* Pitman, London (1980).

[2] A note on optimal domains in a reaction-diffusion problem (to appear).

Bandle, C. and Mossino, J.

[1] Application du réarrangement à une inéquation variationelle. *C.R. Acad. Sc. Paris, 296* (1983), 501-504.

Bandle, C., Sperb, R.P. and Stakgold, I.

[1] Diffusion and reaction with monotone kinetics. *Nonlinear Analysis Th., Meth. and Appl. 8* (1984), 321-333.

Bandle, C. and Stakgold, I.

[1] Isoperimetric inequalities for the effectiveness in semilinear parabolic equations. *ISNM 71* (1984) 289-295.

Banks, H.T.

[1] *Modeling and Control in the Biomedical Sciences.* Lec. Notes in Biomathematics 6, Springer-Verlag (1975).

Baras, P. and Pierre, M.

[1] Singularités éliminables pour des équations Semi-linéaires (to appear in *Ann. Inst. Fourier (Grenoble)).*

Barbu, V.

[1] *Nonlinear Semigroups and Differential Equations in Banach Spaces,* Noordhoff, Leyden (1976).

Bear, J.

[1] *Dynamics of Fluids in Porous Media,* Elsevier, New York (1972).

Benilan, Ph.

[1] *Equations d'évolution dans un espace de Banach quelconque et applications.* Doctoral Thesis, Orsay (1972).

[2] Principe du maximum et perturbation d'operateurs accretifs dans $L^1(\Omega)$ (unpublished).

[3] Sur un problème d'évolution non monotone dans $L^2(\Omega)$. *Publications Mathem. de l'Univ. de Besançon* (1975/76).

[4] Operateurs accrêtifs et semi-groupes dans les espaces L^p $(1 \leq p \leq \infty)$, in *Functional Analysis and Numerical Analysis,* H. Fujita, Ed. Japan Society for the Promotion of Science, Tokyo, (1978) 15-53.

[5] *Evolution Equations and Accretive Operators.* Lecture notes (taken by S. Lenhart) University of Kentucky (1981).

[6] Abstract semi-linear elliptic operators (unpublished).

Benilan, Ph. and Berger, J.

[1] Estimation uniforme de la solution de $u_t = \Delta\phi(u)$ et charactérisation de l'effet régularisant. *C.R. Acad. Sc. Paris, 300* (1985), 573-576.

Benilan, Ph., Brezis, H. and Crandall, M.G.

[1] A semilinear equation in L^1. *Ann. Scu. Norm. Sup. Pisa, 2* (1975), 523-555.

Benilan, Ph., and Catté, F.

[1] Équation d'évolution du type $(du/dt) + \max A_i u = f$ par la théorie des semi-groupes non linéaires dans L^∞. *C.R. Acad. Sc. Paris, 295* (1982), 447-450.

Benilan, Ph., Crandall, M.G. and Pazy, A.

[1] *Nonlinear Evolution Equations Governed by Accretive Operators* (to appear).

Benilan, Ph. and Ha, K.

[1] Equation d'évolution du type $(du/dt) + \beta\partial\phi(u) \ni 0$ dans $L^\infty(\Omega)$. *C.R. Acad. Sci. Paris, 281* (1975), 947-950.

Benilan , Ph. and Picard, C.

[1] Quelques aspects non linéaires du principe du maximum. In Lect. Notes in Math. Springer-Verlag, Berlin, 713 (1979), 1-37.

Benilan, Ph. and Touré, H.

[1] Sur l'équation générale $u_t = \phi(u)_{xx} - \psi(u)_x + v$. *C.R. Acad. Sc. Paris, 299* (1984), 919-922.

Bensousson, A., Brezis, H. and Friedman, A.

[1] Estimates on the free boundary for quasi variational inequalities. *Comm. Partial Diff. Equations, 2* (1977), 297-321.

Bensousson, A. and Friedman, A.

[1] On the support of the solution of a system of quasi-variational inequalities. *J. Math. Anal. Appl. 65*, (1978), 660-674.

Bensoussan, A. and Lions, J.L.

[1] *Applications des Inégalites Variationelles en Controle Stochastique.* Dunod, Paris (1978).

Berestycki, H. and Lieb, E.

[1] To appear.

Berestycki, H. and Lions, P.L.

[1] Une méthode locale pour l'existence de solutions positives de problèmes semilineaires dans R^n. *J. Analyse Mathématique, 38* (1980), 144-187.

Berkovitz, L. and Pollard, H.

[1] A non classical variational problem arising from an optimal filter problem. *Arch. Rat. Mech. Anal. 26* (1967), 281-304.

Bernis, F.

[1] Compactness of the support in convex and nonconvex fourth order elasticity problems. *Nonlinear Analysis, Th., Meth. and Appl. 6* (1982), 1221-1243.

[2] Compactness of the support for some nonlinear elliptic problems of arbitrary order in dimension n. *Comm. Partial Diff. Equations, 9*, (1984), 271-312.

[3] Extinction of the solutions of some quasilinear elliptic problems of arbitrary order. (To appear in *Proc. Symp. Pure Math.*, F.E. Browder, ed., *44*, AMS, 1985).

[4] Asymptotic rates of decay for some nonlinear ordinary differential equations and variational problems of arbitrary order. *Ann. Fac. des Sciences de Toulouse, 6* (1984), 121-151.

Berryman, J.G. and Holland, C.J.

[1] Stability of the separable solution for fast diffusion *Arch. Ration. Mech. Anal. 74* (1980), 379-388.

Bertsch, M., De Mottoni, P. and Peletier, L.A.

[1] Degenerate diffusion and the Stefan problem. *Nonlinear Analysis, Th., Meth. and Appl. 8* (1984), 1311-1336.

Bertsch, M., Kersner, R. and Peletier, L.A.

[1] Positivity versus localization in degenerate diffusion equations (to appear).

Bertsch, M. and Rostamian, R.

[1] The principle of linearized stability for a class of degenerate diffusion equations (to appear).

Bidaut-Veron, M.F.

[1] Propriété de support compact de la solution d'une équation aux derivées partielles non linéaire d'ordre 4. *C.R. Acad. Sci. Paris, 287* (1978), 1005-1008.

[2] Variational inequalities of order 2m in unbounded domains, *Nonlinear Anal. Th. Meth. and Appl. 6*, (1982), 253-269.

[3] On the solutions of some nonlinear elliptic equations of order 2m. In *Contributions to Nonlinear Partial Differential Equations*, C. Bardos, A. Damlamian, J.I. Diaz and J. Hernandez, ed. Pitman London (1983), 69-82.

Boccardo, L. and Dacorogna, B.

[1] A characterization of pseudo-monotone differential operators in divergence form (to appear).

Boccardo, L. and Giachetti, D.

[1] Strongly nonlinear unilateral problems. *Appl. Math. Opt. 9* (1983), 291-301.

[2] Alcune osservazioni sulla regolarità delle soluzioni di problemi fortemente non lineari e applicazioni (to appear).

Boccardo, L., Murat, F. and Puel, J.P.

[1] Existence de solutions faibles pour des équations elliptiques quasi-linéaires à croissance quadratique. In *Nonlinear Partial Differential Equations and their Applications*, vol. IV, H. Brezis and J.L. Lions, ed., Pitman, London (1983), 1973.

[2] Resultats d'existence pour certains problemes elliptiques quasi-lineaires. *Ann. Scuola Norm. Sup. Pisa, 11* (1984), 213-235.

Bouillet, J.E. and Gomes, S.M.

[1] An equation with a singular nonlinearity related to diffusion problems in one dimension, (to appear in *Quart. of Appl. Math.*)

Brauner, C.M., Eckhaus, W., Garbey, M. and van Harten, A.

[1] A nonlinear, singular perturbation problem with some unusual features (to appear).

Brauner, C.M. and Nicolaenko, B.

[1] Sur des problèmes aux valeurs propres non linéaires qui se prolongent en problèmes à frontière libre. *C.R. Acad. Sc. Paris, 288* (1979), 125-127.

[2] On nonlinear eigenvalue problems which extend into free boundaries problems, *Proc. of the Conference on Nonlinear Eigenvalue Problems and Bifurcation*. Lecture Notes in Math. 782, Springer-Verlag, Berlin (1980), 61-100.

[3] Internal layers and free boundary problems. In *Boundary and Interior Layers*, J.J.H. Miller, Ed. Boole Press, Dublin (1980), 50-61.

Brezis, H.

[1] Equations et inéquations non linéaires dans les espaces vectoriels en dualité, *Ann. Institut Fourier (Grenoble), 18* (1968), 115-175.

[2] Seuil de regularité pour certains problèmes unilateraux, *C.R. Acad. Sc. Paris, 273* (1971).

[3] Monotonicity methods in Hilbert spaces and some applications to non-linear partial differential equations. In *Contributions to Nonlinear Functional Analysis*, E. Zarantonello, ed. Academic Press, New York (1971), 101-156.

[4] Integrales convexes dans les espaces de Sobolev, *Israel J. Math. 13* (1972), 9-23.

[5] Problèmes Unilateraux, *J. Math. Pures et Appl. 51* (1972) 1-168.

[6] *Operateurs Maximaux Monotones et Semi-groupes de Contractions dans les Espaces de Hilbert.* Notes de Matematica 5, North-Holland, Amsterdam (1973).

[7] Solutions of variational inequalities with compact support. *Uspekhi Mat. Nauk. 129* (1974), 103-108.

[8] A new method in the study of subsonic flows. In *Partial Differential Equations and Related Topics,* J. Goldstein, ed. Lecture Notes in Math. 476, Springer Verlag (1975).

[9] Some variational problems of the Thomas-Fermi type, in *Variational Inequalities and Complimentarity Conditions,* R.W. Cottle, F. Gianessi and J.L. Lions, eds, Wiley-Interscience, New York (1980), 53-73.

[10] Semilinear equations in R^N without condition at infinity (to appear).

Brezis, H. and Browder, F.

[1] Strongly nonlinear elliptic boundary value problems. *Ann. Scuola Norm. Pisa,* (1978), 587-603.

[2] Some properties of higher order Sobolev spaces *J. Math. Pures et Appl. 61* (1982), 245-259.

Brezis, H. and Duvaut, G.

[1] Ecoulements avec sillages autour d'un profile symmetrique sans incidence. *C.R. Acad. Sci. Paris, 276* (1973), 815-878.

Brezis, H. and Evans, L.C.

[1] A variational inequality approach to the Bellman-Dirichlet equation for two elliptic operators. *Arch. Rational Mech. Anal. 71* (1979), 1-13.

Brezis, H. and Lieb, E.

[1] Minimum action solutions of some vector field equations, *Commun. Math. Phys. 96* (1984), 97-113.

Brezis, H. and Oswald, L.

[1] Remarks on sublinear elliptic equations (to appear).

Brezis, H. and Stampacchia, G.

[1] Sur la regularité de la solution d'inéquations elliptiques, *Bull. Soc. Math. Fr. 96* (1971), 254-265.

[2] Une nouvelle méthode pour l'étude d'écoulements stationnaires. *C.R. Acad. Sci. Paris, 276* (1973), 129-132.

[3] The hodograph method in fluid dynamics in the light of variational inequalities, *Arch. Rational Mech. Anal. 61* (1976), 1-18.

Brezis, H. and Strauss, W.

[1] Semilinear second-order elliptic equations in L^1, *J. Math. Soc. Japan*, *25* (1973), 565-590.

Brezzi, F. and Caffarelli, L.

[1] Convergence of the discrete free boundaries for finite element approximations. *RAIRO Numer. Anal.* *17* (1983), 385-395.

Browder, F.E.

[1] *Nonlinear Operators and Nonlinear Equations of Evolution in Banach Spaces*. Proc. Sym. Pure Math. 28 part 2, A.M.S. Providence (1976).

[2] Pseudo-monotone operators and nonlinear elliptic boundary value problems on unbounded domains. *Proc. Nat. Acad. Sci.* *74* (1977), 2659-2661.

Brunovsky, P. and Mallet-Pavet, J.

[1] Switchings of optimal controls and the equation $y^{(4)} + |y|^{\alpha}\text{sign}y = 0$, $0 < \alpha < 1$ (to appear).

Các, N.P.

[1] Nonlinear elliptic boundary value problems for unbounded domains, *J. Diff. Eqns.* *45* (1982), 191-198.

[2] On some quasilinear elliptic boundary value problems with conditions at infinity, *J. Diff. Eqns.* *52* (1984), 342-355.

Caffarelli, L.A.

[1] The regularity of free boundaries in higher dimensions. *Acta Math.*, *139* (1977), 155-184.

[2] Further regularity for the Signorini problem. *Commun. Part. Diff. Equ.* *4* (1979), 1067-1076.

[3] Compactness methods in free boundary problems, *Commun. Part. Diff. Equ.* *5* (1980), 427-448.

[4] A remark on the Hausdorff measure of a free boundary and the convergence of coincidence sets, *Boll. Unione Mat. Ital.* *18* (1981), 1297-1299.

Caffarelli, L.A. and Friedman, A.

[1] Convexity of solutions of semilinear elliptic equations (to appear).

Caffarelli, L.A. and Spruck, J.

[1] Convexity properties of solutions to some classical variational problems. *Commun. Partial Diff. Equ.* *7* (1982), 1337-1379.

Canturia, T.A.

[1] On singular solutions of nonlinear systems of ordinary differential equations. In *Differential Equations*, M. Farkas, ed. North Holland, Amsterdam (1977), 107-109.

Carrillo, J.

[1] Sobre un problema con frontera libre. *Actas IV Congreso de Ec. Difer. y Aplicaciones*, Univ. de Sevilla (1981) 329-338.

Carrillo, J. and Chipot, M.

[1] On the dam problem. *J. Diff. Eq. 45* (1982), 234-271.

[2] On some nonlinear elliptic equations involving derivatives of the nonlinearity (to appear in *Proc. Royal Soc. Edinburgh*).

Chang, K.C.

[1] Free boundary problems and the set-valued mappings, *J. Diff. Eq. 49* (1983), 1-28.

Chiti, G.

[1] Norme di Orlicz delle soluzioni di una classe di equazioni ellittiche. *Boll. Unione Mat. It. 16-A* (1979), 178-185.

Cohen, D. and Laetsch, T.

[1] Nonlinear boundary value problems suggested by chemical reactor theory. *J. Diff. Eq. 7* (1970), 217-226.

Conrad, F., Issard-Roch, F., Brauner, C.M. and Nicolaenko, B.

[1] Nonlinear eigenvalue problems in elliptic variational inequalities: a local study. *Commun. Par. Diff. Equ. 10* (1985), 151-191.

Cortazar, C.

[1] The application of dissipative operators to nonlinear diffusion equations. *J. Diff. Eq. 47* (1983), 1-23.

Courant, P. and Hilbert, D.

[1] *Methods of Mathematical Physics*. Vol. 2: *Partial Differential Equations*, Interscience, New York (1962).

Crandall, M.G. and Liggett, T.

[1] Generation of semi-groups of nonlinear transformations on general Banach spaces. *Amer. J. Math. 93* (1971), 265-298.

Crandall, M.G. and Rabinowitz, P.H.

[1] Some continuation and variational methods for positive solutions of nonlinear elliptic eigenvalue problems, *Arch. Rat. Mech. Anal. 58* (1975), 207-218.

Crandall, M.G., Rabinowitz, P.H. and Tartar, L.

[1] On a Dirichlet problem with a singular nonlinearity, *Comm. Part. Diff. Eq.* *2* (1977),

Crandall, M.G. and Tartar, L.

[1] Some relations between nonexpansive and order preserving mappings. *Proc. Am. Math. Soc.* *78* (1980), 385-390.

Crooke, P.S. and Sperb, R.P.

[1] Isoperimetric inequalities in a class of nonlinear eigenvalue problems, *SIAM J. Math. Anal.* *9* (1978), 671-681.

Dal Passo, R.

[1] Multiplicity and stability of equilibrium solutions of a one dimensional fast diffusion problem. *Bolletino U.M.I.* (6), *3B* (1984), 1-24.

De Giorgi, E.

[1] Sulla differnziabilità e l'analiticitá delle estremali degli integrali multipili regolari, *Mem. Accad. Sc. Torino, Cl.Sc. Fis. Mat.*, *3* (1957), 25-43.

De Mottoni, P., Schiaffino, A. and Tesei, A.

[1] Attractivity properties of nonnegative solutions for a class of non-linear degenerate problems. *Ann. Mat. Pura. Appl.* *136* (1984), 35-48.

Dellacherie, C.

[1] *Ensembles analytiques. Capacités, Measures de Haussdorff.* Lect. Notes in Math. 295. Springer-Verlag, Berlin (1972).

Descombes, R.

[1] *Integration*, Hermann, Paris (1972).

De Thelin, F.

[1] Local regularity properties for the solutions of a nonlinear partial differential equation, *Nonlinear Analysis, Th. Meth. and Appl.* *6* (1982), 839-844.

Diaz, G.

[1] Estimation de l'ensemble de coincidence de la solution des problèmes d'obstacle pour les équations de Hamilton-Jacobi-Bellman. *C.R. Acad. Sc. Paris*, *290* (1980), 587-591.

[2] Fully nonlinear inequalities and certain questions about their free boundary, *Ann. Fac. des Sciences Toulouse*, *4* (1982), 29-44.

[3] Accion optima en una ecuacion de la programacion dinamica (to appear in *Rev. Real Acad. Ciencias, Madrid*).

[4] Some properties of solutions of degenerate second order P.D.E. in non divergence form (to appear in *Applicable Analysis*).

[5] Fully nonlinear equations in $\mathbb{R}^N$ without conditions at infinity (to appear in *Manuscripta Mathematica)*.

Diaz, G. and Diaz, J.I.

[1] Finite extinction time for a class of nonlinear parabolic equations. *Commun. Part. Diff. Eq. 4* (1979), 1213-1231.

[2] Remarks on the Monge-Ampere equation (to appear).

Diaz, J.I.

[1] Soluciones con soporte compacto de problemas unilatera mixtos. *Rev. Real Acad. Ciencias, Madrid, 69* (1975), 611-616.

[2] Soluciones con soporte compacto para ciertos problemas semilineales. *Collectanea Math. 30* (1979), 3-41.

[3] Localizacion de fronteras libres en inecuaciones variacionales estacionarias dadas por obstaculos. *Actas III Cong. Ec. Dif. y Aplic. Univ. de Santiago de Compostela* (1980), 97-108.

[4] *Tecnica de supersoluciones locales para problemas estacionarios no lineales*. Memory XVI of the Real Acad. Ciencias, Madrid, (1982).

[5] On a fully nonlinear parabolic equation and the asymptotic behaviour of its solutions, *J. Math. Anal. and Appl. 95* (1983), 144-168.

[6] Applications of symmetric rearrangement to certain nonlinear elliptic equations with a free boundary, in *Nonlinear Differential Equations*, J. Hale and P. Martinez, eds. Pitman, London (1985), 155-181.

[7] *Nonlinear Partial Differential Equations and Free Boundaries, Vol. 2. Parabolic and Hyperbolic Equations* (in preparation).

[8] Elliptic and parabolic quasilinear equations giving rise to a free boundary: the boundary of the support of the solutions (to appear in *Proc. Symp. Pure Math.*, F.E. Browder, ed., *44*, AMS, 1985).

Diaz, J.I. and Dou, A.

[1] Sobre flujos subsonicos alrededor de un obstaculo simetrico. *Collectanea Mathematica, 3* (1983), 142-160.

Diaz, J.I. and Hernandez, J.

[1] On the existence of a free boundary for a class of reaction-diffusion systems, *SIAM J. Math. Anal. 5* (1984), 670-685.

Diaz, J.I. and Herrero, M.A.

[1] Propriétés de support compact pour certaines équations elliptiques et paraboliques nonlinéaires. *C.R. Acad. Sc. Paris 286,* (1978), 815-817.

[2] Estimates on the support of the solutions of some nonlinear elliptic and parabolic problems. *Proc. Royal Soc. Edinburgh 89A,* (1981), 249-258.

Diaz, J.I. and Jimenez, R.

[1] Comportamiento en el contorno de la solucion del problema de Signorini. *Actas V Congr. Ec. Dif. y Aplic. Univ. de Zaragoza* (1984), 308-314.

[2] Aplicacion de la teoria no lineal de semigrupos a un operador pseudo-diferencial. *Actas VI Cong. Ec. Dif. y Apl.* Univ. de Granada (1984), 137-142.

Diaz, J.I. and Kawohl, B.

[1] Convexity and starshapedness of solutions of some nonlinear parabolic equations (to appear).

Diaz, J.I. and Kersner, R.

[1] Non existence d'une des frontières libres dans une équation degénérée en théorie de la filtration. *C.R. Acad. Sci. Paris, 296* (1983), 505-508.

Diaz, J.I. and Pierre, M.

[1] Unpublished work.

Diaz, J.I. and Saa, J.E.

[1] Uniqueness of solutions of nonlinear diffusion equations with a possible source term (to appear).

Diaz, J.I. and Veron, L.

[1] Existence theory and qualitative properties of the solutions of some first order quasilinear variational inequalities. *Indiana Univ. Math. J. 32* (1983), 319-361.

[2] Compacité du support des solutions d'équations quasilinéaires elliptiques ou paraboliques, *C.R. Acad. Sci. Paris 297* (1983), 149-152.

[3] Local vanishing properties of solutions of elliptic and parabolic quasi-linear equations (to appear in *Trans. AMS 2* (1985)).

Di Benedetto, E.

[1] $C^{1+\alpha}$ local regularity of weak solutions of degenerate elliptic equations. *Nonlinear Analysis, Th. Meth. and Appl. 7* (1983), 827-850.

Di Blasio, G.

[1] Perturbations of second order elliptic operators and semi-linear evolution equations. *Nonlinear Analysis, Th. Meth. and Appl.* *1* (1977), 293-304.

Donato, P. and Giachetti, D.

[1] Quasilinear elliptic equations with quadratic growth in unbounded domains (to appear).

Donato, P., Migliaccio, L. and Schianchi, R.

[1] Semilinear elliptic equations in unbounded domains of R^n. *Proc. Royal Soc. Edinburgh,* *88A* (1981), 109-119.

Duel, J. and Hess, P.

[1] A criterion for the existence of solutions of nonlinear elliptic boundary value problems. *Proc. Royal Soc. Edinburgh.* *74A* (1974-75), 49-54.

Duff, G.F.D.

[1] A general integral inequality for the derivative and an equimeasurable rearrangement. *Canad. J. Math.* *28* (1976), 793-804.

Duvaut, G. and Lions, J.L.

[1] *Les Inéquations en Mécanique et en Physique.* Dunod, Paris (1972).

Elliot, C.M. and Ockendon, J.R.

[1] *Weak and Variational Methods for Moving Boundary Problems.* Research Notes in Math. 59, Pitman, London (1982).

Evans, L.C.

[1] A convergence theorem for solutions of nonlinear second-order elliptic equations. *Indiana Univ. Math. J.* *27* (1978), 875-887.

[2] On solving certain nonlinear partial differential equations by accretive operator methods. *Israel J. Math.* *86* (1980), 225-247.

[3] Classical solutions of the Hamilton-Jacobi-Bellman equation for uniformly elliptic operators. *Trans. Amer. Math. Soc.* *275* (1983), 245-255.

Evans, L.C. and Friedman, A.

[1] Optimal stochastic switching and Dirichlet problem for the Bellman equation, *Trans. A.M.S.* *253* (1979), 365-389.

Evans, L.C. and Lions, P.L.

[1] Resolution des équations de Hamilton-Jacobi-Bellman. *C.R. Acad. Sci. Paris,* *290* (1980), 1049-1052.

Evans, L.C. and Kner, B.

[1] Instantaneous shrinking of the support of non-negative solutions to certain nonlinear parabolic equations and variational inequalities. *Illinois. J. Math. 23* (1979), 153-166.

Falconer, K.J.

[1] *The Geometry of Fractals Sets,* Cambridge Univ. Press, Cambridge (1985).

Fattorini, H.O.

[1] *The Cauchy Problem.* Encyclopedia of Math. and its Applications Vol. 18, Addison-Wesley, Massachusetts (1983).

Federer, H.

[1] *Geometric Measure Theory.* Springer-Verlag, Berlin (1969).

Fleming, W. and Rishel, R.

[1] An integral formula for total gradient variation, *Arch. Math. 11* (1960), 218-222.

Fraenkel, L. E. and Berger, M.S.

[1] A global theory of steady vortex rings in an ideal fluid, *Acta Math. 132* (1974), 13-51.

Frank, L.S. and Wendt, W.D.

[1] On a elliptic operator with discontinuous nonlinearity. *J. Diff. Eq. 54* (1984), 1-18.

Friedman, A.

[1] Boundary behaviour of solutions of variational inequalities for elliptic operators, *Arch. Rational Mech. Anal. 27* (1969), 95-107.

[2] *Partial Differential Equations of Parabolic Type.* Prentice-Hall, Englewood Cliffs, New Jersey (1964).

[3] *Variational Principles and Free Boundary Problems.* Wiley, New York (1982).

Friedman, A. and Lions, P.L.

[1] The optimal strategy in the control problem associated with the Hamilton-Jacobi-Bellman equation. *SIAM J. Control and Optimization 18* (1980), 191-198; Corrigendum, same journal *20* (1982), 153-154.

Friedman, A. and Phillips, D.

[1] The free boundary of a semilinear elliptic equation. *Trans. Am. Math. Soc. 282* (1984), 153-182.

Fucik, S. and Kufner, A.

[1] *Nonlinear Differential Equations.* Elsevier (1980).

Gagneux, G.

[1] *Approximations Lipschitziennes de la Funtion Signe et Schémas Semi-discrèts Implicites de Problèmes Quasilinéaires Dégénérés.* Publ. Math. Fac. des Sciences de Besançon (1984).

Gallouët, Th. and Morel, J.M.

[1] Resolution of a semilinear equation in L^1. *Proc. Royal Soc. of Edinburgh*, *96A* (1984), 275-288.

[2] On some semilinear problems in L^1. *Boll. Un. Math. It. 6-3A* (1984),

[3] The equation $-\Delta u + |u|^{\alpha-1}u = f$ for $0 < \alpha < 1$ (to appear).

Gariepy, R. and Pierre, M.

[1] Unpublished work.

Giaquinta, M.

[1] *Multiple integrals in the calculus of variations and nonlinear systems.* Annals of Math. Studies, Princeton Univ. Press, 105 (1983),

Giaquinta, M. and Giusti, E.

[1] Sharp estimates for the derivatives of local minima of variational integrals. *Boll. U.M.I. 6* 3-A (1984), 239-248.

Gidas, B., Ni, W.M., and Nirenberg, L.

[1] Symmetry and related properties via the maximum principle. *Comm. Math. Phys. 68* (1979), 209-243.

Gilbarg, D. and Trudinger, N.S.

[1] *Elliptic Partial Differential Equations of Second Order.* Second Edition, Springer-Verlag, Berlin (1983).

Glowinski, R. and Marroco, A.

[1] Sur l'approximation, par éléments finis d'ordre un ... *RAIRO 9* (1975), 41-76.

Giusti, E.

[1] *Minimal Surfaces and Functions of Bounded Variation.* Birkhäuser, Boston (1984).

Gossez, J.P.

[1] Nonlinear elliptic boundary-value problems for equations with rapidly or slowly increasing coefficients, *Trans. A.M.S. 190* (1974), 163-205.

Grun-Rehome, M.

[1] Caractérisation du sous-différentiel d'intégrandes convexes dans les espaces de Sobolev, *J. Math. Pures et Appl, 56* (1977), 149-156.

Gurney, W.S.C. and Nisbet, R.M.

[1] The regulation of inhomogeneous populations, *J. Theoret. Biol.* *52* (1975), 441-457.

Gurtin, M.E. and MacCamy, R.C.

[1] On the diffusion of biological populations, *Math. Biosci.* *33* (1977), 35-49.

Hardy, G.H., Littlewood, J.E. and Polya, G.

[1] Some simple inequalities satisfied by convex functions, *Messenger Math.* *58* (1929), 145-152.

[2] *Inequalities.* Cambridge Univ. Press, Cambridge (1934).

Hartman, P. and Stampacchia, G.

[1] On some nonlinear elliptic differential-functional equations, *Acta Math.* *115* (1966), 271-310.

Haugazeau, Y.

[1] Sur des inequations variationnelles, *C.R. Acad. Sci. Paris*, *265* (1967), 95-98.

Hernandez, J.

[1] Some free boundary problems for predator-prey systems with nonlinear diffusion, (to appear in *Proc. Symp. Pure Math.*, F.E. Browder ed., 44, AMS, 1985).

Herrero, M.A. and Vazquez, J.L.

[1] Asymptotic behaviour of solutions of a strongly nonlinear parabolic problem. *Ann. Fac. Sci. Toulouse Math.* *3* (1981), 113-127.

Hess, P.

[1] Nonlinear elliptic problems in unbounded domains, In *Theory of Nonlinear Operators, Constructive Aspects*, Akademie Verlag, Berlin (1978), 105-110.

Hestenes, M. and Redheffer, R.

[1] On the minimization of certain quadratic functionals, I and II. *Arch. Rat. Mech. Anal.* *56* (1974), 1-14 and 15-33.

Hilden, K.

[1] Symmetrization of functions in Sobolev spaces and the isoperimetric inequality. *Manuscr. Math.* *18* (1976), 215-235.

Hopf, E.

[1] Elementare Bemerkungen über die Lösungen partieller Differentialgleichungen zweiter Ordnung vom elliptischen Typus, *Sitz. Ber. Preuss. Akad. Wissensch. Berlin. Math-Phys. Kl.* *19* (1927), 147-152.

Hummel, R.A.

[1] The hodograph method for convex profiles. *Ann. Sov. Norm. Sup. Pisa, 9* (1982), 341-364.

Il'in, A.M. Kalashnikov, A.S. and Oleinik, O.A.

[1] Linear equations of the second order of parabolic type, *Russian Math. Surveys 17* (1962), 1-143.

Kamin, S. and Rosenau, Ph.

[1] Nonlinear thermal evolution in an inhomogeneous medium. *J. Math. Phys. 23* (1982), 1385-1390.

Kato, T.

[1] Schrödinger operators with singular potentials, *Israel J. Math. 13* (1972), 135-148.

Kawohl, B.

[1] Starshapedness of level sets of solutions for the obstacle problem and for the capacitory potential problem. *Proc. Am. Math. Soc. 89* (1983), 637-640.

[2] A geometric property of level sets of solutions to semilinear elliptic Dirichlet problems. *Appl. Anal. 16* (1983), 229-234.

[3] Geometrical properties of level sets of solutions to elliptic problems (to appear in *Proc. Symp. Pure Math.* F.E. Browder ed. *44*, AMS, 1985).

[4] Starshaped Rearrangement and Applications. LCDS Report 83-20 (1983).

[5] On the Convexity and Symmetry of Solutions to an Elliptic Free Boundary Problem. LCDS Report 84-20, (1984).

[6] When are solutions to nonlinear elliptic boundary value problems convex? (to appear in *Comm. Partial Diff. Eq.*).

[7] *On Rearrangements, Symmetrization and Maximum Principles.* Lecture Notes in Math. Springer-Verlag, (1985).

Keller, H.

[1] Positive solutions of some nonlinear eigenvalue problems. *J. Math. Mech. 19* (1969), 279-296.

Kiguradze, I.T.

[1] The problem of oscillation of solutions of nonlinear differential equations. *Diff. Eq. 1* (1965), 773-782.

[2] On the oscillatory and monotone solutions of ordinary differential equations, *Arch. Math. (Brno) 14* (1978), 21-44.

Kinderlehrer, D.

[1] Variational inequalities and free boundary problems, *Bull. Am. Math. Soc.* *84* (1978), 7-26.

[2] The smoothness of the solution of the boundary obstacle problem. *J. Math. Pures et Appl.* *60* (1981), 193-212.

Kinderlehrer, D. and Stampacchia, G.

[1] A free boundary problem in potential theory. *Ann. Inst. Fourier (Grenoble)*, *25* (1975), 323-344.

[2] *An Introduction to Variational Inequalities and Their Applications.* Academic Press, New York (1980).

Knerr, B.

[1] The porous medium equation in one dimension. *Trans. Amer. Math. Soc.* *234* (1977), 381-415.

Korevaar, N. and Lewis, J.L.

[1] Convex solutions to certain elliptic p.d.e's have constant rank Hessians (to appear).

Krilov, N.V.

[1] *Controlled Diffusion Processes.* Springer, New York (1981).

Lacey, A.A., Ockendon, J.R. and Tayler, A.B.

[1] 'Waiting-time' solutions of a nonlinear diffusion equation, *SIAM J. Appl. Math.* *42* (1982), 1252-1264.

Ladyzhenskaya, O.A. and Ural'tseva, N.N.

[1] *Linear and Quasilinear Elliptic Equations.* Academic Press, New York (1968).

Laetsch, T.

[1] Uniqueness for sublinear boundary value problems, *J. Diff. Eq.* *13* (1973), 13-23.

Landes, R.

[1] Quasilinear elliptic operators and weak solutions of the Euler equations, *Manuscripta Math.* *27* (1979), 47-72.

Langenbach, A.

[1] *Monotone Potentialoperatoren in Theorie und Anwendung.* Springer, Berlin (1977).

Langlais, M. and Phillips, D.

[1] Stabilization of solutions of nonlinear and degenerate evolution equations, *Nonlinear Analysis, Th. Meth. and Appl.* *9* (1985), 321-334.

Lasry, J.M. and Lions, P.L.

[1] Équations elliptiques non linéaires avec conditions aux limites infinies et contrôle stochastique avec contraintes d'état. *C.R. Acad. Sc. Paris, 299* (1984), 213-216.

Lê, C.H.

[1] *Etude de la classe des operateurs m-accrétifs de* $L^1(\Omega)$ *et accrétifs dans* $L^\infty(\Omega)$. These. Univ. Paris VI (1977).

Lewy, H.

[1] On the coincidence set in variational inequalities, *J. Diff. Geom. 6* (1972), 497-501.

Lieb, E.

[1] Existence and uniqueness of the minimizing solution of Choquard's non-linear equation. *Studies Appl. Math. 57* (1977), 93-105.

Lieb, E. and Simon, B.

[1] The Thomas-Fermi theory of atoms, molecules and solids, *Adv. Math. 23* (1977), 22-116.

Lions, J.L.

[1] *Quelques Méthodes de résolution de problèmes aux limites non linéaires.* Dunod, Paris (1969).

[2] Partial differential inequalities. *Russian Math. Surveys, 27* (1972),91-161.

Lions, P.L.

[1] Quelques remarques sur la symmetrization de Schwartz. In *Nonlinear Partial Differential Equations and their Applications.* Vol. I.H. Brezis and J.L. Lions, eds. Pitman, London (1980), 308-319.

[2] Two geometrical properties of solutions of semilinear problems. *Appl. Anal. 12* (1981), 267,272.

[3] Hamilton-Jacobi-Bellman equations and the optimal control of stochastic systems. *Proc. Int. Congr. Math. Warsaw* (1983).

Lions, P.L. and Nisio, M.

[1] A uniqueness result for the semigroup associated with the Hamilton-Jacobi-Bellman operator. *Proc. Japan Acad. 58* (1982), 273-276.

Luning, C.D. and Perry, W.L.

[1] Positive solutions of negative exponent generalized Emden-Fowler boundary value problems. *SIAM J. Math. Anal. 12* (1981), 874-879.

[2] An iterative method for solution of a boundary value problem in non-Newtonian fluid flow, *J. Non-Newtonian Fluid Mech, 15* (1984), 145-154.

Maderna, C.

[1] Optimal problems for a certain class of non linear Dirichlet problems. *Suppl. Bull. U.M.I., I* (1980), 31-34.

Maderna, C. and Salsa, S.

[1] Symmetrization in Neumann problems. *Appl. Anal. 9* (1979), 247-256.

[2] Some special properties of solutions to obstacle problems, *Rend. Sem. Mat. Univ. Padova, 71* (1984), 121-129.

Marcellini, P. and Sbordone, C.

[1] Semicontinuity problems in the calculus of variations, *Nonlinear Analysis Th. Meth. and Appl.* Vol. (1980), 241-257.

Martinson, L.K. and Pavlov, K.B.

[1] The effect of magnetic plasticity in non-Newtonian fluids, *Magnit. Gidrodinamika, 3* (1969), 69-75.

[2] The problem of the three-dimensional localization of thermal perturbations in the theory of non-linear heat conduction. *Zh. vychisl. Mat. mat. Fiz. 12* (1972), 1048-1053.

Mignot, F. and Puel, J.P.

[1] Sur une classe de problèmes non linéaires avec non linéarité positive croissante, convexe. *Proc. Conference Held in Roma*, Pitagora, Roma (1978), 45-72.

Misiti, M. and Guyot, J.

[1] Global solutions for a class of singular free boundary problems (to appear).

Moet, H.J.K.

[1] Asymptotic analysis of the free boundary in singularly perturbed elliptic variational inequalities, In Lecture Notes in Math. 942, Springer-Verlag, Berlin (1982), 1-18.

Morrey, C.B., Jr.

[1] *Multiple Integrals in the Calculus of Variations.* Springer-Verlag, Berlin (1966).

Mossino, J.

[1] A priori estimates for a model of Grad-Mercier type in plasma confinement. *Applicable Anal., 13* (1982), 185-207.

[2] *Inegalites Isoperimetriques et Applications en Physique.* Herman, Paris, (1984).

[3] A generalization of the Payne-Rayner isoperimetric inequality. *Boll. UMI Z-A* (1983), 335-342.

Munroe, M.E.

[1] *Measure and Integration*, Second Edition, Addison-Wesley, Reading, Massachusetts (1971).

Murthy, M.K. and Stampacchia, G.

[1] Boundary value problems for some degenerate elliptic operators, *Ann. Mat. Pura Appl. 80* (1968).

Nagai, T.

[1] Estimates for the coincidence set of solutions of elliptic variational inequalities, *Hiroshima Math. J. 9* (1979), 335-346.

Namba, T.

[1] Density-dependent dispersal and spatial distribution of a population, *J. Theoret. Biol. 86* (1980), 351-363.

Newman, W.I. and Sagan, C.

[1] Galactic civilisations: population dynamics and interstellar diffusion. *Icarus* (in press).

Niezgodgka, M.

[1] Stefan-like problems, In *Free Boundary Problems: Theory and Applications*, A. Fasano and M. Primicerio, eds, Research Notes in Math. 79, Pitman, London (1983), 321-347.

Nirenberg, L.

[1] On elliptic partial differential equations, *Ann. Scuola Norm. Sup. Pisa 13* (1959), 115-162.

Oden, J.T.

[1] Existence theorems and approximations in nonlinear elasticity. In *Nonlinear Equations in Abstract Spaces*. E. Lakshmikantham, ed, North-Holland (1978).

Okubo, A.

[1] *Diffusion and Ecological Problems: Mathematical Models*, Biomathematics, *10*, Springer-Verlag (1980).

Osserman, R.

[1] On the inequality $\Delta u \geqq f(u)$, *Pacific J. Math. 7* (1957), 1641-1647.

Pascali, D. and Sburlan, S.

[1] *Nonlinear Mappings of Monotone Type*, Sijthoff and Noordhoff, Alphen aan den Rijn (1978).

Pazy, A.

[1] The Lyapunov method for semigroups of nonlinear contractions in Banach spaces, *J. d'Analyse Math.* *40* (1981) 241-262.

[2] *Semigroups of Linear Operators and Applications to Partial Differential Equations*, Springer-Verlag, New York (1983).

Payne, L.E.

[1] Some applications of "best possible" maximum principles in elliptic boundary value problems. In *Partial Differential Equations and Dynamical Systems*, W.E. Fitzgibbon III. Ed. Pitman, London (1984), 286-313.

Peletier, L.A. and Serrin, J.

[1] Uniqueness of non-negative solutions of semilinear equations in $\mathbb{R}^N$ (to appear).

Peletier, L.A. and Tesei, A.

[1] Global bifurcation and attractivity of stationary solutions of a degenerate diffusion equation (to appear).

[2] Diffusion in inhomogeneous media: localization and positivity (to appear).

Pelissier, M.C.

[1] *Sur quelques problèmes non lineaires en glaciologie*, Thèse, Publ. Math. d'Orsay, *110* (1975).

Perriot, A.

[1] Utilisation de la théorie des semi-groupes à l'étude d'un problème d'évolution sur une partie du bord d'un ouvert, *Publ. Math. Fac. Sci. Besançon* (1984), 1-26.

Perry, W.L.

[1] A monotone iterative technique for solution of pth order ($p < 0$) reaction-diffusion problems in permeable catalysis. *J. Comput. Chemistry*, *5* (1984), 353-357.

Phillips, D.

[1] A minimization problem and the regularity of solutions in the presence of a free boundary, *Indiana Univ. Math. J.* *32* (1983), 1-17.

[2] Hausdorff measure estimates of a free boundary for a minimum problem. *Commun. Part. Diff. Eq.* *8* (1983), 1409-1454.

Pohozaev, S.I.

[1] Eigenfunctions of the equation $\Delta u + \lambda f(u) = 0$, *Soviet Doklady*, *165* (1965), 1408-1410.

Polya, G. and Szegö, G.

[1] *Isoperimetric Inequalities in Mathematical Physics*, Ann. Math. Stud. *27*, Princeton Univ. Press, Princeton (1952).

Pozio, M.A. and Tesei, A.

[1] Support properties of solutions for a class of degenerate parabolic problems (to appear).

[2] Degenerate parabolic problems in population dynamics (to appear in *Japan J. Appl. Math.*).

Protter, M.H. and Weinberger, H.F.

[1] *Maximum Principles in Differential Equations*. Prenctice Hall., Englewood Cliffs, New Jersey (1967).

Puel, J.P.

[1] Existence, comportement à l'infini et stabilité dans certains problèmes quasilinéaires elliptiques et paraboliques d'ordre 2, *Ann. Sculoa Norm. Sup. Pisa*, *3* (1976), 89-119.

Redheffer, R.

[1] On a nonlinear functional of Berkovitz and Pollard, *Arch. Rational Mech. Anal.* *50* (1973), 1-9.

[2] Nonlinear differential inequalities and functions of compact support, *Trans. Amer. Math. Soc.* *220* (1976), 133-157.

Riesz, F. and Nagy, B.Sz.

[1] *Functional Analysis*, Frederick Ungar, New York (1955);

Roach, G.F.

[1] *Green Functions: Introduction Theory with Applications*, Van Nostrand, London (1970).

Rodrigues, J.F.

[1] Further remarks on the stability of the obstacle problem (to appear in *Comm. Part. Diff. Eqs.).*

Rogers, C.A.

[1] *Hausdorff Measures*, Cambridge Univ. Press, Cambridge (1970).

Sakaguchi, S.

[1] Starshaped coincidence sets in the obstacle problem, *Ann. Sc. Norm. Sup. Pisa*, *11* (1984), 123-128.

Sato, K.

[1] On the generators of non-negative contraction semigroups in Banach lattices, *J. Math. Soc. Japan*, *20* (1968), 431-436.

Schatzman, M.
[1] Stationary solutions and asymptotic behaviour of a quasilinear degenerate parabolic equation. *Indiana Univ. Math. J. 33* (1984), 1-30.
Schoemaver, M.
[1] A monodimensional model for fracturing, In *Free Boundary Problems: Theory and Applications*. A. Fasano and M. Primicerio, eds. Research Notes in Math. 79, Pitman, London (1983), 701-711.
Schwartz, L.
[1] *Theorie des Distributions*, Hermann, Paris (1965).
Serrin, J.
[1] Local behavior of solutions of quasi-linear elliptic equations. *Acta. Math. 111* (1964), 247-302.
Shamir, E.
[1] Regularity of mixed second order elliptic problems, *Israel J. Math. 6* (1968), 150-168.
Shimborski, E.
[1] Encadrement d'une frontière libre relative à un problème d'hydraulique, *Boll. U.M.I. 12* (1975) 66-67.
[2] Variational methods applied to the study of symmetric flows in laval nozzles, *Commun. Part. Diff. Eq. 4* (1979), 41-77.
[3] Variational inequalities arising in the theory of two dimensional potential flows, *Nonlinear Analysis, Th. Meth. and Appl. 5* (1981), 433-444.
Simon, J.
[1] Regularité de la solution d'une équation non linéaire dans $\mathbb{R}^N$. In *Journées d'Analyse Non linéaire*, Ph. Benilan and J. Robert, eds. Lecture Notes in Math. 665, Springer-Verlag, Berlin (1978), 205-227.
Smoller, J. and Wasserman, J.
[1] Global bifurcations steady-state solutions, *J. Diff. Eqns. 39* (1981), 269-290.
Sperb, R.
[1] *Maximum Principles and Their Applications*, Acad. Press, New York (1981).
Spruck, J.
[1] Uniqueness in a diffusion model of population biology, *Commun. Part. Diff. Eqns. 8* (1983), 1605-1620.

Stampacchia, G.

[1] Some limit cases of L^p-estimates for solutions of second order elliptic equations, *Commun. Pure Appl. Math. 16* (1963), 505-510.

[2] *Equations Elliptiques du Second Ordre à Coefficients Discontinus*, Presse de l'Univ. de Montréal, Montréal (1966).

Stakgold, I.

[1] *Green's Functions and Boundary Value Problems*, Wiley, New York (1979).

[2] Estimates for some free boundary problems, in *Ordinary and Partial Differential Equations*, W.N. Everitt and B.D. Sleeman, eds. Lecture Notes in Math. 846, Springer-Verlag (1981).

Stredulinsky, E.W.

[1] *Weighted Inequalities and Degenerate Elliptic Partial Differential Equations*, Lecture Notes in Math. 1074, Springer-Verlag, Berlin (1984).

Stuart, C.A.

[1] Maximal and minimal solutions of elliptic differential equations with discontinuous non-linearities, *Math. Z. 163* (1978) 239-249.

Svec, M.

[1] Behaviour of nonoscillatory solutions of some nonlinear differential equations, *Acta Math. Univ. Comenianae 39* (1980), 115-129.

[2] Some problems of nonlinear differential equations with quasiderivatives (to appear).

Talenti, G.

[1] Best constant in Sobolev inequality, *Ann. Mat. Pura Appl. IV.Ser. 110* (1976), 353-372.

[2] Elliptic equations and rearrangements, *Ann. Sc. Norm. Sup. Pisa, 3* (1976), 697-718.

[3] Nonlinear elliptic equations, rearrangements of functions and Orlicz spaces, *Ann. Mat. Pura Appl. IV, 120* (1977), 159-184.

[7] Linear elliptic pde's: level sets, rearrangements and a priori estimates of solutions (to appear in *Bol. U.M.I*).

Tartar, L.

[1] Equations with Order Preserving Properties, MRC. Technical Report #1580, Madison (1976).

Tepper, D.F.

[1] Free boundary problem, *SIAM J. Math. Anal. 5* (1974), 841-846.

[2] On a free boundary problem, the starlike case, *SIAM J. Math. Anal. 6* (1975) 503-505.

Tolksdorff, P.

[1] Regularity for a more general class of quasilinear elliptic equations, *J. Differ. Eqns. 50* (1983), 1627-1652.

Trombetti, G. and Vazquez, J.L.

[1] A symmetrization result for elliptic equations with lower-order terms (to appear).

Trudinger, N.S.

[1] Fully nonlinear, uniformly elliptic equations under natural structure conditions, *Trans. Am. Math. Soc. 278* (1983), 751-769.

Uhlenbeck, K.

[1] Regularity for a class of nonlinear elliptic systems, *Acta Math., 138* (1977), 219-240.

Vazquez, J.L.

[1] An a priori interior estimate for the solutions of non linear problems representing weak diffusion, *Nonlinear Analysis, Th. Meth. and Appl. 9* (1981), 95-103.

[2] Monotone perturbations of the Laplacian in $L^1(\mathbb{R}^N)$, *Israel J. Math. 43* (1982), 255-272.

[3] Symmetrisation pour $u_t = \Delta\phi(u)$ et applications, *C.R. Acad. Sci, Paris, 295* (1982), 71-74.

[4] On a semilinear equation in $\mathbb{R}^2$ involving bounded measures, *Proc. Royal Soc. Edinburgh, 95A* (1983), 181-202.

[5] A strong maximum principle for some quasilinear elliptic equations (to appear).

Veron, L.

[1] Coercivité et propriétés regularisantes des semi-groups non linéaires dans les espaces de Banach, *Ann. Fac. des Sci. Toulouse, 1* (1979), 171-200.

[2] Equations d'évolution semi-linéaires du second ordre dans L^1, *Rev. Roumaine Math. Pures Appl. 27* (1982), 95-123.

[3] Singularités éliminables d'équations elliptiques non linéaires, *J. Diff. Eqns. 41* (1981), 87-95.

Volquer, R.E.

[1] Nonlinear flow in porous media by finite elements. *J. Hydraulics Division Proc. Am. Sc. Civil Eng. 95* (1969), 2093-2113.

Weinberger, H.F.

[1] Symmetrization in uniformly elliptic problems, in *Studies in Math. Analysis and Related Topics*, Univ. of Calif. Press, Stanford (1962), 424-428.

Weitsman, A.

[1] Spherical symmetrization in the theory of elliptic partial differential equations, *Comm. Part. Diff. Eqns, 8* (1983), 545-561.

Wolanski, N.I.

[1] Flow through a porous column. *J. Math. Anal. and Appl. 109* (1985), 140-159.

Wong, J.S.W.

[1] On the generalized Emden-Fowler equation, *SIAM Review, 17* (1975), 339-360.

Yamada, N.

[1] Estimates on the support of solutions of elliptic variational inequalities in bounded domains, *Hiroshima Math. J. 9* (1979), 7-16.

[2] Estimates on the support of solutions of elliptic variational inequalities in bounded domains, II. *Funckial Ekvac.* (1980).

Zeldovich, Y.B. and Raizer, Y.P.

[1] *Physics of Shock Waves and High-Temperature Hydrodynamic Phenomena*, Academic Press, New York (1969).

Index